T0291087

SUSTAINABLE BLUE REVOLUTION IN INDIA: WAY FORWARD

This book is conceptualized by two authors, which is a unique blend of rich experience from academics and industry. It is an endeavour to provide comprehensive information on several dimensions of blue revolution in a structured form. Material provided in the book has been gathered from several relevant published sources and views expressed are based on practical field experience of the authors. Blue revolution would be one of the big game changers for the Indian economy. The subject of sustainable development of fisheries sector being very vast, concerted efforts have been made to accommodate all the relevant elements. Very little reading material with proper analysis is currently available and this book is expected to bridge the gap and project way-forward to achieve sustainable development of fisheries and aquaculture in India under the blue revolution.

Dr (Mrs) Latha Shenoy is retired Principal Scientist from the ICAR-Central Institute of Fisheries Education, Mumbai. She is recipient of number of gold medals consistently for her brilliant academic performance. She has close to four decades of experience in Fisheries education, research and human resource development.

Dr. (Prof) Shridhar Rajpathak is Director of Garware Technical Fibres Ltd, one of the leading companies in the field of technical textiles, and a top manufacturer and exporter in the field of fishing nets. Dr Rajpathak has worked for this organisation for over 45 years. He also worked as Director Garware Bestretch Ltd, a leading manufacturer of elastic rubber tapes and sheets. Management strategy, export promotion, international business, indirect taxes, policy advocacy and other commercial activities are his core areas of expertise.

SUSTAINABLE BLUE REVOLUTION IN INDIA: WAY FORWARD

Dr. Latha Shenoy
Retired Principal Scientist,
ICAR-Central Institute of Fisheries Education,
Versova, Mumbai-400061, Maharashtra, India
Dr. Shridhar Rajpathak
Director, Garware Technical Fibers Ltd.,
MIDC, Chinchwad,
Pune-411019, Maharashtra, India

CRC Press is an imprint of the
Taylor & Francis Group, an **informa** business

NARENDRA PUBLISHING HOUSE
DELHI (INDIA)

First published 2021
by CRC Press
2 Park Square, Milton Park, Abingdon, Oxon, OX14 4RN

and by CRC Press
6000 Broken Sound Parkway NW, Suite 300, Boca Raton, FL 33487-2742

CRC Press is an imprint of Informa UK Limited

British Library Cataloguing-in-Publication Data
A catalogue record for this book is available from the British Library

Library of Congress Cataloging-in-Publication Data
A catalog record has been requested

ISBN: 978-1-032-02463-9 (hbk)
ISBN: 978-1-003-18344-0 (ebk)

Dedicated to all the
blue revolution enthusiasts

CONTENTS

Central Agricultural University
Lamphel Pat, Imphal-795004
Manipur

Prof. S. Ayyappan
Chancellor

July 14, 2020

FOREWORD

After successful green and white revolutions, India, endowed with both aquatic resources, marine and freshwater, as also rich biodiversity, is in the throes of a blue revolution. Several new programmes were initiated under the *Neeli Kranti* Mission, 2015, to ensure sustainable fisheries development in the country. Recognising it as the gateway for untapped potential in the field of fishing and aquaculture, *Pradhan Mantri Matsya Sampada Yojana* (PMMSY) was launched in 2020, with a funding of over Rs 20,000 crore, to give a further boost fisheries sector. Fisheries and aquaculture have all the components and potentials to provide for the shift from 'starch-to-protein' in Indian diet, as well as contribute to doubling fishers' and farmers' income over the next five years. This is duly echoed in the Vision statement of PMMSY as *'Ecologically healthy, economically viable and socially inclusive fisheries sector that contributes towards economic prosperity and well-being of fishers and fish farmers and other stakeholders, food and nutritional security of the country in a sustainable and responsible manner'.*

In this context, there is a need for a comprehensive reference book on sustainable fisheries and aquaculture in the country. As if to address this lacuna, the present volume on 'Sustainable Blue Revolution in India: Way Forward' has been penned by Prof. Latha Shenoy and Prof. Shridhar Rajpathak. The book covers a wide range of subjects related to Sustainable development of fisheries and aquaculture including SDGs, Overview of global and Indian fisheries and aquaculture production, Seafood trade & exports, Policy initiatives & legislation, Impact of climate change, Priorities of the Government of India, Schemes for achieving blue revolution, Blue economy and Innovative concepts & approaches, including counter measures for the recent outbreak of Covid-19 pandemic.

Mobile: 0091-9582898989; e-mail: sayyappan1955@gmail.com
Residence: 106, SankalpBasant, # 40 & 41, Akkamahadevi Road, Industrial Suburb, Mysuru South, Mysuru 570008, Karnataka, India

Central Agricultural University
Lamphel Pat, Imphal-795004
Manipur

I compliment Prof. (Dr) Latha Shenoy, Former Principal Scientist, ICAR-Central Institute of Fisheries Education, Mumbai, with over 36 years of experience in fisheries education and research; and Prof. (Dr) Shridhar Rajpathak, Director, Garware Technical Fibres Ltd. having more than 45 years industry experience, for their relentless efforts in planning and preparation of this comprehensive and exhaustive book. Having known both the authors for a long time,

I appreciate this academia-industry partnership enriching the volume. I am sure the book will benefit various stakeholders of fisheries sector including academicians, researchers, students, teaching faculty, Government officials, Policy makers, NGOs, fisher community, fish processors/exporters, entrepreneurs and other fisheries practitioners alike.

(S. Ayyappan)

Dr. S Ayyappan, is Chancellor, Central Agricultural University, Imphal, Manipur and Chairman, Karnataka Science and Technology Academy, Bangalore, Karnataka and former Secretary DARE & DG, ICAR.

Mobile: 0091-9582898989; e-mail: sayyappan1955@gmail.com
Residence: 106, SankalpBasant, # 40 & 41, Akkamahadevi Road, Industrial Suburb, Mysuru South, Mysuru 570008, Karnataka, India

PREFACE

The contents of the book have been structured under 13 chapters. Chapter 1 titled "Overview of global fisheries and aquaculture" highlights the present status of global fisheries and aquaculture, global export and import of fisheries products. In the backdrop of ecolabelling being acknowledged globally as a market-based tool to promote sustainable use of fisheries resources, information on major ecolabels in wild catch fisheries and aquaculture and guidelines are covered in this chapter. Chapter on "Overview of Indian fisheries and its role in Indian economy" presents the status of fish production in India, Indian seafood exports and imports, modern seafood trade in India, retail seafood market and guidelines for seafood retailing. Development of innovative solutions is the need of the hour to address the growing global demand for seafood without causing any damage to environment and ecosystems. Role of industry in promoting sustainability in fishing and aquaculture has been elaborated.

Marine fisheries are faced with challenges such as overfishing, destructive fishing, bycatch and discards, ghost fishing. Issues and measures are discussed in the chapter titled "Issues in Indian marine fisheries and measures for sustainable development". The status of FAO CCRF in the Indian context, its implementation and compliance in Indian marine fisheries, major causes of non-compliance and mitigation measures are elaborated. Chapter on aquaculture deals with common aquaculture practices, culture and seed production of major cultivable species, Institutional framework of aquaculture in India, new approaches to fish farming, Feed Innovation, and new product developments.

Salient features of Regulations / Restrictions / Prohibitions in the Maritime States of India, National Marine Fisheries Regulation and Management policy, Challenges in implementation, Aquaculture Legislation in India, Draft National Inland Fisheries and Aquaculture Policy, 2019, National Mariculture Policy, 2019 (Revised Draft) are covered in the chapter on Regulations for fisheries and aquaculture in India. Chapter on "FAO structure and its contribution to India's fisheries development priorities"covers Government's fisheries development priorities, Contribution of FAO through its Country Programming Framework (CPF) for India, Structure and role of FAO in fisheries and aquaculture development.

The 2030 Agenda for sustainable development recognizes that eradicating poverty in all its forms and dimensions is the greatest global challenge and an indispensable requirement for sustainable development. Information on Sustainable Development Goals and in particular Conservation and sustainable use of the oceans, seas and marine resources (SDG14), global efforts towards achieving SDG 14 and Indian scenario of goal 14 and Indian initiatives towards achieving SDGs has been provided under a separate chapter. The chapter on "Overview of recently launched Pradhan Mantri Matsya Sampada Yojana (PMMSY) in India" covers vision, objectives, strategies and funding pattern for achieving blue revolution. Role of National Fisheries Development Board (NFDB) in achieving blue revolution, Central Sector assistance schemes, centrally sponsored schemes, Linkages with other schemes, Implementing Agencies and Fisheries and Aquaculture Infrastructure Development Fund (FIDF) have been included.

Overall adverse impact of global warming on sustainability of fish resources, Impacts of climate change on marine capture fisheries and aquaculture, small scale fishers, major challenges to fishing communities posed by climate change, impact of climate-driven disasters, adaptation to climate change are described in the chapter on "Climate change and impact on fisheries and aquaculture". Aquaculture in the context of blue dimensions, Green House Gas (GHG) emissions in marine capture fisheries and aquaculture and contribution of aquaculture to climate change are also dealt with.

Components of blue economy and innovative concepts such as development of plant and cell-based seafood to provide safe, nutritious, tasty, delicious, easily available and affordable seafood product to consumers are covered under separate chapters. Chapter on Impact of Covid-19 on fisheries and aquaculture covers wide range of impacts of the pandemic, role of FAO, initial assessment of impact on global fisheries & aquaculture, global/regional initiatives amid COVID-19, ongoing coping measures and potential adaptation strategies and standard operating procedures imposed by GoI. Concepts, approaches and technologies which can be adopted to bring about blue revolution in the country have been suggested in the chapter "Way forward".

Firstly, we are extremely grateful to Dr S. Ayyappan, Chancellor, Central Agricultural University, Imphal, Manipur for readily agreeing to pen Foreword for this book. We express our profuse thanks to all our colleagues and friends in ICAR, CIFE, Mumbai and sister Fisheries institutes in ICAR, Department of Fisheries, Ministry of Fisheries, Animal Husbandry and Dairying, and M/S Garware Technical Fibres Ltd, Pune, Maharashtra for their help at different times. Valuable technical help from Mr. Amod Ashok Salgaonkar, Global Seafood Business,

Marketing Sustainability Professional (International Advisory Board: Friend of the Sea) is thankfully acknowledged. Our special thanks to family members and well-wishers for their support in making this book a reality. We express our sincere thanks to Narendra Publishing House, N. Delhi for readily accepting to publish this book.

We hope that this book will be useful to professionals and people associated with fisheries and aquaculture.

Latha Shenoy
Shridhar Rajpathak

CHAPTER 1

OVERVIEW OF GLOBAL FISHERIES AND AQUACULTURE

Present status of global fisheries and aquaculture production

Global total capture fisheries production, as derived from the FAO capture database, was 90.9 million tonnes in 2016. Total marine catches by China, by far the world's top producer, were stable in 2016. As in 2014, Alaska pollock *(Theragra chalcogramma)* surpassed anchoveta as the top species in 2016, with the highest catch since 1998. Total global catch in inland waters was 11.6 million tonnes in 2016.

Global aquaculture production (including aquatic plants) in 2016 was 110.2 million tonnes. The contribution of aquaculture to the global production of capture fisheries and aquaculture combined has risen continuously, reaching 46.8 percent in 2016, up from 25.7 percent in 2000. The growth of farming of fed-aquatic animal species has outpaced the farming of unfed species in world aquaculture. In 2016, aquaculture was the source of 96.5 percent by volume of the total 31.2 million tonnes of wild-collected and cultivated aquatic plants combined.

The most recent official statistics indicate that 59.6 million people were engaged in the primary sector of capture fisheries and aquaculture in 2016, with 19.3 million people engaged in aquaculture and 40.3 million people engaged in fisheries. In 2016, 85 percent of the global population engaged in the fisheries and aquaculture sectors was in Asia, followed by Africa (10 percent) and Latin America and the Caribbean (4 percent).

It is estimated that in 2016, overall, women accounted for nearly 14 percent of all people directly engaged in the fisheries and aquaculture primary sector, as compared with an average of 15.2 percent across the reporting period 2009–2016.

WORLD FISHERIES AND AQUACULTURE PRODUCTION AND UTILIZATION (MILLION TONNES)[a]

Category	2011	2012	2013	2014	2015	2016
Production						
Capture						
Inland	10.7	11.2	11.2	11.3	11.4	11.6
Marine	81.5	78.4	79.4	79.9	81.2	79.3
Total capture	**92.2**	**89.5**	**90.6**	**91.2**	**92.7**	**90.9**
Aquaculture						
Inland	38.6	42.0	44.8	46.9	48.6	51.4
Marine	23.2	24.4	25.4	26.8	27.5	28.7
Total aquaculture	**61.8**	**66.4**	**70.2**	**73.7**	**76.1**	**80.0**
Total world fisheries and aquaculture	**154.0**	**156.0**	**160.7**	**164.9**	**168.7**	**170.9**
Utilization[b]						
Human consumption	130.0	136.4	140.1	144.8	148.4	151.2
Non-food uses	24.0	19.6	20.6	20.0	20.3	19.7
Population (*billions*)[c]	7.0	7.1	7.2	7.3	7.3	7.4
Per capita apparent consumption (*kg*)	18.5	19.2	19.5	19.9	20.2	20.3

[a] Excludes aquatic mammals, crocodiles, alligators and caimans, seaweeds and other aquatic plants.
[b] Utilization data for 2014–2016 are provisional estimates.
[c] Source of population figures: UN, 2015e.

Source: The State of Fisheries and Aquaculture in the World, 2018

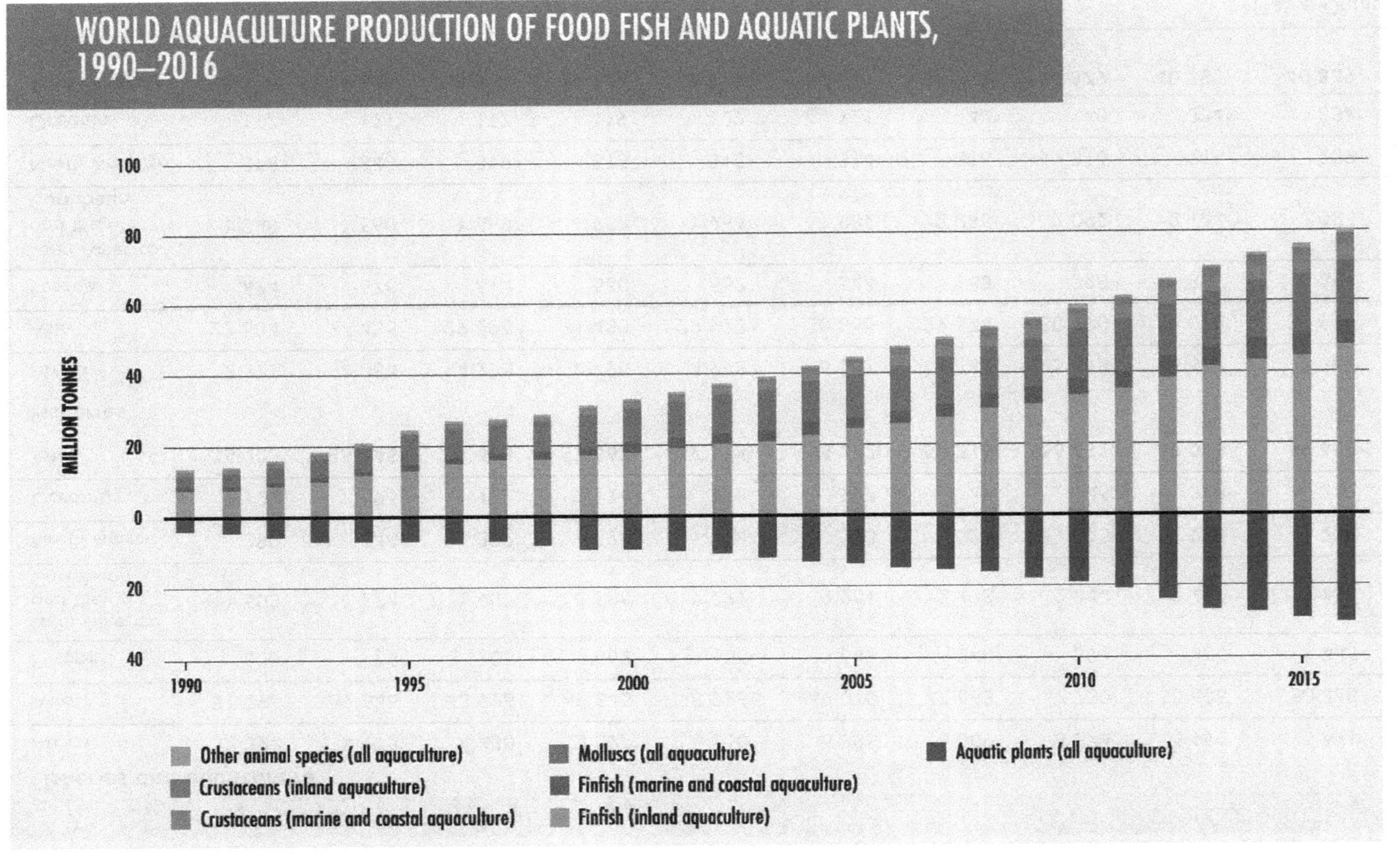

Source: The State of Fisheries and Aquaculture in the World, 2018

WORLD EMPLOYMENT FOR FISHERS AND FISH FARMERS BY REGION (thousands)

Region	1995	2000	2005	2010	2011	2012	2013	2014	2015	2016
Fisheries and aquaculture										
Africa	2 392	4 175	4 430	5 027	5 250	5 885	6 009	5 674	5 992	5 671
Asia	31 296	39 646	43 926	49 345	48 926	49 040	47 662	47 730	50 606	50 468
Europe	530	779	705	662	656	647	240	394	455	445
Latin America and the Caribbean	1 503	1 774	1 907	2 185	2 231	2 251	2 433	2 444	2 482	2 466
North America	382	346	329	324	324	323	325	325	220	218
Oceania	121	126	122	124	128	127	47	46	343	342
Total	**36 223**	**46 845**	**51 418**	**57 667**	**57 514**	**58 272**	**56 716**	**56 612**	**60 098**	**59 609**
Fisheries										
Africa	2 327	4 084	4 290	4 796	4 993	5 587	5 742	5 413	5 687	5 367
Asia	23 534	27 435	29 296	31 430	29 923	30 865	29 574	30 190	32 078	31 990
Europe	474	676	614	560	553	544	163	328	367	354
Latin America and the Caribbean	1 348	1 560	1 668	1 937	1 966	1 982	2 085	2 092	2 104	2 085
North America	376	340	319	315	315	314	316	316	211	209
Oceania	117	121	117	119	122	121	42	40	334	334
Total fishers	**28 176**	**34 216**	**36 304**	**39 157**	**37 872**	**39 411**	**37 922**	**38 379**	**40 781**	**40 339**

[Table Contd.

Contd. Table]

WORLD EMPLOYMENT FOR FISHERS AND FISH FARMERS BY REGION (thousands)

Region	1995	2000	2005	2010	2011	2012	2013	2014	2015	2016
Aquaculture										
Africa	65	91	140	231	257	298	267	261	305	304
Asia	7 762	12 211	14 630	17 915	18 373	18 175	18 088	17 540	18 528	18 478
Europe	56	103	91	102	103	103	77	66	88	91
Latin America and the Caribbean	155	214	239	248	265	269	348	352	378	381
North America	6	6	10	9	9	9	9	9	9	9
Oceania	4	5	5	5	6	6	5	6	9	8
Total fish farmers	**8 049**	**12 632**	**15 115**	**18 512**	**19 015**	**18 861**	**18 794**	**18 235**	**19 316**	**19 271**

Source: The State of Fisheries and Aquaculture in the World, 2018

The total number of fishing vessels in the world in 2016 was estimated to be about 4.6 million, unchanged from 2014. The fleet in Asia was the largest, consisting of 3.5 million vessels, accounting for 75 percent of the global fleet. Globally, the number of engine-powered vessels was estimated to be 2.8 million in 2016. In 2016, about 86 percent of the motorized fishing vessels in the world were in the length overall (LOA) class of less than 12 m, the vast majority of which were undecked, and those small vessels dominated in all regions. The fraction of fish stocks that are within biologically sustainable levels has exhibited a decreasing trend, from 90.0 percent in 1974 to 66.9 percent in 2015.

In 2015, maximally sustainably fished stocks (formerly termed fully-fished stocks) accounted for 59.9 percent and underfished stocks for 7.0 percent of the total assessed stocks. In 2015, among the 16 major statistical areas, the Mediterranean and Black Sea (Area 37) had the highest percentage (62.2 percent) of unsustainable stocks, closely followed by the Southeast Pacific 61.5 percent (Area 87) and Southwest Atlantic 58.8 percent (Area 41).

In 2016, of the 171 million tonnes of total fish production, about 88 percent or over 151 million tonnes were utilized for direct human consumption. In 2016, the greatest part of the 12 percent used for non-food purposes (about 20 million tonnes) was reduced to fishmeal and fish oil (74 percent or 15 million tonnes). Live, fresh or chilled is often the most preferred and highly priced form of fish and represents the largest share of fish for direct human consumption, 45 percent in 2016, followed by frozen (31 percent), prepared and preserved (12 percent) and cured (dried, salted, in brine, fermented smoked) (12 percent). In 2016, about 35 percent of global fish production entered international trade in various forms for human consumption or non-edible purposes.

The share of global fish production being exported as fish and fish products for human consumption has shown an upward trend, from 11 percent in 1976 to 27 percent in 2016. During the same period, world trade in fish and fish products also grew significantly in value terms, with exports rising from USD 8 billion in 1976 to USD 143 billion in 2016, at an annual growth rate of 8 percent in nominal terms and 4 percent in real terms.

China is the main fish producer and since 2002 has also been the largest exporter of fish and fish products, although they represent only 1 percent of its total merchandise trade. In per capita terms, food fish consumption has grown from 9.0 kg in 1961 to 20.2 kg in 2015, at an average rate of about 1.5 percent per year. Globally, fish and fish products provide an average of only about 34

KEEPING MOMENTUM TO ACHIEVE THE 2030 AGENDA

DELIVERABLES

2030: Increased economic benefits to SIDS and LDCs from sustainable use of marine resources (SDG target 14.7)

2025: Marine pollution significantly reduced (SDG target 14.1)

Fish mainstreamed into food security and nutrition policy by end of UN Decade of Action on Nutrition

2020: Marine ecosystems sustainably managed (SDG target 14.2)

An end to overfishing and IUU fishing (SDG target 14.4) and subsidies that contribute to them (SDG target 14.6), for earliest possible restoration of fish stocks

At least 10 percent of coastal and marine areas conserved (SDG target 14.5 and Aichi target 11)

2016: PSMA enters into force; data exchange operational at national, regional and international levels

2030

FAO Committee on Fisheries (COFI) every two years

2022: International Year of Artisanal Fisheries and Aquaculture (IYAFA)

2018: First International Day for the Fight Against IUU Fishing (every 5 June)

2017, 2020: UN Ocean Conferences

2016–2025: UN Decade of Action on Nutrition

2016: First Global Integrated Marine Assessment: World Ocean Assessment I

UN ACTIVITIES: RAISING AWARENESS, PROMOTING ACTION

Source: The State of Fisheries and Aquaculture in the World, 2018

calories per capita per day. However more than as an energy source, the dietary contribution of fish is significant in terms of high-quality, easily digested animal proteins and especially in fighting micronutrient deficiencies. A portion of 150 g of fish provides about 50 to 60 percent of an adult's daily protein requirement. Europe, Japan and the United States of America together accounted for 47 percent of the world's total food fish consumption in 1961 but only about 20 percent in 2015. Of the global total of 149 million tonnes in 2015, Asia consumed more than two-thirds (106 million tonnes at 24.0 kg per capita).

Global export and import of fisheries Products

Export Status

Asia is the largest region of seafood export in the world with a trade value of 59.2 US billion dollars in 2017. Its estimated trade value may increase and reach 61.1 US billion dollars in 2018 and decrease again to forecasted trade value of 59.4 US billion dollars in 2019. Europe is another major region with seafood export trade value of 55 US billion dollars in 2017-18. China alone has contributed about 14.76% to the total global export of seafood valuing 23.1 US billion dollars in 2017. Norway had seafood export worth of 11.3 US billion dollars with contribution of 7.22% to the global exports of seafood. Among the top five seafood exporters of the world; Vietnam is the only country which showed consistent positive estimated & forecasted growth in 2018 & 2019. Vietnam exported seafood products worth 8.5 US billion dollars in 2017 and estimated export trade of 9 US billion dollars with forecasted export trade of 9.5 US billion dollars in 2019. Ecuador is one of the fast-growing seafood exporters. India is one of the largest exporters of seafood products and exported seafood commodities worth 7.2 US billion dollars in 2017. India's share in global seafood export market was 4.60% in 2017.

Import Status

The import of seafood trade globally was worth 146.3 US billion dollars in 2017. Europe was one of the largest importers of seafood that consumed seafood commodities worth 61.3 US billion dollars. China, USA and Japan imported seafood products worth of 15.9, 21.6 & 15 US billion dollars respectively in 2017. China appears to be one of the promising destinations of seafood trade as forecasted figures show that it may touch exports worth 22.2 US billion dollars in 2019 compared to 15.9 US billion dollars in 2017.

Region-wise Status of Export/Import of Seafood Commodity

Sr No	Region Name	Exports			Imports		
		2017 US $ billion	2018 estim. US $ billion	2019 f'cast US $ billion	2017 US $ billion	2018 estim. US $ billion	2019 f'cast US $ billion
1	ASIA	59.2	61.1	59.4	48.6	54.6	57
2	AFRICA	7.2	7.3	7.2	5	5.4	5.5
3	CENTRAL AMERICA	2.8	3	2.7	2	1.9	1.8
4	SOUTH AMERICA	16.5	18.1	18.4	2.9	3	2.9
5	NORTH AMERICA	12.5	12.1	11.8	24.6	25.6	25.5
6	EUROPE	55	57.9	57.6	61.2	65.3	62.3
7	OCEANIA	3.3	3.3	3.4	2	2	1.9
	WORLD TOTAL	**156.5**	**162.9**	**160.5**	**146.3**	**157.9**	**157**

Country-wise Status of Export/Import of seafood commodities

Sr No	Country Name	Exports			Imports		
		2017 US $ billion	2018 estim. US $ billion	2019 f'cast US $ billion	2017 US $ billion	2018 estim. US $ billion	2019 f'cast US $billion
1	European Union	35.5	37.2	36.2	55.8	59.3	56.5
2	China	23.1	24.2	22.7	15.9	19.9	22.2
3	Norway	11.3	12	11.9	1.2	1.3	1.2
4	Vietnam	8.5	9	9.5	1.7	1.9	2
5	India	7.2	6.9	6.6	0.1	0.1	0.1
6	Chile	6	6.8	6.7	0.4	0.4	0.4
7	Thailand	6	6	5.7	3.5	3.9	3.7
8	United States of America	6.1	6	5.5	21.6	22.6	22.4
9	Canada	5.4	5.4	5.5	2.9	3	3.1
10	Ecuador	4.6	4.9	5.4	0.1	0.1	0.2
11	Russian Federation	4.5	4.5	5.4	2	2.2	2
12	Indonesia	4.2	4.5	4.4	0.4	0.4	0.4
13	Japan	2	2	1.8	15	15.3	15.5
14	Korea, Republic of	1.7	1.7	1.8	5.1	5.9	5.7

Source: tables derived from data of Food Outlook, Nov 2019

Global shrimp trade

Shrimp is one of the high- value and fastest trading seafood products globally and Indian fish economy is also largely dependent on export of shrimp. Farmed shrimp production in Asia in 2019 was likely to be lower than in 2018. In Latin America, increased production was expected in Ecuador. Shrimp imports were disappointing in the United States of America and the EU28 but strong demand from China kept the international shrimp trade stable in 2019.

Supply

Asian shrimp farmers remained conservative during the main aquaculture season between April and September of 2019, amid continued low market prices in the international trade. In India, where shrimp aquaculture is mainly export-oriented, production forecast for 2019 suggested a 30–40 percent decrease in comparison with 2018. In the main aquaculture region, Andhra, the often-unsuccessful price negotiations between farmers and processors/exporters, resulted in a much lower production in 2019. In Odisha, cyclone and floods disrupted farmed shrimp production during the second half of the year and the region of Tamil Nadu was affected by the unusual and extreme hot weather in 2019. Production trend in Gujarat and West Bengal remained moderate but insufficient to offset the falling supplies in the southern farming regions.

Farmed shrimp production in Ecuador continued to grow, which became evident in its increased export trade. The overall supply of sea-caught shrimp in Argentina was 16 percent below (compared to 2018) during the first six months of 2019. Subsequently, the Federal Fisheries Council of Argentina announced in September 2019 an early closure of the shrimp fishing season effective 15 October 2019.

International Trade

During the second half of 2019, international shrimp trade escaped another market crash supported by strong imports by China. However, the three other large traditional markets, the United States of America, the EU28 and Japan posted negative import growth during this period. Closely following the United States of America, China became the second largest shrimp importer in the world market during the first half of the year. Therefore, China is now the world's number one market for shrimp. China produces about one million tonnes of farmed shrimp annually and about less than 20 percent of those are exported. Farmers in Asia, as well as in Latin America, benefited much from the strong import growth in China.

In East Asia, stable local demand and firm prices of head-on shrimp also absorbed fresh shrimp in the regional markets.

Exports

Overall exports of shrimp declined from most countries in Asia, due to lower import demand in the United States of America, EU28 and Japan, although exports increased to China. Interestingly, Ecuador emerged as the top shrimp exporter, replacing India, during the first half of the year in 2019, which could be linked to its increasing production of farmed shrimp. For the first time in recent years, India reported a negative growth in shrimp exports.

Imports

Imports of the United States of America reduced during the first half of 2019. The trend was similar in most EU28 markets. In Asia, China's dominance in the import market was very strong. Imports in Viet Nam were nearly 60 percent below the previous year for the same period, following the large direct imports by China and stringent control by the Chinese authority on unreported border trade with Viet Nam. In the Middle East, shrimp imports also increased in the Gulf Cooperation Council (GCC) markets, as shown by the shrimp exports from India, the main supplier.

Market Situation

Overall, 2018 had been a year of further expansion for the fisheries and aquaculture sectors, with production, trade and consumption reaching historical peaks. The growth in production was due to an increase in capture fisheries (mainly of anchoveta in South America) and the continued expansion of aquaculture production, at 3-4% a year. Fish prices grew during the first part of 2018, driven by demand growing faster than supply for a number of key species, and weakened over the rest of the year due to increased supply and softening consumer demand in the United States and some European markets. The aggregate FAO Fish Price Index reached a record high in March 2018 (165 from a base of 100 in 2002-04), and then started to slightly decline. However, fish prices remained above 2017 levels for most species and products. These high prices, combined with sustained trade volumes, resulted in the value of total trade in fish and fish products peaking at USD 166 billion in 2018, more than a 7% increase compared with the previous year.

World top exporters and importers of shrimp (all types) (January-June)

Exporters	2018	2019	% change 2018/19
	(1 000 tonnes)		
Ecuador	242.7	315.1	+29.8
India	293.5	284.9	-2.9
Viet Nam(e)	140.0	120.0	-14.3
Indonesia	95.1	94.2	-0.9
China	89.9	74.9	-16.6
Thailand	78.2	76.9	-1.6
Argentina	63.7	53.3	-16.4

Importers	2018	2019	% change 2018/19
	(1 000 tonnes)		
EU28	365.3	358.7	-1.9
USA	303.3	301.5	-0.6
China	100.0	285.0	+185.8
Viet Nam(e)	220.0	90.5	-58.9
Japan	94.0	93.4	-0.7
Rep. of Korea	35.8	37.6	+5.4
Canada	23.2	24.3	+4.7

Source: *National data. Note: (e) Estimate*

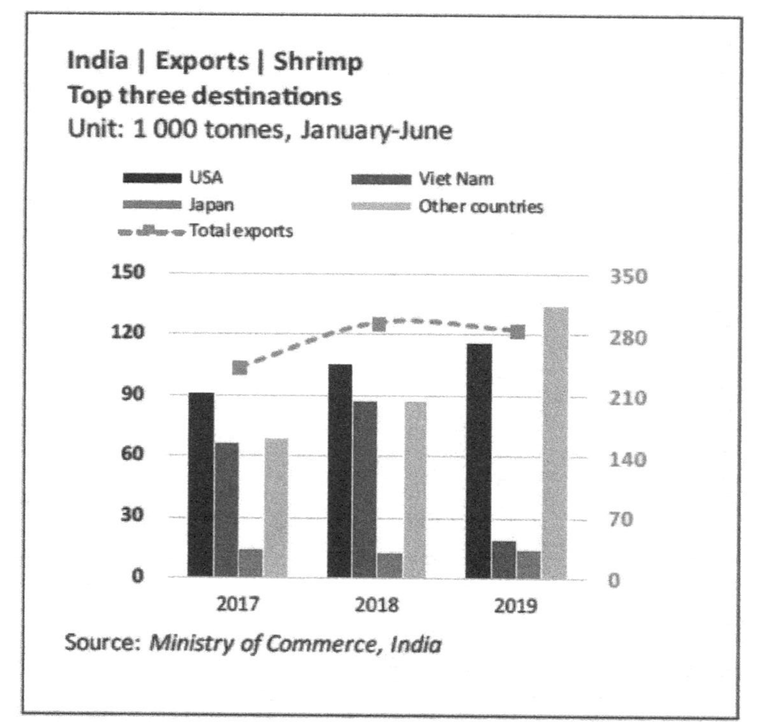

Source: *Ministry of Commerce, India*

Source: GLOBEFISH HIGHLIGHTS ISSUE 4, OCTOBER 2019

Global Fish Economy

Global fish production recorded a 3.4 percent decline in capture fisheries production in 2019 offset by a 4 percent increase in aquaculture harvests. Cephalopods and cod were among the wild stocks for which supplies have been tight. Anchoveta production was also lower in the first fishing season in Peru from April to July 2019. Meanwhile, the growth trajectory of the aquaculture sector remains steady. Supplies of the major farmed finfish species increased in 2019 by a forecasted increase in supplies of the major farmed finfish species – salmon, tilapia, pangasius, seabass and seabream. However, shrimp production in Asia was expected to drop sharply, particularly in India. Both aquaculture and capture harvests have been affected by higher water temperatures in 2019. Pressured by unfavourable macro-economic developments, in particular slower economic growth prospects and trade tensions, global trade in fish and fish products contracted in 2019, projected to drop by 1.2 percent in volume and 1.4 percent in value.

Although global per capita fish consumption continues to grow at around half a percent per year, world trade in fish and fish products was expected to contract in 2019 in volume and in US dollar terms. This loss of momentum was in line with forecasts issued by the World Trade Organization (WTO) for the lowest growth in total merchandise trade in a decade due to various economic headwinds. Many large economies were now on the brink of recession as a result of persisting trade tensions and further Brexit delays, leading to a general weakening of demand. The US dollar has generally been stronger in 2019 and this inevitably accentuates declines in the value of trade as expressed in other currencies.

The tariffs introduced by both China and the United States of America have negatively impacted bottom lines all along the supply chain for a number of heavily traded species, including lobster and tilapia. For other commodities, such as bivalves and small pelagics, the impact of trade tensions has been lower and the demand outlook is more positive. Aside from the direct impact of the tariffs on US-China trade flows, the wider geopolitical uncertainty is also translating into an increasingly cautious decision-making environment for seafood businesses, consumers and investors alike. From a somewhat more positive perspective, the potentially permanent transformations that are taking place in key markets such as those for cephalopods, lobster, groundfish and tilapia may boost aggregate demand in the longer term and foster new trading relationships as Chinese exporters seek out alternative markets and US buyers look for new suppliers.

After reaching record in 2018, The FAO Fish Price Index has fallen by 2.1 percent between January and September in 2019, compared with the same period in 2018 primarily due to price declines for many important farmed species, including

World Seafood Market Scenario

	2014	2015	2016	2017	2018 estim.	2019 f'cast (Nov)	Change 2019 over 2018	Change 2017 over 2014
Million Tonnes (Live weight) % %								
WORLD BALANCE								
Production	167.2	169.2	170.9	172.6	177.7	177.8	0.1	3.2
Capture Fisheries	93.4	92.6	90.9	92.5	94.5	91.3	-3.4	-1
Aquaculture	73.8	76.6	80	80.1	83.2	86.5	4	8.5
Trade Value (exports US $ Billion)	148.3	133.2	142.5	156.5	162.9	160.5	-1.5	5.5
Trade Volume (Live Weight)	60	59.6	59.5	64.9	65.1	64.3	-1.2	8.2
Total Utilization	167.2	169.2	170.9	172.6	177.7	177.8	0.1	3.2
Food	146.3	148.8	151.2	153.4	155.7	158.2	1.6	4.9
Feed	15.8	15.1	14.6	14.6	17.5	15	-14.3	-7.6
Other Uses	5.1	5.2	5.1	4.7	4.6	4.6	0	-7.8
SUPPLY AND DEMAND INDICATORS								
Per caput food consumption								
Food Fish (Kg/Year)	20.1	20.2	20.3	20.3	20.4	20.5	0.5	1
Capture Fisheries (Kg/Year)	10	9.8	9.5	9.7	9.5	9.3	-2.1	-3
Aquaculture (Kg/Year)	10.1	10.4	10.7	10.6	10.9	11.2	2.8	5

Source: Table derived from various reports of Food Outlook

shrimp, salmon, pangasius and tilapia, a consequence of both increased supplies and faltering demand. Prices were also weaker for canned tuna, with limited prospects for recovery although more positive indications are reported for fresh tuna.

Based on year-to-date performance, imports into the United States of America were expected to fall marginally in 2019, while Japanese import growth remains slow but positive. Price declines for multiple commodities imported into the EU28 will contribute to an estimated 2.8 percent fall in import value in 2019, a reversal of positive indications in 2017. Declines in imports were also projected for Latin America and most emerging economies in Asia, but China is the notable exception. The forecasted 12 percent increase in imports into China is somewhat unusual considering the broader trends and is explained largely by significant increases in shrimp imports from Ecuador and India. This is likely to be related to the Chinese government's crackdown on illegal (unreported) trade through Viet Nam.

On the export side, the challenging trade environment, combined with a drop in shrimp production in Asia and generally weaker prices, is contributing to lower export revenues for many large producers. India, Indonesia, the Philippines and Thailand were all expected to see a decline in export value in 2019. The world's largest seafood exporter, China, shall also see a decrease in export revenues due in large part to contractions in trade with the United States of America across multiple commodity categories.

Projection highlights of global fish production, consumption and prices in the coming decade

The report of OECD-FAO Agricultural Outlook 2019-28 highlighted projections regarding global fish production, consumption and prices for the coming decade (OECD-FAO,2019).

Production

The quantity of fish produced at the global level is projected to continue growing (1.1% p.a.), but at a slower rate than observed over the previous decade (2.4% p.a.). Among the major contributors to the slowdown, there are the impacts of the China's 13th five-year plan (2016-2020), affecting both capture and aquaculture production, and the downwardly revised Chinese production data since 2009. The relative and growing importance of aquaculture should continue and the average growth in this sector (2.0% p.a.) is expected to be the principal driver of growth in total fish production at the global level. By 2028, aquaculture is projected to produce substantially more fish than the capture sector (8.0 Mt). While inadequate governance and stock depletion of some fisheries will continue to be a concern at the global level, the quantity of fish produced by capture fisheries is projected to increase slightly over the outlook period (0.2% p.a.), partly supported by expectations that improved management conditions in several regions will continue to pay dividends.

The total quantity of fish produced at the world level is projected to be 196.3 Mt by 2028, an increase of 14% relative to the base period (average of 2016-18) and an additional 24.1 Mt of fish and seafood in absolute terms. While the total quantity being produced continues to increase, both the rate and absolute level of growth continue to fall. In absolute terms, growth in total fish and seafood production over this outlook period is projected to be 51% of that observed over the previous decade, when annual world production was 32.2 Mt higher by the final year.

The majority of growth in world fish and seafood production will continue to come from the aquaculture sector, where output is projected to increase by an average of 2 Mt per year to 102.2 Mt in 2028, an increase of 28% over the outlook period. While breaking the 100 Mt threshold for the first time in 2027 will be a milestone event for the aquaculture sector, the annual rate of aquaculture growth is expected to continue to slow over the next decade, projected at less than half that observed in the previous one (2.0% vs 4.6%). This is largely a consequence of how China's current five-year plan is expected to constrain output growth of its aquaculture industry. China's aquaculture production is expected to grow by 24% in the next decade, halving from the 54% increase in the previous decade. China accounted for 59% of global aquaculture production in the base period (average of 2016-18) and this is projected to fall to 57% by 2028, despite aquaculture's contribution to total Chinese fish and seafood production increasing from 75% to 82% in the same period, as levels of capture fisheries production fall (-14%). At the world level, the anticipated lower productivity gains in aquaculture, as a consequence of environmental regulations and a reduction in the availability of optimal production locations, will also contribute to lower production growth.

In contrast to the relative plateau of recent years, some growth in capture fisheries production is projected, resulting in world capture production at the end of the outlook period being 94.2 Mt, about 1.7 Mt larger than the average of 2016-18 and an increase of 1.9% compared to the base period. This increase is despite the anticipated reduction in capture fisheries production from China and is driven by expectations associated with better management in some regions (e.g. North and Latin America, Europe) and the relatively high price of fish driving demand.

At a country level, the greatest absolute increases in capture production are expected to occur in the Russian Federation and Indonesia. As a consequence of the overall increase in capture production, aquaculture is now not expected to surpass total capture production (including that utilised for non-food uses) until 2022, an event previously anticipated to occur in 2021. Assumed *El Niño* events in 2021 and 2026 result in world capture production falling in both years, as these periodic environmental events have a substantial impact on the pelagic fisheries of South American countries.

At the species group level, all forms of aquaculture production will continue to increase, but rates of growth will be uneven across groups and the importance of different species, in terms of quantities produced at the world level, will change as a consequence. By 2028, carp and molluscs, are projected to remain the most significant aquaculture groups and will together account for 55% of total production by 2028, 35.8% and 19.2% respectively. The dominance of these groups continues to diminish though, especially for molluscs, having slowly fallen from a combined peak of 77% in the mid-1990s, as production growth in other species has outpaced them. This pattern will continue over the next decade, with tilapia followed by catfish and pangassius (part of other freshwater and diadromous fish) projected to experience the highest rate of growth, at 3.4% p.a.

After a decade of little growth in fish oil production and falling levels of fishmeal production, both are expected to increase over the outlook period by 3.9% in 2028 compared to the base period in the case of fish oil and 10.6% in the case of fishmeal. This is despite the share of capture production that is reduced into fishmeal and fish oil not being expected to move much from its current level of around 16%. Growth is instead expected to be a consequence of the ongoing increase in the proportion of fishmeal and fish oil being sourced from fish waste, their relative higher prices; and the expected small increase in capture fisheries production. The proportion of fishmeal being produced from waste is projected to increase from 25% in 2018 to 31% by 2028, while for fish oil it is projected to increase from 35% to 40%.

Consumption

Fish is a versatile and heterogeneous commodity covering a wide variety of species. It can be prepared and consumed in many different ways and forms for either food or non-food uses. Marked differences also exist in how fish is utilised, processed and consumed within and between continents, regions and countries. The bulk of the utilisation of fish production is in the form of products for human consumption and this share is projected to grow from 89% in the base period

(2016-18) to 91% by 2028. Overall, the amount of fish for human consumption is projected to increase by 25 Mt by 2028, reaching 178 Mt. This represents an overall increase of 16% compared to the average for 2016-18, a slower pace when compared to the 32% growth experienced in the previous decade. This slowdown mainly reflects the reduced amount of additional production available, a deceleration in population growth and saturated demand in some countries, particularly developed ones, where food fish consumption is projected to show little growth (+0.6% p.a. by 2028).

It is projected that fish production will predominantly continue to be consumed as food (178 Mt in 2028), with only 9.4% utilised for non-food uses (mainly as fishmeal and fish oil). The share of fish for human consumption originating from aquaculture is projected to increase from 52% (average 2016-18) to 58% in 2028. The slowdown in world fish production growth means global food fish consumption is projected to increase by only 1.3% p.a., a substantial decline when compared to the 2.7% p.a. growth rate witnessed over the previous decade. World per capita apparent fish food consumption is projected to reach 21.3 kg per capita in 2028, up from 20.3 kg per capita in 2016-18. While Sub-Saharan Africa is expected to decline slightly in per capita terms, or remain static in the case of Africa overall, Latin America and Europe should show the highest growth rates.

Growth in demand will stem mostly from developing countries (in particular in Asia), which are expected to account for 93% of the increase in consumption and to consume 81% of the fish available for human consumption in 2028 (vs 79% in 2016-18). Overall, Asia is projected to consume 71% (or 126 Mt), of the total food fish, while the lowest quantities will be consumed in Oceania and Latin America. Asia will also continue to dominate growth in consumption, accounting for 71% of the additional fish consumed by 2028. This growth, in particular in eastern (minus Japan) and south-eastern Asia, will be driven by a combination of further increases in domestic production, in particular from aquaculture, rising income and increased commercialisation and a large, growing and increasingly urban population, which will push the intake of animal proteins, including fish, at the expense of foods of vegetal origin. Being the largest fish producer, China will remain by far the world's largest fish consuming country, projected to consume about 36% of the global total in 2028, with per capita consumption reaching about 44.3 kg compared with 39.3 kg in the base period.

Overall, growth in demand is also expected to be fuelled by ongoing changes in dietary trends, which should continue towards a greater variety in food choice along with increased health, nutrition and diet concerns. Fish, being a concentrated source of protein and of many other essential fatty acids and micronutrients, plays a particular role in this regard by providing a valuable and nutritious contribution to a diversified and healthy diet. Trade is expected to continue playing a major role as expanding the commercialisation of fish which will further help to reduce the impact of geographical location and limited domestic production, broadening the markets for many species and offering wider choices to consumer. Imports are projected to represent up to 69% of the food fish consumption in Europe and up to 71% in North America.

Per capita fish consumption should increase in all continents, except Africa where consumption is expected to remain static. A small decline is expected to occur in Sub-Saharan Africa. Despite an expected overall increase in total food fish supply (+30% for Africa and +31% for Sub-Saharan Africa compared to the base period), obtained through increased production and imports, it will not be sufficient to outstrip similar growth rates of the African population, with a consequent static or declining per capita fish consumption. This stagnation of per capita fish consumption for Africa as a whole and the decline for the Sub-Saharan region, raises an alarm in terms of food security, considering that the highest prevalence of undernourishment in the world is in Africa and that the food security situation has recently worsened, especially in parts of sub-Saharan Africa. Even if current per capita fish consumption in Africa is lower than the world average, fish plays a major role in the region, providing valuable micronutrients and proteins, at higher levels than the world average in the case of proteins.

Just under 10% of fish production is projected to be utilised for non-food uses. The greatest amount will be used to produce fishmeal and fish oil, and about 2% for ornamental fish, culturing, fingerlings and fry, bait, pharmaceutical inputs, or as direct feed for aquaculture, livestock and other animals. Fishmeal and fish oil can be processed from whole fish, fish trimmings or other fish by-products resulting from processing. Both fishmeal and fish oil are mainly used as animal feed in aquaculture and livestock breeding (in particular pigs), and as dietary supplements for human consumption and ingredients in the food industry. Currently about 70% of fishmeal is used as feed in aquaculture and this share increases to 75% in the case of fish oil. Furthermore, about 7% of fish oil is utilised as dietary supplement. No major significant changes are expected in the next decade, except for a potential increase in the share of fish oil being utilised as dietary supplement, that usually obtain higher prices.

Due to limited increase in production volumes, together with high prices and innovation efforts, fishmeal and fish oil will be more frequently used as strategic ingredients to enhance growth at specific stages of fish or livestock production, as they are considered the most nutritious and most digestible ingredients for fish and livestock breeding. Their inclusion rates in compound feeds for aquaculture have shown a clear downward trend, as they are used more selectively and are substituted by lower priced oilseed meal. By 2028, the oilseed meal used in aquaculture is projected to reach almost 9 Mt in 2028, compared to 4.4 Mt for fishmeal. Being by far the main aquaculture producer, China will continue to be the main consumer of fishmeal, accounting for about 38% of total consumption, while Norway will remain the main consumer of fish oil due to its salmon industry.

Fish and fish products (fish for human consumption and fishmeal) are amongst the most traded food items in the world. By 2028, export volumes of fish and fish products are projected to account for about 36% of total production (31% excluding intra-EU trade). World trade of fish for human consumption is projected to continue growing over the coming decade (+1.1 p.a.) but at a slower rate than in the past decade (+1.9% p.a.), reflecting the slowdown in production growth. The long-term trend, which has seen Asian countries steadily increasing their proportion of world trade in fish for human consumption, is projected to continue with 52% of world exports by 2028, compared with 49% in 2016-18. After experiencing a downward trend over the previous decade, world fishmeal trade is expected to grow over the outlook period, boosted by higher fishmeal production as greater quantities are recovered from fish waste and capture production increases slightly.

Fish prices

Fish prices are expected to remain relatively flat in real terms over the duration of the outlook period, as production constraints prevent prices falling to the same extent as expected in potential meat substitutes, such as poultry. Growth rates are projected to be within +/-1% p.a. in all cases; slightly negative for capture species, the world price for traded fish, and fishmeal; and slightly positive for aquaculture species and fish oil. When compared to the previous decade (2009-18), annual rates of price growth are all projected to be either lower or remain negative. In the wake of the higher price plateau observed for many agricultural products over the previous decade, prices of fishmeal and fish oil in particular have been historically high, a situation that is expected to continue into the foreseeable future. Growth in the weighted average real price of aquaculture species outpaces that of the price of low protein feeds such as maize. This is potentially a positive sign for profitability as low protein feeds represent a major input in the production of many aquaculture species. In nominal terms, fish prices are expected to gradually increase in all cases.

Ecolabelling in Seafood Sector

Seafood Trade

Seafood is one of the most important commodities in terms of value traded globally. In 2014, the value of the seafood economy was estimated at US$140 billion (Rabobank, 2015), with both the primary and secondary seafood sectors supporting an estimated 10 to 12 per cent of the world's population (FAO,2014b). The two main systems of production for seafood products are wild fish harvesting and aquaculture. Strategic development of the seafood economy, therefore, represents a major opportunity for securing more sustainable livelihoods. The dynamic context of seafood production systems represents an invaluable opportunity to transition to systems that foster and promote social, economic and environmental sustainability over the long term. Markets for sustainable fisheries products were an estimated US$12.9 billion in 2013.

Fisheries trade is a major economic driver in many developing nations, accounting, in some instances, for more than half of the total value of traded commodities (FAO, 2014). The global fisheries market is also volatile and therefore difficult to predict. Demand and supply factors, along with cost of production and transportation, as well as the value and supply of substitutes like meat and feed all influence fish prices and overall trade values. Fish prices have increased over the past decade (FAO,1998). The aggregate FAO Fish Price Index reached a record high in October 2013 (FAO, 2014).

Aquaculture is contributing to a growing share of international trade in fishery commodities for high-value species such as salmon, sea bass, sea bream, shrimp and prawns, bivalves and other molluscs, as well as low-value species such as tilapia, catfish (including pangasius) and carp (FAO, 2014). Developing countries typically have a fish trade surplus. Approximately 30 per cent of their total fish production is exported to the United States, Japan and the European Union (FAO, 2014) and is mostly made up of high-value species like shrimp and prawns, lobster, and tuna (Pérez-Ramírez, Phillips, Lluch-Belda, & Lluch-Cota, 2012).

What is Ecolabel?

Ecolabelling is a market-based tool to promote the sustainable use of natural resources. Ecolabels are seals of approval given to products that are deemed to have fewer impacts on the environment than functionally or competitively similar products. The ecolabel itself is a tag or label placed on a product that certifies that the product was produced in an environmentally friendly way. The label provides information at the point of sale that links the product to the state of the resource and/or its related management regime. Behind the label is a certification process. Organizations developing and managing an ecolabel set standards against which applicants wishing to use the label will be judged and, if found to be in compliance, eventually certified. The parent organization also markets the label to consumers to ensure recognition and demand for labelled products. The theory is that ecolabels provide consumers with sufficient information to enable them to recognize and choose environmentally friendly products.

A range of ecolabelling and certification schemes exists in the fisheries sector, with each scheme having its own criteria, assessment processes, levels of transparency and sponsors. What is covered by the schemes can vary considerably: bycatch issues, fishing methods and gear, sustainability of stocks, conservation of ecosystems, and even social and economic development. The sponsors or developers of standards and certification schemes for fisheries sustainability also vary: private companies, industry groups, NGOs, and even some combinations of stakeholders. A few governments have also developed national ecolabels. The first fisheries ecolabelling initiatives appeared in the early 1990s and were largely concerned with incidental catch, or bycatch, during fishing.

Major Ecolabels in Wild Catch Fisheries and Aquaculture

Eco-labelling in the seafood sector has evolved considerably from its humble roots of single-issue tuna labels in the 1970s. With the growth in consumer awareness of sustainability issues, retailers and manufacturers serving developed country markets have increasingly recognized value in affiliation with one or another sustainability standard. Standard compliant seafood production has grown consistently and dramatically as a percentage of global seafood production over the past decade. By 2015, certified production had reached 23 million metric tons, accounting for 14 per cent of global seafood production, up from 0.5 million metric tons (or 0.5 per cent of global) in 2003. From 2008 to 2015, certified seafood production grew at an annual rate of 30 per cent, over 10 times faster than total seafood production. Eighty per cent of certified seafood comes from certified wild catch production, reflecting the longer history of certification in wild catch markets but also the primacy of sustainability challenges in wild catch production due to issues related to stock management, which, to date, has been the primary driver behind seafood certification.

Two initiatives, FOS (Friend of the Sea) and the MSC (Marine Stewardship Council), dominate certification for wild catch markets, each accounting for 10 per cent of total wild catch seafood. As a consequence, these two initiatives also lead as a portion of global seafood production (including aquaculture) with FOS accounting for 6.2 per cent and the MSC accounting for 5.7 per cent of total seafood production (although of all the standards covered, only FOS and Naturland operate in both wild catch and aquaculture). GLOBALG.A.P., the leading aquaculture certification scheme, by contrast, accounted for 3.0 per cent of the global aquaculture market and 1.3 per cent of the global seafood market (2015).

Key Statistics: Certified Wild Catch and Aquaculture Production, 2015

Major international stanards	ASC, GAA, ChinaG.A.P., GLOBALG.A.P., MSC. Natural and Organic
Standard-complaint production	23 million mt (14% of global production, 58% from developing countries)
Top 5 standard-compliant producers and proportion of total	Peru (25%), United States (15%), Norway (11%), Chile (8%), Russia (6%) Total combined proportion: 65%
Top 5 standard-compliant species groups and proportion of total	Anchoveta (29%), cod (16%, including Alaska pollock, salmon (15%), mackerel (4%) Total combined proportin = 72%

Source: State of Sustainability Initiatives Review, Blue Economy (by International Institute of Sustainable Development), 2016

As of 2015, certified seafood made up more than 14 per cent of global seafood production. MSC and FOS certified production accounted for virtually all certified wild catch and for 80 per cent of global certified seafood. Six aquaculture certifications accounted for 20 per cent of certified seafood in 2015.

Historically, wild catch fisheries have provided the vast majority of seafood products available on global markets. At the international level, two certification systems, FOS and the MSC, compete for global market share, with each initiative accounting for roughly 50 per cent of total certified wild catch by 2015. These two initiatives alone certified 18.6 million metric tons of wild catch seafood, accounting for 20 per cent of total wild catch production and 80 per cent of the total certified seafood market. Total certified wild catch production has been growing at an annual rate of 36 per cent (2003–2015), significantly outpacing the relative stagnant growth across global wild catch markets over the same period. Although wild catch fisheries are present in most countries and 57 countries had some level of production certified under a sustainability standard in 2015, 70 per cent of certified wild catch production was sourced from 5 countries, with Peru and the United States accounting for 50 per cent of total certified wild catch. China, on the other hand, which accounts for 17 per cent of the global wild catch supply, is notably absent from the list of suppliers of certified wild catch production, with the exception of 60,000 metric tons of MSC certified yesso scallops, certified in 2015.

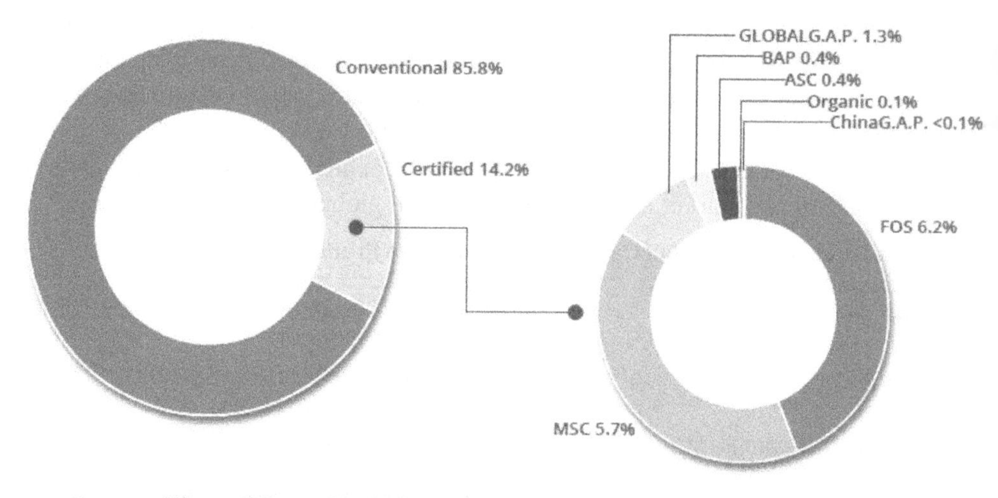

Data years: ASC, 2015; BAP, 2013; ChinaG.A.P., 2013; Conventional, 2013; FOS, 2014; GLOBALG.A.P., 2015; MSC, 2015; Organic, 2013.
Sources: FAO Fishstat, 2015; ASC, BAP, MSC, FOS, Naturland, GLOBALG.A.P., FiBL, ChinaG.A.P., personal communication, 2015.

Source: State of Sustainability Initiatives Review, Blue Economy (by International Institute of Sustainable Development), 2016

Majority of MSC-certified production being sourced from developed countries and the majority of FOS-certified production being sourced from developing countries.

Key Statistics: Wild Catch Production 2015

Global production	92. 6 million mt
Top 5 producers and proportion of total	China (17%), Indonesia (7%), Peru (69%) United States (6%), India (5%) Total combined proportion: 41%
Top 5 species groups produced and proportion of total	Anchoveta (9%), tuna (6%), cod (6%, 3% of which is Alaska pollock, sardines (4%), shrimp/prawns (4%) Total combined proportion: 29%
Major international standards	Friend of the Sea, Marine Stewardship Council
Standard-compliant production	18.6 million mt (20% of global production)
Top 5 standard-compliant producers	Peru (31%), Unites Stated (19%), Norway (8%), Russia (6%), Chile (6%) Total combined proportion: 70%
Top 5 standard-compliant species groups	Anchoveta (36%), cod (19% of which is Alska pollack, (tuna (10%, mackerel (5%, salmon (4%) Total combined proportion: 74%
Retail value of compliant production	US$7.9 billion

Source: State of Sustainability Initiatives Review, Blue Economy (by International Institute of Sustainable Development), 2016

Certified wild catch accounted for 20 per cent of global wild catch in 2015, with FOS and MSC certifying nearly equal portions of total certified production.

Certified Catch as Portion of Total Wild Catch

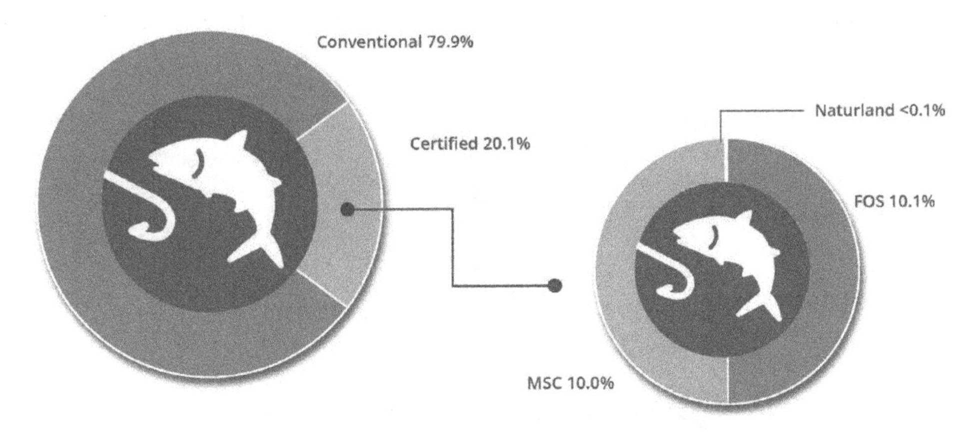

Source: State of Sustainability Initiatives Review, Blue Economy (by International Institute of Sustainable Development), 2016

Certified wild catch is growing substantially faster than conventional wild catch production. FOS has grown five times as fast as MSC over the last seven years. By 2015 the total production volumes of the two initiatives converged at just over 9 million metric tons, growing at a rate of around 6 per cent per annum (2014–2015).

The largest single source of certified wild catch is anchoveta, primarily destined for fish meal markets. Cod, tuna and salmon are the main certified wild catch species destined for retail markets. Overall certified wild catch production is concentrated in fewer species than global production as a whole, with the top five species groups accounting for 74 per cent of total certified wild catch.

In 2015, the MSC certified just over 9 million metric tons. The MSC has experienced rapid and consistent growth over the past seven years, with an annual average growth rate of 18 per cent and a reported retail value of US$4.5 billion in 2015.

FOS now operates as one of the most diversified seafood labelling initiatives certifying both aquaculture and wild catch fisheries. FOS production has grown at a rate of 91 per cent per annum between 2008 and 2015, reaching 9.3 million metric tons of FOS-certified wild catch seafood in 2015 (5.7 per cent of global; 10.1 per cent of total wild catch) and making it the single largest source of certified wild catch on the global market. FOS has certified the entire production of Peruvian and Chilean anchovies, which at a combined total of about 6 million metric tons of production per year accounts for about half of the world's fish meal production (Eurofish,2012). As a result, Peruvian fish meal exported to China, which at half a million metric tons per year (Rabobank, 2015) represents one of the largest trade flows in the entire seafood industry, is now almost entirely FOS certified.

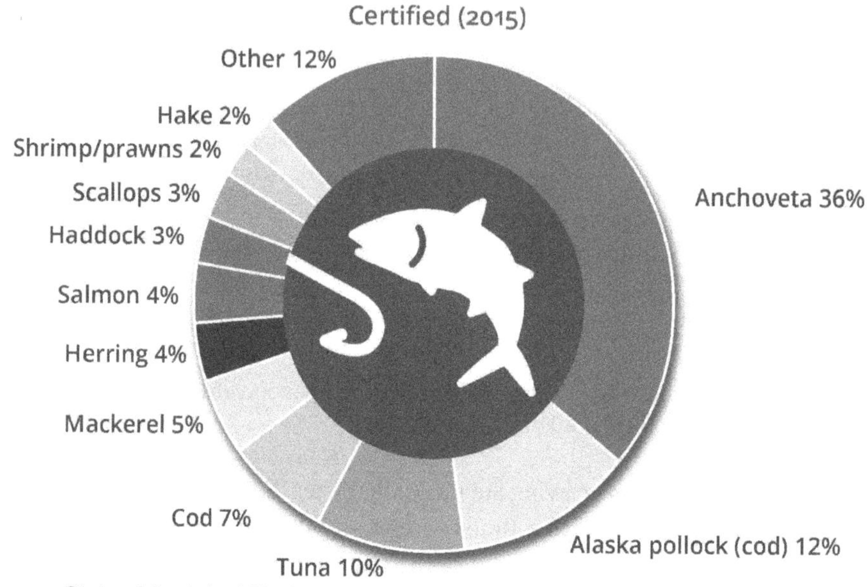

Source: State of Sustainability Initiatives Review, Blue Economy (by International Institute of Sustainable Development), 2016

Certified aquaculture production has grown exponentially, at an average rate of 76 per cent per year from 2003 to 2015, significantly outpacing the growth of conventional aquaculture. Six certification initiatives dominate global supply for certified aquaculture, supplying 4.5 million metric tons and accounting for 6 per cent of global aquaculture supply in 2015.

Key Statistics: Aquaculture Production 2015

Global production	70.2 million mt
Top 5 producers and proportion of total	China (60%), India (6%), Indonesia (5%), Vietnam (5%), Bangladesh (3%) Total combined proportion: 79%
Top 5 global species groups and proportion of total	Carp (39%), clams (8%), tilapia (7%), oysters (7%), shrimp/prawns (6%) Total combined proportion: 67%
Major international standards	ASC, GAA BAP, FOS, GLOBALG.A.P., Organic

[Table Contd.

Contd. Table]

Standard-compliant production	4.5 million mt (6% of total)
Top 5 standard-compliant producers	Norway (25%), Chile (19%), Spain (9%), Vietnam (8%), Italy (7%) Total combined proportion: 68%
Top 5 standard-compliant species groups	Salmon (56%), pangasius (10%), mussels (8%), tilapia (8%), shrimp/prawns (6%) Total combined proportion: 88%

Source: State of Sustainability Initiatives Review, Blue Economy (by International Institute of Sustainable Development), 2016

Certified aquaculture accounted for just over 6 per cent of total aquaculture production in 2015. GLOBALG.A.P. accounted for almost half of all certified aquaculture production, while BAP, ASC and FOS shared near-equal portions of the remainder.

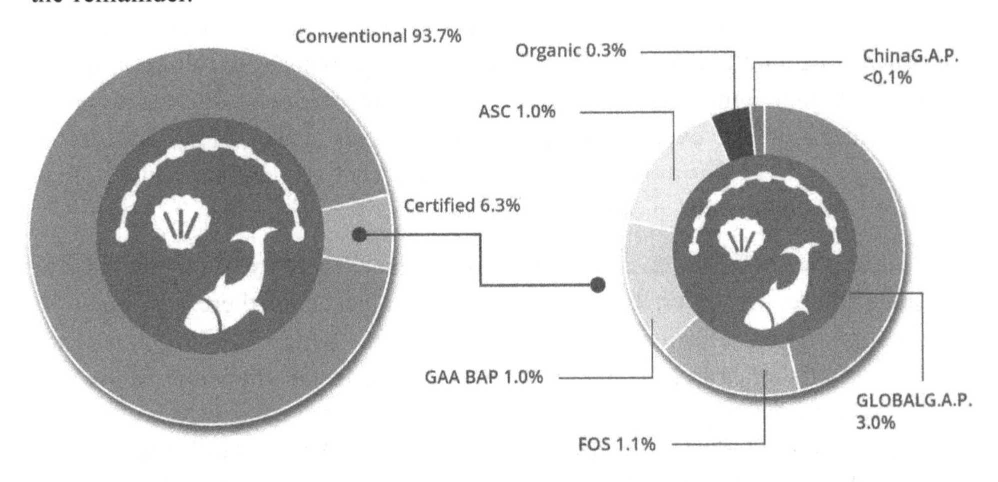

Source: State of Sustainability Initiatives Review, Blue Economy (by International Institute of Sustainable Development), 2016

Certified aquaculture is primarily aimed at a limited number of species with high commercial value. The largest single source of certified aquaculture in 2015 was salmon, accounting for 56 per cent of the global total. With only seven species groups accounting for more than 97 per cent of the global total. Certified aquaculture production is significantly more concentrated than global aquaculture production as a whole. Certified aquaculture displays a focus on a more limited number of species groups than wild catch certified production.

Seven species groups (salmon, pangasius, mussels, tilapia, shrimp/prawns, trout and sea bream) account for 97 per cent of certified aquaculture production, with salmon alone accounting for over half of total certified production. Initiatives serving the aquaculture sector tend to focus on the certification of a few species. The six leading certified species groups account for 23 per cent of conventional production, pointing to the particularities of the certified market. Notably, as the most widely produced aquaculture species, accounting for 39 per cent of global production, carp has no significant certified volumes.

GLOBALG.A.P. (Global Partnership for Good Agricultural Practices) aquaculture standards were first launched at the global level in 2004. GLOBALG.A.P. is by far the world leader in terms of volume of aquaculture certified. In 2015, the standard reported an estimated 2.1 million metric tons of reported compliant production, accounting for approximately 3 per cent of global aquaculture production. Between 2008 and 2015 GLOBALG.A.P. also reported significant growth at 29 per cent per annum.

FOS (Friend of the Sea), with about 700,000 metric tons certified in 2014 and 750,000 metric tons certified in 2015, represents the second-largest source of certified aquaculture as well as the second-fastest growing aquaculture initiative, with an average annual growth rate of 47 per cent between 2008 (its founding year) and 2015. The overwhelming majority (47 per cent) of FOS aquaculture production in 2014 was mussels. Trout and Arctic char account for another 35 per cent of FOS aquaculture production.

So far, the ecolabels have been related to exclusive activity of capture fisheries or aquaculture. However, fisheries is a large subject and there are many co-sectors associated with it. Friend of the Sea (FOS) which is one of the major ecolabels in fisheries and aquaculture of the world has taken the initiative in recent years and have developed many innovative sustainable standards. FOS is a Member Associate of INFOFISH. Its criteria follow the UN FAO - Guidelines for the Eco-labelling of Fish and Fishery Products from Marine Capture Fisheries. In particular, only products from stocks which are not overexploited can be certified.

Friend of the Sea under its international standards since long have covered majority areas as suggested by the Govt of India in development of sustainability, traceability standards and certification for fisheries/aquaculture, sustainable shipping, sustainable restaurants etc. that may prove useful for domestic sector development.

Friend of the Sea's mission, in line with the United Nations 2030 Sustainable Development Goals, is to protect the oceans by means of promoting sustainable fisheries, aquaculture and shipping. Friend of the Sea certification requirements have been recognized as top performers by several international benchmark studies. (Pew Charitable Trusts, Sea Ecology, Food & Water Watch, University of Tier/Froese and Proelss*). The European Commission lists Friend of the Sea among the main certification ecolabels for sustainable seafood in its "How do you choose your fish?" report.

ASC (Aquaculture Stewardship Council) certified production grew from 88,096 metric tons to 688,138 metric tons, making it the fastest-growing initiative in recent years, with a growth rate of 98 per cent per annum (2012–2015). ASC production to date has primarily focused on three species groups (salmon, tilapia and pangasius), which alone accounted for over 90 per cent of ASC certified production in 2015.

FAO in Relation to Ecolabels in Fisheries

FAO has developed three major instruments which are given below.

Guidelines for the Ecolabelling of Fish and Fishery Products from Inland Capture Fisheries

The Guidelines for the Ecolabelling of Fish and Fishery Products from Inland Capture Fisheries are of a voluntary nature. They are applicable to ecolabelling schemes that are designed to certify and promote labels for products from well-managed inland capture fisheries and focus on issues related to the sustainable use of fisheries resources. The guidelines refer to principles, general considerations, terms and definitions, minimum substantive requirements and criteria, and procedural and institutional aspects of ecolabelling of fish and fishery products from inland capture fisheries.

Guidelines for the Ecolabelling of Fish and Fishery Products from Marine Capture Fisheries

The Guidelines for the Ecolabelling of Fish and Fishery Products from Marine Capture Fisheries are of a voluntary nature. They are applicable to ecolabelling schemes that are designed to certify and promote labels for products from well-managed marine capture fisheries and focus on issues related to the sustainable use of fisheries resources. The guidelines refer to principles, general considerations, terms and definitions, minimum substantive requirements and criteria, and procedural and institutional aspects of ecolabelling of fish and fishery products from marine capture fisheries.

FAO Technical Guidelines on Aquaculture Certification

The Technical Guidelines on Aquaculture Certification have been developed by FAO upon the request of its Members attending the 3rd Session of the Committee on Fisheries (COFI) Sub-Committee on Aquaculture, held in India from 4-8 September 2006, through a consultative process with major stakeholders from governments, industry and civil society. The guidelines were approved by the 29th Session of COFI, held in Rome from 31 January to 4 February 2011. The guidelines provide advice on developing, organizing and implementing credible aquaculture certification schemes, which are viewed as potential market-based tools for minimizing negative impacts and increasing societal and consumer benefits and confidence in the process of aquaculture production and marketing.

National Sustainable Certification/Ecolabel for Fisheries/ Aquaculture in India

India has vast resources of fisheries and is one of the largest producers of seafood commodity in the world. India follows all the regulations as set by importing countries while exporting its seafood products and many companies follow voluntary ecolabel schemes during production subject to demand from buyer countries. Considering the requirement of a national sustainable certification/ecolabel for fisheries, a brainstorming session under the convenorship of Dr Iddya Karunasagar was held in 2015 with participation of representatives from QCI, BIS, MPEDA, CAA, ICAR, NFDB, different fisheries colleges and private sector. Recommendations made towards development of National Aquaculture Certification scheme in India were published by NAAS (NAAS,2015). Further, activities regarding certification, accreditation, traceability and labelling have been incorporated under the recently launched Pradhan Mantri Matsya Sampada Yojana (PMMSY),2020.

Certification, Accreditation, Traceability and Labelling under PMSSY, 2020

Background and broad activities covered under this sub-component of the scheme

A system of seed and feed certification and accreditation needs to be setup for finfish and shellfish. Emerging concerns of presence of antibiotics and residues in shrimp needs to be effectively addressed in order to ensure that the marine exports continue to show a sustained double-digit growth. A system of end to end traceability in fish needs to be urgently put in place including use of Block chain technology. Under PMMSY, special focus will be given for establishing a comprehensive traceability and labelling system using IT applications wherever required. Certification of aquaculture inputs including seed and feed, accreditation of production units like brood banks, farms, hatcheries, supporting extension systems, etc. will be supported. Any other need-based activity and infrastructure related to certification, accreditation, traceability and labelling in fish will be supported. The sub-component "Certification, Accreditation, Traceability and Labelling" has been covered with 100% central funding under PMMSY launched by Govt of India in 2020.

Aspects of accreditation are dealt by National Accreditation Board for Certification Bodies (NABCB). NABCB may approve certain Inspection Bodies (IBs)/ Certifying Bodies (CB) under Quality Management System (QMS), for fish/ shrimp hatcheries and feed mill with the following objectives:

- For setting quality standards for shellfish / finfish hatcheries/Feed mills in India and ensuring that their production process conforms to norms of quality seed/feed
- Economically empower the hatchery owners/ feed mill
- To ensure the availability and supply of quality fish/shrimp seed and feed to all farmers at a reasonable price
- To keep traceability of broodstocks and documentation of seed production in case of hatcheries and traceability of raw material, process documentation for feed mill

The Accreditation and Certification systems will be made mandatory to all hatcheries of Finfish/Shellfish (Shrimp, Crab etc) and feed mills in India – under both private and public sector that undertake breeding of fish/ shrimp hatcheries and feed mill. Department of Fisheries would also work out suitable model for traceability and labelling for hatcheries, seed farm, fish/shrimp farm/capture fish etc. Terms & Conditions of the sub-component under this scheme will be implemented on the basis of DPR/Self Contained Proposal recommended by CAC and approved by DoF. Detailed operational guidelines for this sub-component would be worked out and issued in due course.

Some Innovative Sustainable Standards Related to Fisheries & Allied Industries

Standards for sustainable UV cream is one of the recent innovations in sustainability. In order to minimize the impact of sunscreens on the marine environment, Friend of the Sea (FoS) decided to award as Friend of the Sea certified sunscreen products releasing less than 20% of their chemicals to the water after 80 minutes and thus validating the producer's long-term commitment towards environmental sustainability. The companies should not use chemicals that damage corals and marine life. Also, there should be an evidence of testing from independent laboratories.

Shipping is the biggest mode of transport and having sustainable standards for shipping is requirement of the industry. With 90% of the world's international trade occurring by sea, and the industry accounting for 3 to 4% of global CO_2 emissions, shipping has the potential to reduce its environmental impact. FOS has developed a new certification, with the aim to increase the responsible practices of the shipping industry. The criteria included in the certification scheme range from pollution prevention to fuel efficiency, from waste management to social accountability and legal compliance.

Ornamental species are tropical marine and freshwater species kept at home and public aquaria and include fish, corals, crustaceans, mollusks etc. Friend of the Sea sustainability certification of ornamental species makes consumers aware about conservation measures undertaken by suppliers to reduce their potential environmental impact. The Friend of the Sea certification contributes to conservation and sustainability of aquariums by certifying the animal welfare and promoting the environmental policy. AquaRio - Rio de Janeiro Marine Aquarium, in Brazil, is the first aquarium to obtain this sustainable aquarium certification, which is the first of its kind.

Dolphin and Whale watching represents an important contribution to the economy of many countries in the developed and developing world. It consists of boat trips to see a range of marine life including cetaceans, seals, birds and sharks. The rapid growth of the number of whales watching operators is known to potentially change cetaceans' behaviour and, migratory patterns. Friend of the Sea standard helps operators assess threats, promote best practices and support a responsible and sustainable industry.

The Sustainable Restaurants project has been created by Friend of the Sea to help people find the nearest restaurants serving sustainable certified seafood. The project selects and rewards those restaurants, take-away and catering chains that serve Friend of the Sea certified sustainable seafood products.

Friend of the Sea has established requirements to safeguard beaches. FOS criteria for sustainable beaches require: environmental awareness; proper waste disposal; no disposable plastic; water quality; respect for the natural ecosystem.

Salt can be obtained from the earth or from naturally occurring sea water. Friend of the Sea has created criteria and standard for sustainable salt which is unique and first of its kind in the world.

Conclusion

Global capture fisheries production decreased to 92.5 million tonnes in 2017 from 93.4 million tonnes of 2014. Global aquaculture production has shown 8.5% growth in 2017 over 2014 (73.8 million tonnes aquaculture production in 2014 and 80.1 million tonnes in 2017). The global per capita fish consumption reached 20.3kg/year in 2017 while the global per capita consumption from aquaculture reached 10.6kg/year in 2017 from 10.1kg/year in 2014 which showed a positive trend, maintaining consistency. Asia is the largest region in terms of exports with value worth 59.2 US $ billion while Europe is the largest region in terms of imports with value worth 61.2 US $ billion in 2017. EU excluded (as it's a bunch of countries), China was the world's largest exporter of seafood products in the world with export value of 23.1 US $ billion and USA was the world's largest importer of seafood products with import value of US $ 21.6 US $ billion in 2017.

Projection shows fish production will predominantly continue to be consumed as food (178 Mt in 2028), with only 9.4% utilised for non-food uses (mainly as fishmeal and fish oil) and the share of fish for human consumption originating from aquaculture is projected to increase from 52% (average 2016-18) to 58% in 2028. The total quantity of fish produced at the world level is projected to be 196.3 Mt by 2028, an increase of 14% relative to the base period (average of 2016-18). It is expected that aquaculture will cross 100 Mt threshold for the first time in 2027. In short, aquaculture will remain as one of the main food sectors supplying nutrition to the entire world for many years to come.

In a world in which the demand for fishery products is increasing in leaps and bounds, and the pressure on the natural resources is rising, ecolabelling appears to be a possible way to bring about a greater degree of control in the system. The increasing proportion of aquaculture in the production system for aquatic products is also being addressed by global organizations. Certified seafood products have potentially significant economic and sustainability impacts. The special capabilities of voluntary standards in managing credible traceability and conformity assessment protocols, combined with their ability to promote efficient implementation by leveraging market forces, gives them a special and invaluable role in promoting and verifying policy objectives in a cost-effective manner. In India, activities under Certification, Accreditation, Traceability and Labelling have been incorporated with 100% central funding under the recently launched PMSSY, 2020.

CHAPTER 2

OVERVIEW OF INDIAN FISHERIES AND ITS ROLE IN INDIAN ECONOMY

Present status of Indian fisheries

Fisheries sector plays an important role in the Indian economy. It contributes to the national income, exports, food, nutritional security and employment generation. The sector is important in the socio-economic development providing livelihood to around 1.6 crore number of people. The country is on its way to fully harness the potential of fisheries. Government of India has identified thrust areas to achieve the targeted production with sustainable development of the country in focussed manner. With this mandate, Government of India (GoI) created Ministry of Fisheries, Animal Husbandry and Dairying in 2019 for the sector development. India is endowed with a Coastline of 8118 kms; Exclusive Economic Zone (EEZ) of 2.02 million sq. km; and Continental Shelf Area of 0.53 million sq. km. Total fish Production in 2017-18 was 12.59 million metric tonnes: 3.69 (Marine) and 8.90 (Inland). India exported 1377.24 ('000 tonnes) of fisheries products earning a revenue of 45106.89 Crores in 2017-18. Gross Value Added (GVA) from Fisheries in 2017-18 was 111018 Crores forming 0.92% of total GVA of India.

Fish Production in India

Fish Production in India showed tremendous growth in the past seven decades. The total fish production of India increased from 7.52 lakh tonnes in 1950-51 to 125.9 lakh tonnes in 2017-18. Marine fish production increased from 5.34 lakh tonnes to 36.88 lakh tonnes during this period while the inland fish production increased from 2.18 lakh tonnes to 89.02 lakh tonnes. The percentage contribution of inland fish to the total landings showed significant increase from 29% to 71% while the share of marine fisheries reduced from 71% to 29% in the last several decades. The shift in landings reflect the potential of inland resources in Indian fisheries.

Percentage contribution of Inland and Marine fish production

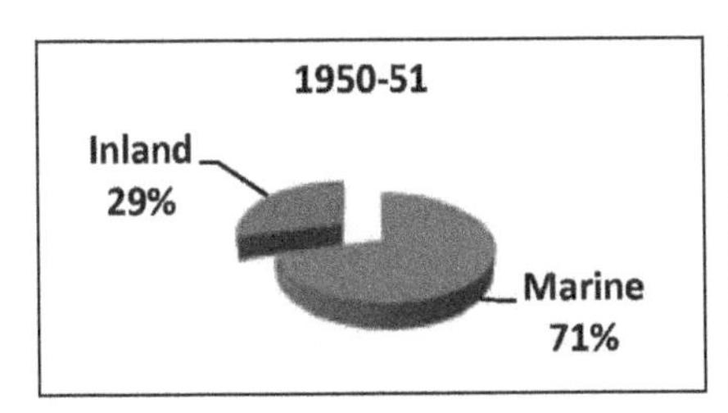

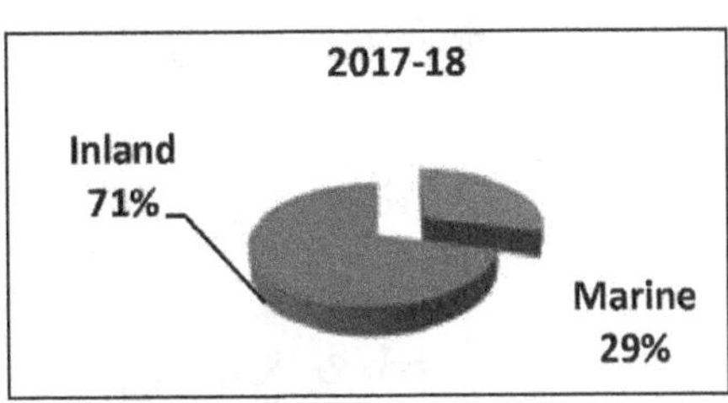

Source: Handbook of Fisheries Statistics, Government of India, 2018

Fish Production of India during 2005-06 to 2017-18 (Marine and Inland)

Fish production of India showed gradual increase during the last decade. However, growth in marine fisheries was minimal while there was significant growth in inland fisheries.

State-wise total fish production

Andhra Pradesh became the highest fish producing state in India with annual production of 34.50 lakh tonnes in 2017-18 followed by West Bengal (17.42 lakh tonnes) and Gujarat (8.35 lakh tonnes). These top three fish producing states contributed together 47.87% to the entire fish production of India.

State-wise (major) percentage contribution to total fish production 2017-18

Sr No	State Name	Fish Production (in lakh tonnes)	% Contribution to India's Fish Production
1	Andhra Pradesh	34.5	27.40%
2	West Bengal	17.42	13.84%
3	Gujarat	8.35	6.63%
4	Odisha	6.85	5.44%
5	Tamil Nadu	6.82	5.42%
6	Uttar Pradesh	6.29	5.00%
7	Maharashtra	6.06	4.81%
8	Karnataka	6.03	4.79%
9	Bihar	5.88	4.67%
10	Kerala	5.63	4.47%

Source: Table derived from Handbook of Fisheries Statistics, Government of India, 2018

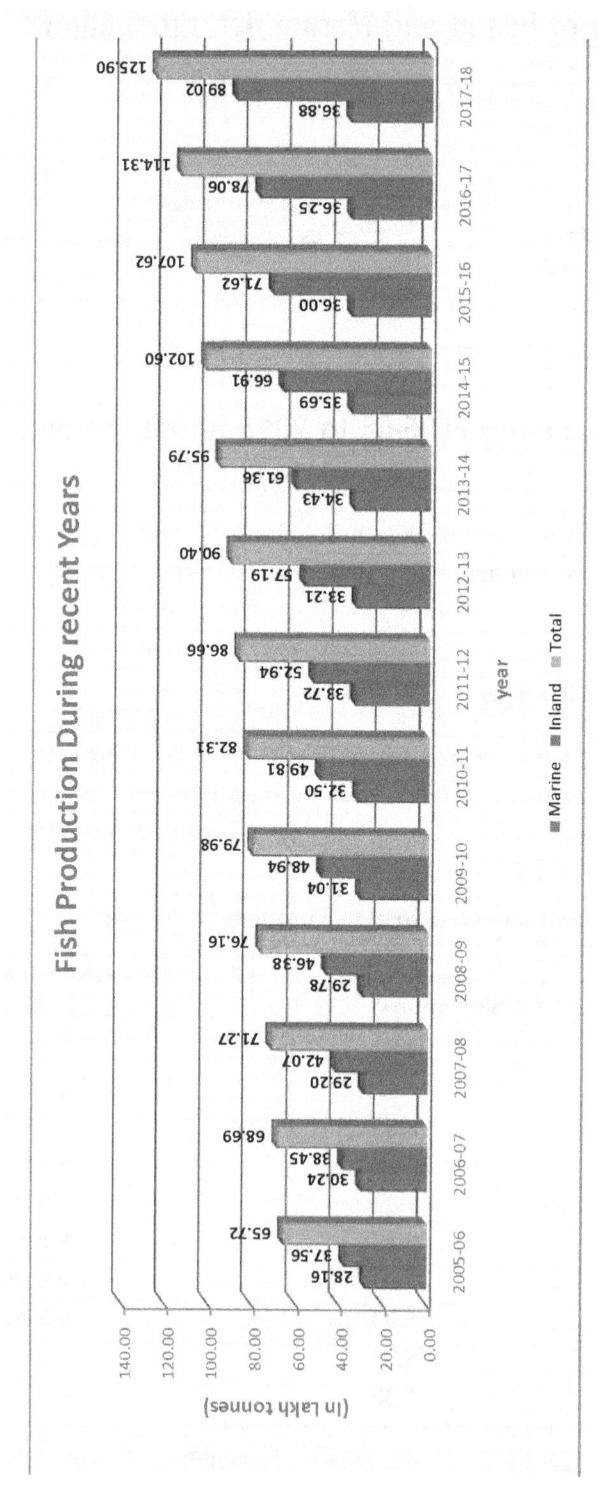

Source: Handbook of Fisheries Statistics, Government of India, 2018

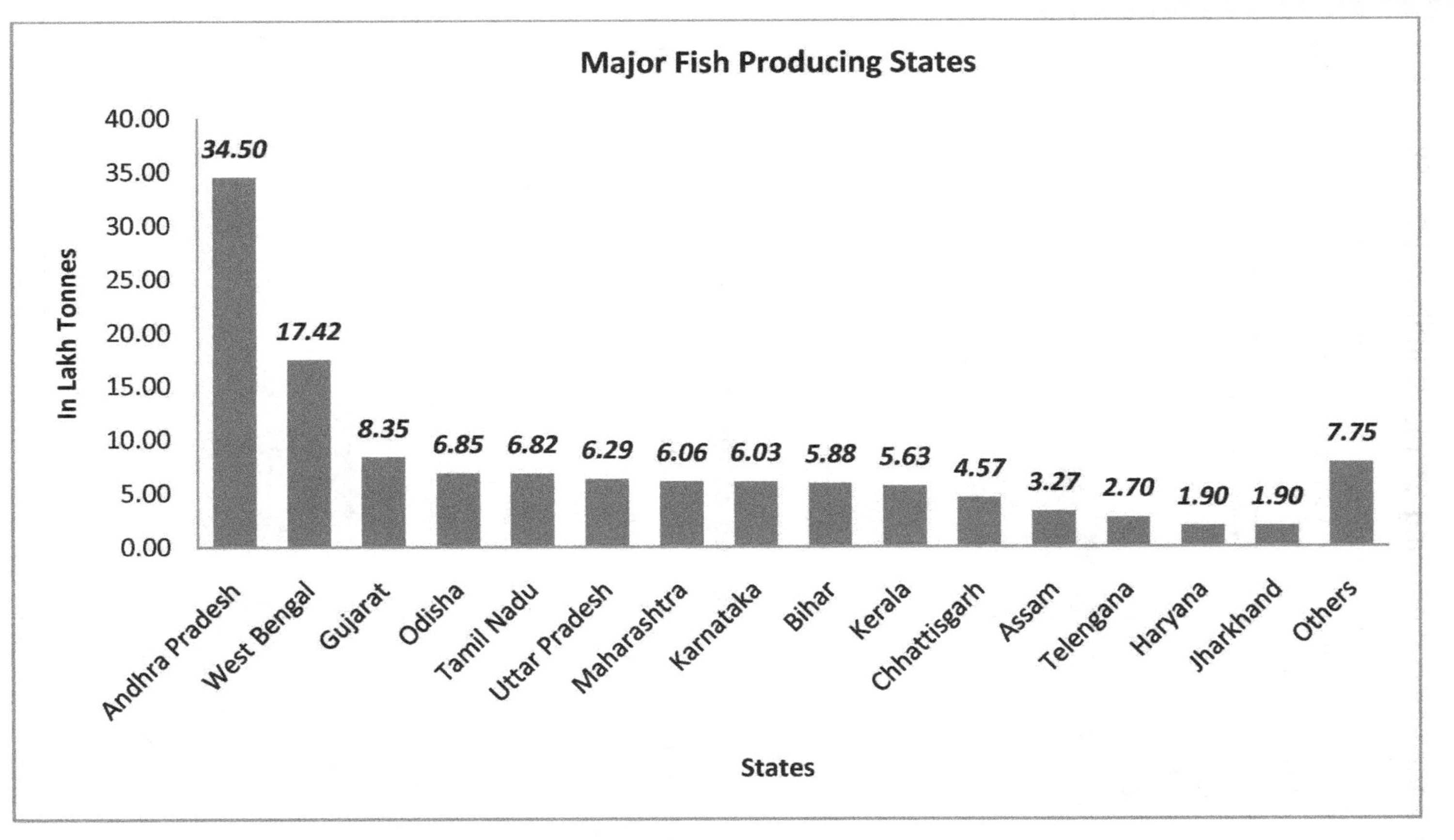

Source: Handbook of Fisheries Statistics, Government of India, 2018

Major Marine Fish Producing States of India in 2017-18

Gujarat showed the highest seafood production in marine sector with production of 7.01 lakh tonnes followed by Andhra Pradesh (6.05 lakh tonnes) and Tamil Nadu (4.97 lakh tonnes).

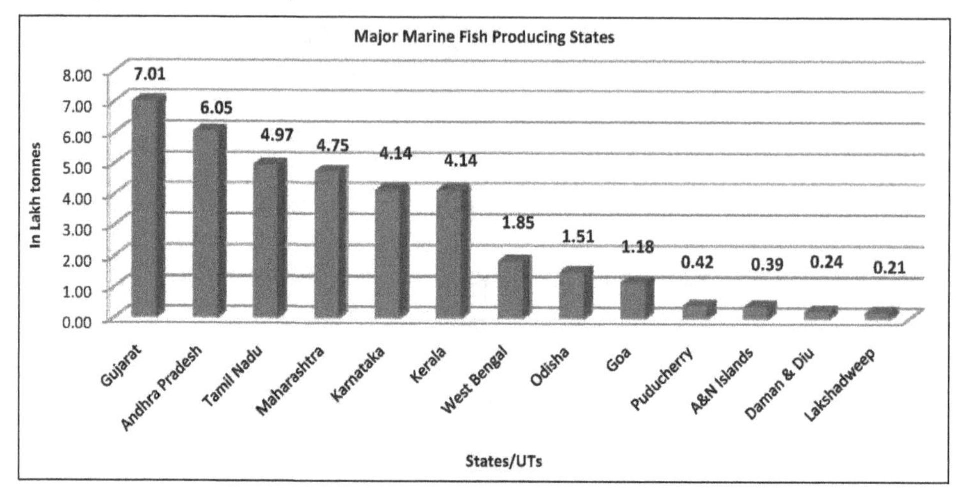

Source: Handbook of Fisheries Statistics, Government of India, 2018

Major Inland Fish Producing States of India in 2017-18

Andhra Pradesh picked up a top slot in inland fish production (28.45 lakh tonnes) followed by West Bengal (15.57 lakh tonnes) and Uttar Pradesh (6.29 lakh tonnes). Bihar produced 5.88 lakh tonnes followed by Odisha 5.34 lakh tonnes.

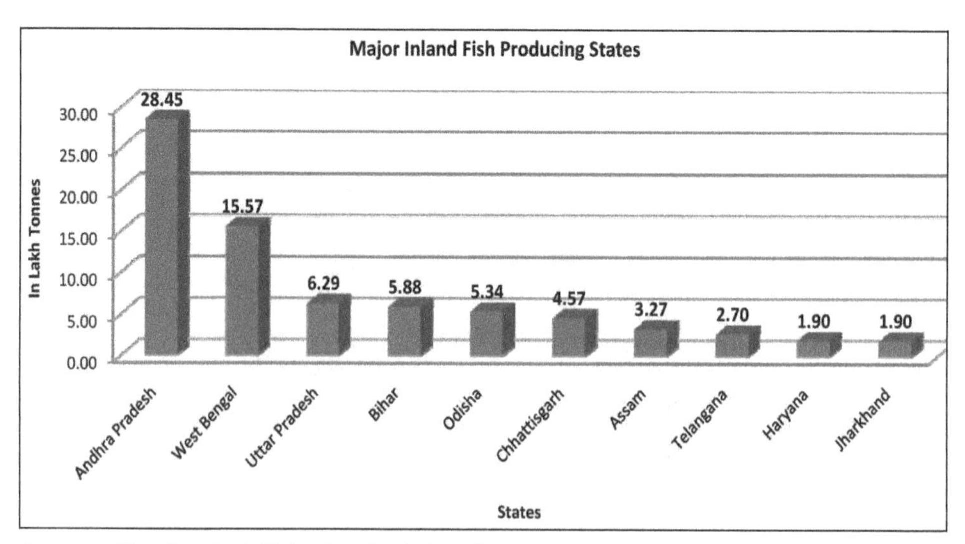

Source: Handbook of Fisheries Statistics, Government of India, 2018

Inland Fisheries Resource Framework of India as of 2017-18

- Total Length of Rivers and Canals: 388957 Kms
- Total Number of Reservoirs in the Country: 9058
- Total Area of Reservoirs in the Country: 3524724 Ha
- Total Area of Tanks and Ponds: 2478263 Ha
- Total Area of Brackish Water: 1160162 Ha
- Total Area of Bheels: 434850 Ha
- Total Area of Oxbow Lakes: 117800 Ha
- Total Area of Derelict Water Bodies: 230136 Ha
- Other Than Lakes and Canals Area: 300724 Ha

Source: DoF Handbook of Fisheries Statistics 2018

Percentage share of disposition of catch

Majority of the catch is marketed fresh (78%) and the remaining as frozen, canned and for miscellaneous purposes.

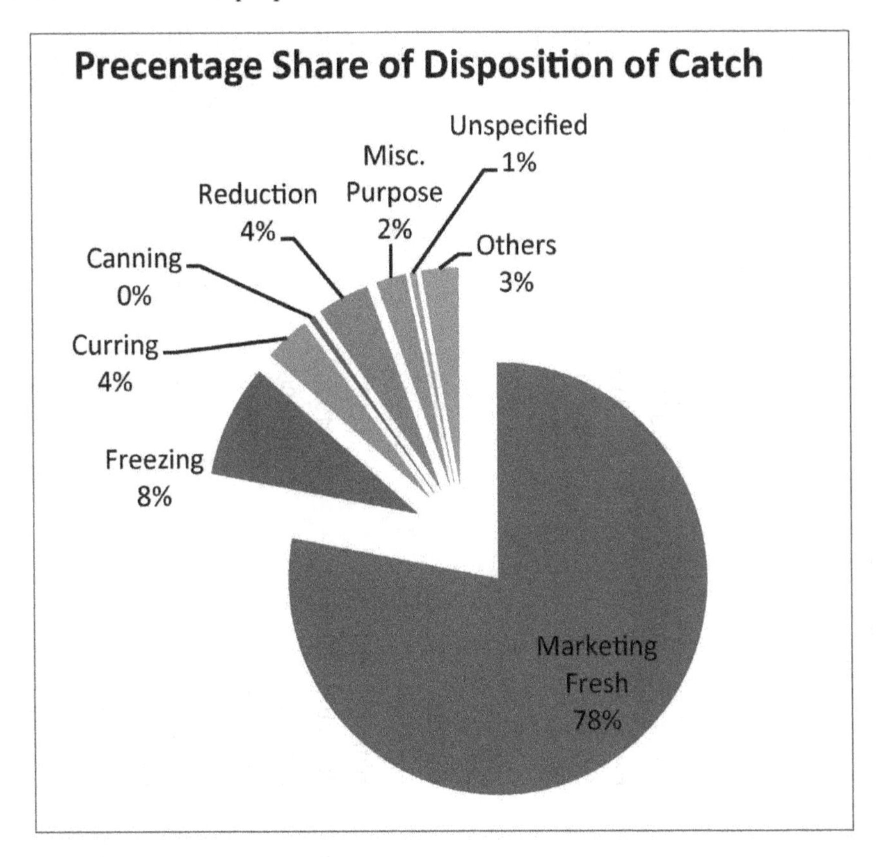

Source: Handbook of Fisheries Statistics, Government of India, 2018

State-wise landings of selected major inland fishes of India (2017) in tonnes

Sr No	State Name	Major Carps (Rohu, Catla, Mrigal)	Minor Carps	Exotic Carps (Common Carp, Silver Carp, Grass Carp)	Murrels	Catfishes (Wallago Attu, Pangassius)	Other FW Fishes	Total
1	Andhra Pradesh	1619710	0	0	62858	433262	312303	2428133
2	Arunachal Pradesh	2010	200	1820	0	0	220	4250
3	Assam	136221	53435	45408	9535	15770	42348	302717
4	Bihar	249449	111998	10182	35636	5091	96725	509081
5	Chhattisgarh	303967	60793	40529	0	0	0	405289
6	Goa	100	0	40	0	603	0	743
7	Gujarat	39388	9	0	1872	6603	36677	84549
8	Haryana	117800	24650	41800	0	3715	2035	190000
9	Himachal Pradesh	11431	0	0	0	0	344	11775
10	Jammu & Kashmir	1323	668	11899	484	698	5618	20690
11	Jharkhand	161800	1520	17200	0	9500	0	190020
12	Karnataka	116582	3763	52775	1882	4704	8468	188174
13	Kerala	323268	30431	60087	7895	18653	10836	451170
14	Madhya Pradesh	103156	10034	18785	481	4627	836	137921
15	Maharashtra	116223	18110	15472	10198	0	14946	174949
16	Manipur	9030	1290	13650	595	97	330	24992
17	Meghalaya	4303	915	3839	209	256	1579	11100
18	Mizoram	1274	1274	1274	1274	1274	1274	7643

[Table Contd.

Contd. Table]

Sr No	State Name	Major Carps (Rohu, Catla, Mrigal)	Minor Carps	Exotic Carps (Common Carp, Silver Carp, Grass Carp)	Murrels	Catfishes (Wallago Attu, Pangassius)	Other FW Fishes	Total
19	Nagaland	4369	58	4308	16	85	156	8991
20	Odisha	323268	30431	60087	7895	18653	10836	451170
21	Punjab	72324	3253	54221	0	5443	2070	137310
22	Rajasthan	39015	9130	4090	0	1800	0	54035
23	Sikkim	0	0	200	0	0	200	400
24	Tamil Nadu	76538	1527	21648	4370	6666	72099	182849
25	Telangana	180560	8905	27489	16809	15790	8554	258107
26	Tripura	46348	7718	15444	1537	3855	2311	77215
27	Uttarakhand	4546	0	0	0	33	0	4578
28	Uttar Pradesh	518000	94000	0	300	11600	4800	628700
29	West Bengal	996229	55365	192975	24600	43840	116725	1429734
30	Andaman & N Islands	95	0	4	0	0	121	220
31	Chandigarh	282817	56564	37709	0	0	0	377090
32	Dadra & Nagar Haweli	108	0	0	0	0	0	108
33	Daman & Diu	0	0	0	0	0	0	0
34	Delhi	690	0	0	0	0	0	690
35	Lakshadweep	0	0	0	0	0	0	0
36	Pondicherry	1322	280	627	306	518	606	3659
	Total	**5863263**	**586321**	**753560**	**188753**	**613136**	**753017**	**8758050**

Source: DoF Handbook of Fisheries Statistics 2018

Major carps are dominating in terms of total landings of inland fish production of India with a contribution of 67% followed by exotic carps (common carp, silver carp, grass carp) with a contribution of 9%. Other varieties include murrels, catfishes such as Wallago attu, pangasius etc. Andhra Pradesh is the top producer of inland fish in India and contributed around 28% to the entire production followed by West Bengal with contribution of around 16%.

Production of cultured shrimps in India (2017-18)

Production of cultured shrimps (2017-18) (in mt)

Sr No	State Name	P Monodon	L Vannamei	Total	% Contribution
1	West Bengal	49319	22191	71510	10.52%
2	Odisha	3887	37229	41116	6.05%
3	Andhra Pradesh	2714	456300	459014	67.50%
4	Tamil Nadu	25	43622	43647	6.42%
5	Kerala	1522	208	1730	0.25%
6	Karnataka	59	1465	1524	0.22%
7	Goa	0	78	78	0.01%
8	Maharashtra	0	6073	6073	0.89%
9	Gujarat	162	55161	55323	8.14%
	Total	**57688**	**622327**	**680015**	**100.00%**

Indian seafood exports

HS Code-wise export of Indian fishery products

India exports seafood commodity through different HS codes, HS Code 03 being the main HS Code dedicated for seafood. India is one of the main seafood exporters in the world. Codes for different varieties of seafood are given below.

1) **HS CODE 301**: LIVE FISH
2) **HS CODE 302**: FISH, FRESH OR CHILLED, EXCLUDING FISH FILLETS AND OTHER FISH MEAT OF HEADING 0304
3) **HS CODE 303**: FISH FROZEN EXCLUDING FISH FILLETS AND OTHER FISH MEAT OF HEADING NO 0304
4) **HS CODE 304**: FISH FILLETS AND OTHER FISH MEAT (WHETHER OR NOT MINCED), FRESH, CHILLED OR FROZEN
5) **HS CODE 305**: FISH DRIED SALTED OR IN BRINE; SMOKED FISH COOKED OR NOT BEFORE OR DURING THE SMOKING PROCESS; FISH MEAL FIT FOR CONSUMPTION

6) **HS CODE 306**: CRSTCNS W/N IN SHL, LIVE, FRSH, CHLD, FRZN, DRDSLTD/IN BRINE; CRSTCNS, IN SHL, CKD BY STMNG OR BOILING,W/N CHLD,FRZN,DRD,SLTD/IN
7) **HS CODE 307**: MOLUSCS W/N SHL, LIVE, FRSH, CHLD, FRZN, DRIED, SLTD/INBRINE; AQUATIC INVRTEBRTS EXCLCRSTCNS AND MOLUSCS LIVE, FRSH, CHLD, FRZ
8) **HS CODE 308**: SEA CUCUMBERS (STICHOPUS JAPONICUS, HOLOTHUROIDEA)

HS Code wise Export of Indian Fishery Products

HS Code 306 is the major sub-group (Crustaceans) under which 31663.50 Crore Rs & 30798.01 Crore Rs worth exports were made from India in 2017-18 & 2018-19 respectively. It is an important sub-group which contributed 71.68% & 70.26% to the total seafood export of India in 2017-18 & 2018-19 respectively. Though it is a strong sub-group, it showed a negative growth of -2.73% for the first time in 2018-19. Other major sub-groups are HS Code 303 (Frozen fish excluding fillets) & HS Code 307 (Molluscs) which contributed more than 10% individually to the total seafood export of India during 2017-18 & 2018-19. The HS Code 302 (fish fresh or chilled excluding fillets) showed substantial growth in 2018-19 (430.06 Crore Rs) with a growth of 37.81% compared to its export in 2017-18 (312.06 Crore Rs). HS Code 305 (fish dried salted) too showed growth rate of 37.07% in 2018-19 (Rs 667.41 Crore Rs) compared to previous year 2017-18 (486.91 Crore Rs). Seafood export contributed 1.90% to the total all-commodity export from India in 2018-19. HS Code 308 (sea cucumbers) has been introduced recently.

HS Code-wise Export of Indian Fisheries Products (in Rs Crores)

Sr No	HS Code	Commodity	2017-2018	2018-2019	% Growth
1	301	LIVE FISH	13.06	10.70	-18.07%
2	302	**FISH, FRESH OR CHILLED**, EXCLUDING FISH FILLETS AND OTHER FISH MEAT OF HEADING 0304	312.06	430.06	37.81%
3	303	**FISH FROZEN EXCLUDING FISH FILLETS** AND OTHER FISH MEAT OF HEADING NO 0304	4727.65	4837.64	2.33%
4	304	**FISH FILLETS AND OTHER FISH MEAT** (WHETHER OR NOT MINCED), FRESH, CHILLED OR FROZEN	1610.42	1848.13	14.76%

[Table Contd.

Contd. Table]

Sr No	HS Code	Commodity	2017-2018	2018-2019	% Growth
5	305	**FISH DRIED SALTED** OR IN BRINE; SMOKED FISH COOKED OR NOT BEFORE OR DURING THE SMOKINGPROCESS; FISH MEAL FIT FOR CONSUMPTIO	486.91	667.41	37.07%
6	306	**CRSTCNS** W/N IN SHL,LIVE,FRSH,CHLD, FRZN,DRDSLTD/IN BRINE;CRSTCNS, IN SHL,CKD BY STMNG OR BOILING, W/N CHLD,FRZN,DRD,SLTD/IN	31663.50	30798.01	-2.73%
7	307	**MOLUSCS** W/N SHL,LIVE,FRSH,CHLD, FRZN, DRIED,SLTD/INBRINE;AQUATIC INVRTEBRTS EXCLCRSTCNSAND MOLUSCS LIVE,FRSH,CHLD,FRZ	5386.18	5224.04	-3.01%
8	308	**SEA CUCUMBERS** (STICHOPUS JAPONICUS, HOLOTHUROIDEA)	6.00	16.18	169.67%
Total	**3**	**FISH AND CRUSTACEANS, MOLLUSCS AND OTHER AQUATIC INVERTABRATES.**	**44175.76**	**43832.19**	**-0.78%**
		India's All Commodities Export	**1956514.00**	**2307726.00**	**17.95%**
		% Share of Fisheries (HS Code 03) to All Commodities Export of India	**2.26%**	**1.90%**	

Source: Data derived from Government of India, Ministry of Commerce and Industry, Department of Commerce; updated as of 16.12.2019

Contribution of Sub Group HS Code to the Main Group HS Code 3 in the Export of Indian Fisheries Products

Sr No	HS Code	Commodity	2017-2018	2018-2019
1	301	**LIVE FISH**	0.03%	0.02%
2	302	**FISH, FRESH OR CHILLED**, EXCLUDING FISH FILLETS AND OTHER FISH MEAT OF HEADING 0304	0.71%	0.98%
3	303	**FISH FROZEN EXCLUDING FISH FILLETS** AND OTHER FISH MEAT OF HEADING NO 0304	10.70%	11.04%
4	304	**FISH FILLETS AND OTHER FISH MEAT** (WHETHER OR NOT MINCED), FRESH, CHILLED OR FROZEN	3.65%	4.22%
5	305	**FISH DRIED SALTED** OR IN BRINE;SMOKED FISH COOKED OR NOT BEFORE OR DURING THE SMOKINGPROCESS; FISH MEAL FIT FOR CONSUMPTION	1.10%	1.52%
6	306	**CRSTCNS** W/N IN SHL,LIVE,FRSH,CHLD,FRZN, DRDSLTD/ IN BRINE; CRSTCNS, IN SHL, CKD BY STMNG OR BOILING,W/ N CHLD, FRZN, DRD,SLTD/IN	71.68%	70.26%

[Table Contd.

Contd. Table]

Sr No	HS Code	Commodity	2017-2018	2018-2019
7	307	**MOLUSCS** W/N SHL, LIVE, FRSH,CHLD, FRZN, DRIED, SLTD/INBRINE;AQUATIC INVRTEBRTS EXCLCRSTCNS AND MOLUSCS LIVE, FRSH, CHLD, FRZ	12.19%	11.92%
8	308	**SEA CUCUMBERS** (STICHOPUS JAPONICUS, HOLOTHUROIDEA)	0.01%	0.04%
Total	**3**	**FISH AND CRUSTACEANS, MOLLUSCS AND OTHER AQUATIC INVERTABRATES.**	100%	100%

Source: Data derived from Government of India, Ministry of Commerce and Industry, Department of Commerce; updated as of 16.12.2019

Indian Seafood Export during 2005-06 to 2017-18

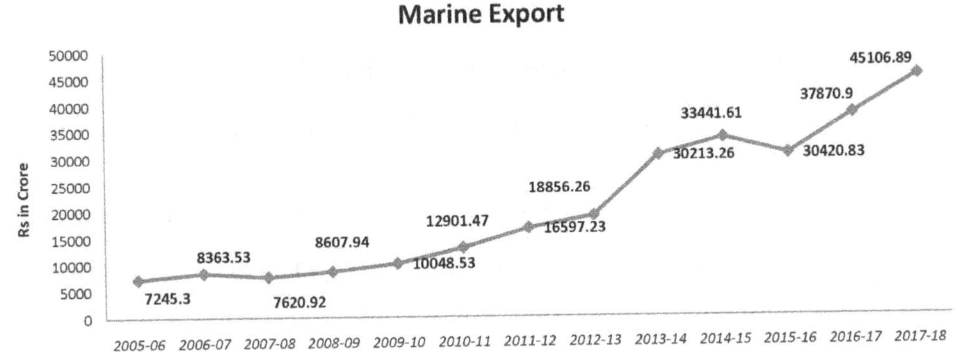

Source: Handbook of Fisheries Statistics, Government of India, 2018

Top ten Export Markets of Indian Fisheries Products

India has exported seafood commodities to 122 countries in 2017-18 which increased to 127 countries in 2018-19.

Top ten Export Markets of Indian Fisheries Products (Value in Rs Crores)

Sr No	Name of the Country	2017-18	2018-19	% Growth
1	U S A	13574.41	14002.58	3.15%
2	VIETNAM SOC REP	11893.64	6849.80	-42.41%
3	CHINA P RP	1043.89	5094.28	388.01%
4	JAPAN	2860.79	2832.07	-1.00%
5	THAILAND	1654.63	2234.46	35.04%

[Table Contd.

Contd. Table]

Sr No	Name of the Country	2017-18	2018-19	% Growth
6	U ARAB EMTS	1250.09	1254.47	0.35%
7	SPAIN	1575.75	1154.22	-26.75%
8	ITALY	1035.85	993.34	-4.10%
9	U K	1107.57	911.87	-17.67%
10	BELGIUM	855.49	784.56	-8.29%
	Total of Top 10	36852.12	36111.65	-2.01%
	India Total Fish Export	44,175.76	43,832.19	-0.78%
	% Contribution of Top 10 to Total Fish Export	**83.42%**	**82.39%**	

Source: Data derived from Government of India, Ministry of Commerce and Industry, Department of Commerce; updated as of 16.12.2019

Countries with Notable Export Growth from India for HS Code 03

The seafood export to some of the countries showed excellent growth; prominent of them are listed below.

Countries with Notable Export Growth from India for HS Code 03 (Value in Rs Crore)

Sr No	Country Name	2017-18	2018-19	% Growth
1	HONG KONG	539.20	683.56	26.77%
2	TAIWAN	373.58	454.77	21.73%
3	KUWAIT	103.43	202.46	95.74%
4	BANGLADESH PR	94.35	124.64	32.10%
5	SOUTH AFRICA	60.85	100.67	65.45%
6	BELARUS	61.79	97.07	57.09%
7	BAHARAIN IS	36.06	84.42	134.10%
8	ALGERIA	47.40	70.44	48.62%
9	MALDIVES	39.42	67.73	71.81%
10	CAMEROON	1.12	52.63	4601.20%
11	PUERTO RICO	29.54	41.54	40.62%
12	NEW ZEALAND	13.39	23.91	78.60%
13	IRAN	9.37	21.55	130.08%
14	NAMIBIA	2.06	20.85	913.86%
15	LIBYA	1.12	14.63	1200.59%

Source: Data derived from Government of India, Ministry of Commerce and Industry, Department of Commerce; updated as of 16.12.2019

Export of HS CODE 303: FISH FROZEN EXCLUDING FISH FILLETS AND OTHER FISH MEAT OF HEADING NO 0304

Top ten Export Markets of Indian Fisheries Products HS Code 303 (Value in Rs Crores)

Sr No	Name of the Country	2017-18	2018-19	% Growth
1	CHINA P RP	287.93	1,860.32	546.10%
2	THAILAND	941.11	1,237.04	31.44%
3	VIETNAM SOC REP	2,580.53	803.51	-68.86%
4	TUNISIA	160.14	161.37	0.77%
5	MALAYSIA	117.01	96.41	-17.61%
6	U S A	49.94	72.77	45.71%
7	CAMEROON	1.11	52.63	4641.44%
8	HONG KONG	53.58	48.19	-10.06%
9	TAIWAN	41.64	40.62	-2.45%
10	ALGERIA	27.75	37.32	34.49%
	Total of Top 10	4260.74	4410.18	3.51%
	India Total Fish Export	44175.76	43832.19	-0.78%
	% Contribution of Top 10 to Total Fish Export	**9.64%**	**10.06%**	

Source: Data derived from Government of India, Ministry of Commerce and Industry, Department of Commerce; updated as of 16.12.2019

Export of HS CODE 306

CRSTCNS W/N IN SHL, LIVE, FRSH, CHLD, FRZN, DRDSLTD/IN BRINE; CRSTCNS, IN SHL,CKD BY STMNG OR BOILING, W/N CHLD, FRZN, DRD, SLTD/IN

Top 10 Export Markets of Indian Fisheries Products HS Code 306 (Value in Rs Crores)

Sr No	Name of the Country	2017-18	2018-19	% Growth
1	U S A	13,155.78	13,496.16	2.59%
2	VIETNAM SOC REP	7,955.19	4,798.03	-39.69%
3	CHINA P RP	640.79	2,887.62	350.63%
4	JAPAN	2,241.22	2,218.91	-1.00%
5	U ARAB EMTS	1,135.23	1,145.56	0.91%
6	U K	1,010.05	822.79	-18.54%
7	BELGIUM	805.84	739.16	-8.27%

[Table Contd.

Contd. Table]

Sr No	Name of the Country	2017-18	2018-19	% Growth
8	CANADA	770.17	701.03	-8.98%
9	NETHERLAND	727.05	620.09	-14.71%
10	RUSSIA	371.59	393.58	5.92%
	Total of Top 10	28812.91	27822.93	-3.44%
	India Total Fish Export	44175.76	43832.19	-0.78%
	% Contribution of Top 10 to Total Fish Export	**65.22%**	**63.48%**	

Source: Data derived from Government of India, Ministry of Commerce and Industry, Department of Commerce; updated as of 16.12.2019

Import of fishery products in India

The contribution of import of fisheries products to the all-commodity import segment of India is very much negligible. Imported seafood is one of the upcoming promising trade of seafood business which is gradually picking up. With increasing number of Indians travelling across the globe, the benefits of eating healthier seafood (e.g. Atlantic salmon) is fast spreading and in turn resulting in growth of the market. Having multibillion-dollar food service, hotel, retail & online industry in India – there is tremendous potential for imported seafood commodity sector. India imported seafood commodity worth of Rs 745 crore in the year 2018-19 with 27.72% growth compared to 2017-18.

HS Code-wise Import of Fisheries Products in India (in Rs Crores)

Sr No	HS Code	Commodity	2017-2018	2018-2019	% Growth
1	301	LIVE FISH	28.76	34.96	21.53
2	302	FISH, FRESH OR CHILLED, EXCLUDING FISH FILLETS AND OTHER FISH MEAT OF HEADING 0304	106.97	109.82	2.67
3	303	FISH FROZEN EXCLUDING FISH FILLETS AND OTHER FISH MEAT OF HEADING NO 0304	69.29	72.55	4.7
4	304	FISH FILLETS AND OTHER FISH MEAT (WHETHER OR NOT MINCED), FRESH, CHILLED OR FROZEN	138.15	190.44	37.84
5	305	FISH DRIED SALTED OR IN BRINE; SMOKED FISH COOKED OR NOT BEFORE OR DURING THE SMOKINGPROCESS;FISH MEAL FIT FOR CONSUMPTIO	28.42	26.51	-6.71

[Table Contd.

Contd. Table]

Sr No	HS Code	Commodity	2017-2018	2018-2019	% Growth
6	306	CRSTCNS W/N IN SHL,LIVE,FRSH,CHLD, FRZN,DRDSLTD/IN BRINE;CRSTCNS,IN SHL, CKD BY STMNG OR BOILING,W/N CHLD, FRZN,DRD,SLTD/IN	187.44	274.09	46.23
7	307	MOLUSCS W/N SHL,LIVE,FRSH,CHLD,FRZN, DRIED,SLTD/INBRINE;AQUATIC INVRTEBRTS EXCLCRSTCNSANDMOLUSCS LIVE,FRSH, CHLD,FRZ	24.34	36.67	50.65
8	308	SEA CUCUMBERS (STICHOPUS JAPONICUS, HOLOTHUROIDEA)		0.05	
Total	3	FISH AND CRUSTACEANS, MOLLUSCS AND OTHER AQUATIC INVERTABRATES.	583.40	745.12	27.72

Source: Data derived from Government of India, Ministry of Commerce and Industry, Department of Commerce; updated as of 16.12.2019

Seafood Imports from top ten Countries (Value in Rs Crores)

Sr No	Name of the Country	2017-18	2018-19	% Growth
1	VIETNAM SOC REP	119.33	167.30	40.20%
2	BANGLADESH PR	136.76	159.35	16.52%
3	U S A	85.55	139.47	63.03%
4	OMAN	5.67	35.11	519.22%
5	MYANMAR	40.00	28.91	-27.73%
6	UNSPECIFIED	94.38	21.95	-76.74%
7	SINGAPORE	11.40	17.95	57.46%
8	NORWAY	15.97	16.40	2.69%
9	U K	15.99	14.46	-9.57%
10	ECUADOR		13.88	
	Total of Top 10	525.05	614.78	17.09%
	India Total Fish Import (in Rs Crore)	583.40	745.12	27.72%
	% Contribution of Top 10 to Total Fish Import	**90.00%**	**82.51%**	

Source: Data derived from Government of India, Ministry of Commerce and Industry, Department of Commerce; updated as of 16.12.2019

Vietnam is the top importer of seafood to India with a value worth 167.30 Crore Rs in 2018-19 followed by Bangladesh with an import value of Rs 159.67 Crore Rs. Oman emerged as the fourth largest importer of seafood to India with growth of 519.22% in 2018-19 and a value of 35.11 Crore Rs against import of 5.67 Crore Rs in 2017-18.

Economic activity in fisheries of India

Per Capita Income, Product and Final Consumption in India

Per Capita Income, Product and Final Consumption of India (2017-18) (Rs)

Population (in million)		1316	
At Current Prices		**At Constant Prices**	
Per Capita GDP	129901	Per Capita GDP	100151
Per Capita GNI	128497	Per Capita GNI	99043
Per Capita NNI	114958	Per Capita NNI	87623
Per Capita GNDI	131580		
Per Capita PFCE	76619	Per Capita PFCE	56364

Source: Table derived from Statement 1.2, Page No 08, National Account Statistics Report 2019, Ministry of Statistics and Programme Implementation (MoSPI), Government of India

Output of economic activity and capital formation by industry of use – Agriculture, Forestry and Fishing in India (at constant 2011-12 prices)

Agriculture, Forestry and Fishing	Value in Rs Crore
Output	2342177
Intermediate consumption	539139
GVA at basic prices	1803039
CFC	137584
NVA at basic prices	1665455
Gross Capital Formation	273755
GVA to Output Ratio	77
GCF to Output Ratio	11.7

Source: Table derived from Statement 1.5, Page No 13, National Account Statistics Report 2019, Ministry of Statistics and Programme Implementation (MoSPI), Government of India

Gross value added by economic activity specific to fisheries and aquaculture in India (at constant 2011-12 prices)

Item	2011-12	2012-13	2013-14	2014-15	2015-16	2016-17	2017-18
Fishing and Aquaculture	68027	71362	76487	82232	90205	99224	111018
Agriculture, Forestry and Fishing	1501947	1524288	1609198	1605715	1616146	1717467	1803039
Total GVA at basic prices	8106946	8546275	9063649	9712133	10491870	11318972	12104165
% Growth YOY	2011-12	2012-13	2013-14	2014-15	2015-16	2016-17	2017-18
Fishing and Aquaculture		4.90%	7.18%	7.51%	9.70%	10.00%	11.89%
Agriculture, Forestry and Fishing		1.49%	5.57%	-0.22%	0.65%	6.27%	4.98%
Total GVA at basic prices		5.42%	6.05%	7.15%	8.03%	7.88%	6.94%

Values are in Crore Rs

Contribution of Fisheries GVA (in %)	2011-12	2012-13	2013-14	2014-15	2015-16	2016-17	2017-18
To Agriculture GVA	4.53	4.68	4.75	5.12	5.58	5.78	6.16
To Total GVA	0.84	0.84	0.84	0.85	0.86	0.88	0.92

Source: Table derived from Statement 1.6 & 1.6A, Page No 21 & 23, National Account Statistics Report 2019, Ministry of Statistics and Programme Implementation (MoSPI), Government of India

Net Value Added (NVA) by Economic Activity of India (at constant 2011-12 prices)

Item	2011-12	2012-13	2013-14	2014-15	2015-16	2016-17	2017-18
Fishing and Aquaculture	60039	62915	67436	72535	79782	88000	98896
Agriculture, Forestry and Fishing	1406268	1421409	1497458	1486314	1491336	1586172	1665454
Total NVA at basic prices	7189771	7535614	7963039	8533489	9220979	9938018	10601202
% Growth YOY	2011-12	2012-13	2013-14	2014-15	2015-16	2016-17	2017-18
Fishing and Aquaculture		4.79%	7.19%	7.56%	9.99%	10.30%	12.38%
Agriculture, Forestry and Fishing		1.08%	5.35%	-0.74%	0.34%	6.36%	5.00%
Total NVA at basic prices		4.81%	5.67%	7.16%	8.06%	7.78%	6.67%

Values are in Rs Crores

Contribution of Fisheries NVA (in %)	2011-12	2012-13	2013-14	2014-15	2015-16	2016-17	2017-18
To Agriculture NVA	4.27	4.43	4.50	4.88	5.35	5.55	5.94
To Total NVA	0.84	0.83	0.85	0.85	0.87	0.89	0.93

Source: Table derived from Statement 1.7, Page No 27, National Account Statistics Report 2019, Ministry of Statistics and Programme Implementation (MoSPI), Government of India

The fishing and aquaculture GVA increased from 68027 Crore Rs in 2011-12 to 111018 Crore Rs in 2017-18 which is almost a growth of 63% in merely 07 years with a base year of 2011-12. Also, the contribution of fisheries GVA not only increased from 4.53% (2011-12) to 6.16% (2017-18) in Agriculture GVA but it also increased from 0.84% (2011-12) to 0.93% (2017-18) to the total GVA of India. This reflects the important role fisheries sector is playing in the Indian economy.

Net Value Added (NVA) is an important term in economic activities. The fishing and aquaculture NVA increased from Rs. 60039 crores in 2011-12 to 98896 crores in 2017-18 showing a substantial growth of 65% in the last 7 years with a base year of 2011-12. At the same time, the percentage contribution of fisheries NVA to agriculture NVA increased from 4.27% (2011-12) to 5.94% (2017-18) & from 0.84% (2011-12) to 0.92% (2017-18) to total NVA of India. Fisheries is one of the consistently fast-growing and high-performing sectors in the Indian economy and continue to play its key role in boosting development in the country.

Modern seafood trade in India

India is the third largest economy in the world in PPP terms (Euromonitor). The retail sector is emerging as one of the largest sectors of the Indian economy & contributes 10% to GDP and 8% to employment generation of India (IBEF, 2019). India ranked 63rd as per the World Bank's 'Doing Business 2020' report and 2nd according to the 2019 Global Retail Development Index.

Modern trade comprises of supermarkets, hypermarkets, large format chain of stores which cater to all kinds of household needs right from food, non-food, apparels etc. It includes both lead retail chains (like Big Bazar, Reliance, Aditya Birla Retail More, D-Mart etc) and cash and carry players like METRO Cash and Carry India Pvt Ltd, Best Price Modern Wholesale, Booker Wholesale, Reliance Cash and Carry etc. Modern Trade has strong presence across many cities of India. Normally retail chains are of B2C concept (Business to Consumer) whereas Cash and Carry segment is of B2B concept (Business to Business). Any individual can purchase from any of the retail stores as per their buying pattern; all kinds of customers are allowed to shop in retail stores where as in case of Cash and Carry segment, one may require to produce documents of business based on which that particular cash and carry store issues a card necessary while buying.

Fresh seafood is one of the important categories in 'perishables' division of any Hypermarket, Supermarket or Cash and Carry chains. Modern chains take extra care to keep the products fresh by following appropriate standards of quality as per norms. Though the price may be higher, buyer may find it worth as it assures good quality, hygiene and traceability of the products. Many of the lead Modern Trade and/or Cash & Carry players follow the HACCP standards which are usually being adopted for export of seafood products. It implies that the Indian consumer is more aware of eating healthy & safe seafood.

Demographically, India is the youngest consumer market with 33% of the population less than 15 years and 50% less than 24 years (Technopak). Customers visit both traditional 'mom and pop' (Kirana) stores and modern stores in the ratio 5:1 time on a weekly basis (India Retailing). Rising urbanization leading to changing lifestyles with less time to prepare food at home has changed the taste and preference of Indian consumers. Increase of tourism in India and international travel by Indians increased the interest of Indians in international food products in addition to innovative advertisements, rise in supermarkets and e-commerce boom, increased awareness among consumers and easy accessibility of products to the consumers.

Retail Landscape and Food Service Industry in India

The Indian retail market is one of the fastest growing markets in the world with the economic growth the country is currently experiencing that is expected to reach •926 billion by 2020. The Food & Grocery (FG) category is expected to capture 66% of the total retail sector by 2020. Within Indian retail, there is a strong distinction between organized and unorganized retailing. The unorganized sector is dominated by the 'Kirana' ("mom-and-pop stores"), general stores, street markets and convenience stores whereas the organized retail encompasses department stores, gourmet stores, supermarkets and hypermarkets and e-commerce retailers. However, the organized retail sector only accounted for 9% of the total retail sector in 2017, valued at •48.6 billion (IBEF)

Modern trade is an important contributor in the retail industry of India. There are many players, some of the known formats for seafood retail comprises of Spencer Retail; Aditya Birla "More" ; Spar; Star Bazaar; Big Bazaar; Best Price Modern Wholesale; and METRO Cash and Carry India Pvt Ltd.

India is booming with food joints and new restaurants are constantly coming up. Systems of online ordering, home delivery, cashback facilities, reward points and heavy discounts have changed the Indian consumer pattern. By 2021, the restaurant sector will contribute almost 2.1% to the nation's GDP witnessing a CAGR of 10%. Delhi, Mumbai, Ahmedabad, Pune, Chennai, Kolkata, Bangalore and Hyderabad together contribute around 42% to the food service market of India. The organized standalone market share in the total food services market is projected to rise to 29% in 2022 from 24% in 2017. In case of independent restaurants (in hotels), the market share is estimated to remain constant at 3% till 2022. However, this segment is still expected to grow at a CAGR of 9% from 2017 to 2022.

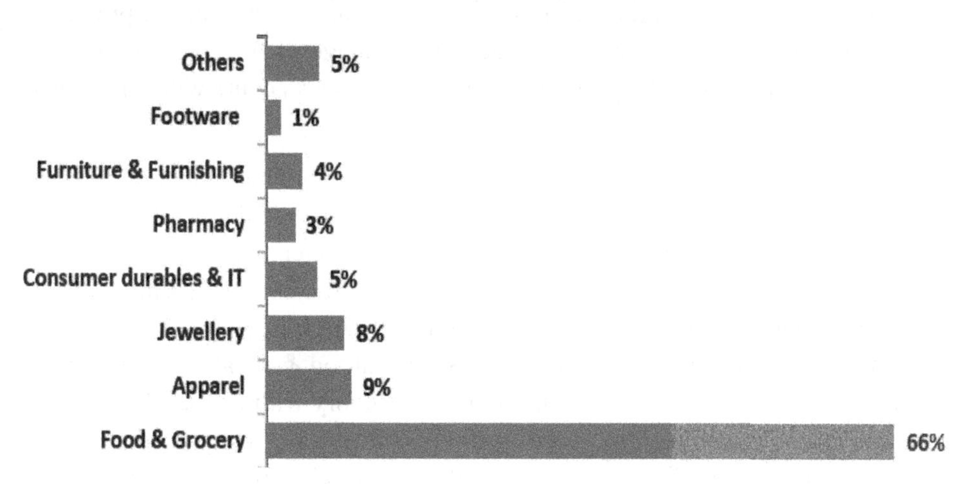

Source: Flanders,2018

Among modern hotels and restaurants, opportunities are typically for foods or ingredients that are not readily available in India. Among four and five-star hotels, casual and fine dining restaurants, imported food and ingredients are typically limited to products that cannot be sourced in India, or products which cannot match the imported quality. Imported products primarily include wine, other alcoholic beverages, dairy products, meat, seafood, fruits, frozen French fries, sauces, seasonings, drink mixes and food ingredients.

Seafood market scenario of India

The retail market of perishables including meat, poultry, fish and seafood, eggs and such other items is valued at Rs. 2,84,688 crores (USD 44.5 billion) in 2017. In the wake of growing non-vegetarianism in India, the perishables market is expected to grow at a healthy CAGR of 15.2 per cent to reach Rs 4,35,238 crores (USD 68 billion) in 2020. Currently, 57 per cent of the perishables market is rural. The demand for perishable foods in the urban market seems to have grown over the years since the urban share in total market has increased from 41 per cent (2014) to 43 per cent (2017). Almost 65 per cent of the combined market share is held by south, east and north-east regions with north contributing the least among all regions.

(**Source:** India Food Report, 2018)

Frozen/Processed Seafood in Modern Trade

Equipped with freezer capacities, modern retailers provide consumers with an option to buy appealing frozen seafood. Packaged and frozen seafood provide a quick and convenient meal option. An added advantage of frozen seafood is the shelf life of the product. Many of the frozen seafood packs come with an average shelf life of one year.

Processed seafood and meats are among the important segments of large packaged food industry in India. Busy lifestyles and convenience of food have supported the growth of processed seafood and meat category. Frozen processed seafood is the fastest growing category. However, sufficient availability of fresh meat & affordable pricing across India make it difficult for companies to convert the mind of the consumer for buying processed seafood & meat at greater level. Domestic brands have shown dominance in the category with negligible presence from international arena. Processed seafood and meat is forecasted to see a constant value CAGR of 11% over 2016-2021. Fresh chilled salmon commodity though negligible is emerging with presence at many of the hypermarkets in metropolitan cities. Breaded fish fillets, fish fingers, prawns, rolls form the major sales in processed seafood category.

Trend is shifting towards use of flexible plastic in frozen processed meat. Major players are targeting family as a customer rather than individual and hence offering packs of 250 gms and upward. Flexible plastic option for packing is slowly increasing in frozen processed seafood. 200 gms flexible plastic packs in frozen processed seafood saw notable growth of 18% whereas 500 gms packs saw strongest growth of more than 25% in 2016. 1 Kg packs of frozen processed seafood have also seen quite good growth in 2016. With good disposable incomes and time constraints of consumer, it is projected to record growth of more than 30% in 1 kg packs in frozen processed seafood continuously for next five years. Zip/press closure flexible plastic could be seen as offering greater convenience to consumers and may be the rising of more gourmet products with this packaging. Top brands in frozen seafood in modern retail include Gadre Marine, Cambay Tiger, Big Sams, Empire, Sumeru, IFB, Buffet etc.

Diagrammatic Representation of 'Ease of Doing Business in India'

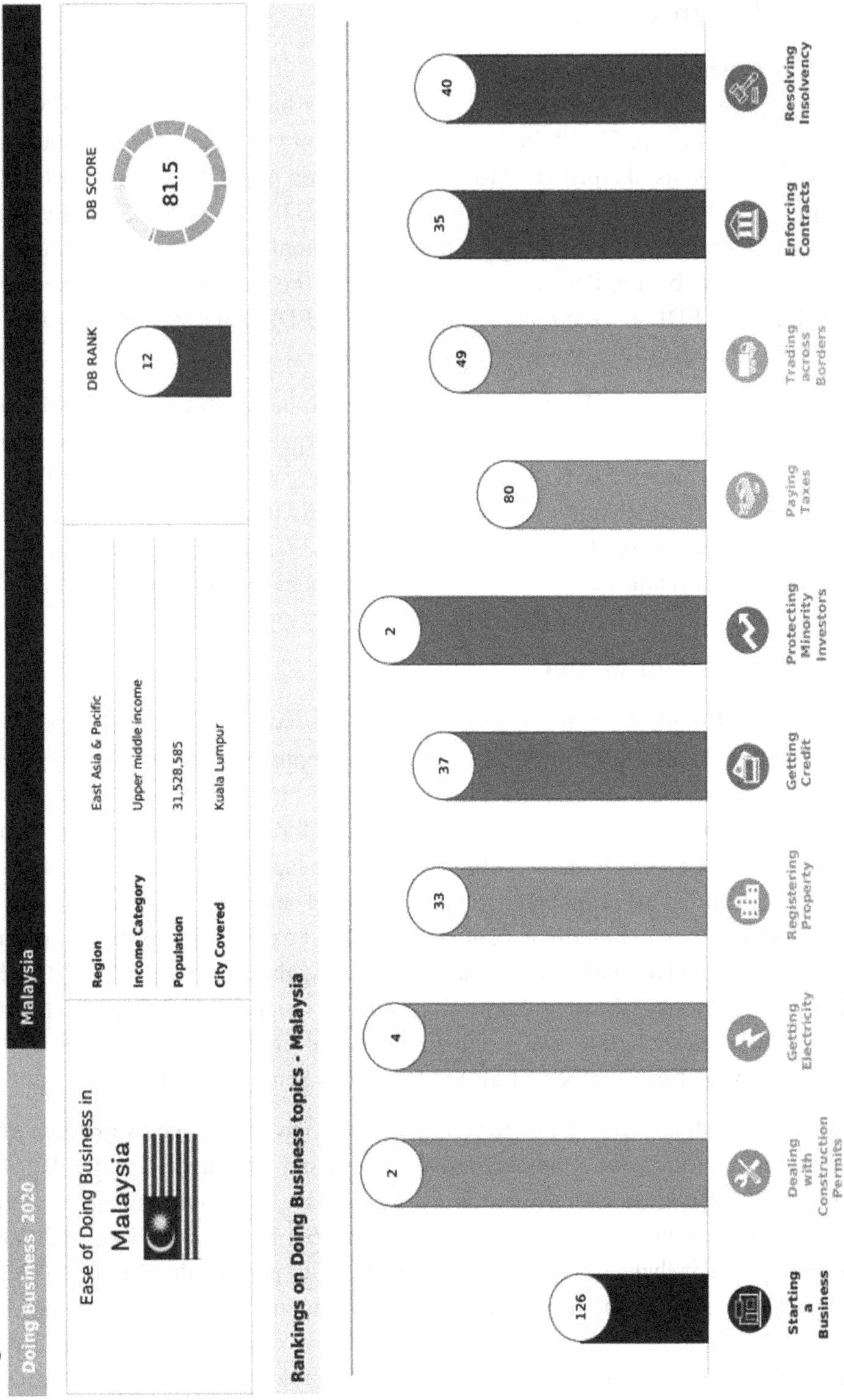

Doing Business 2020 | Malaysia

Ease of Doing Business in
Malaysia

Region	East Asia & Pacific
Income Category	Upper middle income
Population	31,528,585
City Covered	Kuala Lumpur

DB RANK: 12
DB SCORE: 81.5

Rankings on Doing Business topics - Malaysia

Topic	Rank
Starting a Business	126
Dealing with Construction Permits	2
Getting Electricity	4
Registering Property	33
Getting Credit	37
Protecting Minority Investors	2
Paying Taxes	80
Trading across Borders	49
Enforcing Contracts	35
Resolving Insolvency	40

(**Source:** Economic Profile India, Doing Business 2020)

Seafood in an Online Segment

The E-commerce market is expected to reach US $ 200 billion by 2026 from US $ 38.5 billion in 2017. The total internet subscriber base in India stood at 636.73 million subscribers in FY 19. India's internet economy is expected to double from US $ 125 billion as of April 2017 to US $ 250 billion by 2020; majorly backed by e-commerce. In India, 100% FDI is permitted in B2B e-commerce. As per new FDI policy, online entities through foreign investments cannot offer the products which are sold by retailers in which they hold the equity stake. As per new guidelines on FDI in e-commerce, 100 percent FDI under automatic route is permitted in marketplace model of E-commerce.

The online retail market in India is estimated to be worth US $ 17.8 billion in terms of GMV as of 2017. E-retail market is expected to continue its strong growth, by registering a CAGR of over 35% and to reach Rs 1.8 trillion by FY 20. There are lots of opportunities for e-retailers in India to capitalize upon gradually growing internet penetration in India. As of 2016-17, online retail is made up of 1.5% of overall retail market in India and is expected to contribute 5% in 2020.

Inventory Based Model

Inventory led models are those shopping websites where online buyers choose from among the products owned by the online shopping company or shopping website that take care of whole process end-to-end, starting with product purchase, warehousing and ending with product dispatch. Seafood is one of the new entrants in an online category in India. It is challenging to manage the operations for seafood category online. This segment with its additional services and convenience is expected to grow very rapidly in coming years. Big basket is one of the companies which sells almost all major categories of food through its online platform. On the other hand, Licious or Fresh to home are exclusive companies who sell fresh non-veg of all types including seafood. Jalongi is one of the online platforms who dedicatedly sell fresh seafood of all kinds besides some other new entrants.

Operationally, few of the major online players have their own warehouse and processing facilities. The major challenge with online players is to fulfil the requirement in shortest possible time as customers place orders on an online platform for their convenience. One of the major differences between modern trade and online segment is that almost all the online companies provide the seafood in clean and packed manner in the form of fillets or steaks or curry cuts or cubs or portions but not the whole fish. The percentage of returns is an area of concern for few of the online players. Due to improvements in the lifestyle, awareness towards eating hygienic and cleaned seafood, saving on time, increase in spending power, people are attracted towards placing the orders online instead of physically visiting the stores.

Marketplace Model

It's a digital platform for consumers and merchants without warehousing the products. It is a platform where there are many sellers from around the country or even globe who can connect with large number of customers. Marketplace model allow companies (e.g Flipkart) to act as aggregators to third party merchants who can sell via the online firm's platform. The recent development is with online giant Amazon who showed keen interest in food categories in general and fresh categories in particular as the last frontier categories for them as being done globally. Many of the seafood players register as third-party merchants on the marketplace model platforms like Amazon, Flipkart etc.

Daily Chores of Fish Department in typical Modern Trade Store

Every organisation follows its own standards for fish department. However, generally there is a department manager who is responsible for ensuring normal functioning of the department with his/her subordinates and superiors. Department manager manages the regular work flow which defines all the procedures in fish department.

Below is the normal workflow of the daily chores of a typical fish department.

1. Calculation of stocks in hand, sales pattern and forecasted demand will be taken into consideration by department manager based on which the purchase order for the stocks get raised.
2. Fish receiving will happen as per aligned day and time. Quality checks as per set standards of the company will get followed in presence of quality personnel along with fish department staff and goods receiving staff. Only those fish stocks that fulfil the criteria are received and rest get rejected.
3. FIFO (First In First Out) principle gets strictly followed in all retail formats.
4. There is a manual and process set for all the activities to be followed in fish processing area. Department manager monitors all those activities.
5. Presentation of fish merchandise is a key for sale of seafood and more hygienic, appealing and attractive display attracts customers.
6. Stock monitoring and pricing checks will get observed by department manager periodically every day in order to keep exact track of business.

7. Main aim of the department manager is to provide excellent service to the customers. Setting own example by doing and training the staff in the department is crucial to addition of new customers. Customer service is top priority in all modern trade stores.
8. Usually there is no fish return policy across all the formats working in India.

Guidelines for Seafood Retailing in Modern Trade Stores

Guidelines provide assistance and guidance for the seafood safety and quality management by giving practical information and suggestions on how to ensure seafood safety and quality during all stages of retail distribution and sale. There is a public perception that seafood belongs to high risk category. Food safety is of the utmost importance in the mind of the consumer. Therefore, it makes sense for the seafood industry and retail operators to be perceived as providing safe, quality food.

Seafood should always reach the consumer in good condition & it must be handled with care and speed always. Seafood spoilage is the physical and biochemical deterioration or breakdown of tissue in seafood. As seafood spoils, the external appearance undergoes great change, particularly the skin, eyes, gills, flesh and organs. At the same time, the odour changes from an initial fresh odour to a sour fishy smell.

Contamination is the introduction of a contaminant on to or into a food. There are three types of contamination namely bacterial, chemical and physical. Contamination of seafood may result from exposure to the environment (for example, polluted waters) or through direct contact during processing. It is often during processing activities that bacterial contamination of seafood occurs. Even hygienically produced seafood will have some bacteria on it after processing. Seafood spoilage cannot be stopped, but it can be slowed to a minimum. Reducing the temperature of seafood is the single most effective way of slowing spoilage, obtaining maximum shelf life, and preventing food poisoning. Temperature control is critical for controlling the rate of seafood spoilage and preventing bacterial growth. One day of shelf life is lost for every hour fish remains at room temperature (25°C). Seafood at -10°C looks and feels as hard as seafood at -30°C, yet it will deteriorate more rapidly. Causes and effects and mode of control of fish spoilage are given in Annexure 1 and 2 respectively.

People carry bacteria in their gut, nose, mouth, ears, hair and on their skin. These bacteria are quite normal and do not affect us, however, if they are transferred to food they can grow to numbers that can cause food poisoning. *Staphylococcus aureus* (Golden Staph) lives in the hair, ears, nose, armpits, groin, cuts, pimples, and boils as well as under jewellery worn. The intestines of humans can also contain salmonella and *E. Coli*. Bacteria can be transferred to our hands when we use the toilet, scratch our face or sneeze into our hands. Cross contamination of seafood with these bacteria can occur during any handling and preparation activity. Good personal hygiene is therefore essential for the safe handling and preparation of seafood and for preventing bacterial contamination of seafood.

It is a good idea for seafood businesses to document their requirements for personal hygiene and have them accessible to all staff. A sign or poster of the requirements may be useful. It is important to note that visitors and contractors must also abide by the personal hygiene requirements. Good premise-hygiene is essential for producing safe seafood and a good shelf life. Good premise-hygiene includes cleaning and sanitation of seafood preparation surfaces and equipment, controlling pests and vermin (such as rodents, cockroaches, flies) & effective waste disposal. Seafood purchased by a retail store for onward selling to the consumer, must be safe and of sufficient quality to ensure it will not spoil during expected storage times. Good handling and storage practices are needed by the retail store to ensure that the customer receives safe seafood.

Traditionally, food safety has been determined through premises inspection and end product testing. This form of reactive control has proved costly and ineffective. To reverse the increasing incidences of food poisoning, the new proactive approach to managing food safety is the adoption of food safety programs. Developing a food safety program will help a seafood business understand its product and process, and will ensure controls are in place for identified food safety hazards. The proactive approach of Hazard Analysis Critical Control Point (HACCP) has been adopted internationally by regulators and industry. A food safety program (sometimes called a food safety plan) is a written document which details the business activities and responsibilities associated with the production of safe food. It documents staff and operational requirements necessary to minimise the risk of food poisoning. The food safety program will detail how a food business is going to maintain staff and premises hygiene, and manage the safety of seafood from ingredient receipt - during food preparation and storage - through to customer delivery. The identification of potential food safety hazards and their methods of control may be achieved by utilising HACCP principles, or by documenting them into procedures. The decision about quality rests ultimately with the consumer. Quality seafood can be defined as seafood that meets the customer's requirements. Quality is a combination of various properties that influence acceptability. Few of the attributes that determine the quality of retail seafood may include freshness, food safety, size and appearance, texture and taste, nutritional value, cost, packaging service etc.

The importance of seafood display and presentation cannot be over-emphasised. The display should lure the customer to the shop and an interesting presentation attracts customers. Seafood presentation, the image of premises and staff all have an impact on customers. Attributes required are staff knowledge about product (know about today's product), promotion of freshness through presentation, hygienic premises, well-uniformed staff, full display in cabinets, good housekeeping, effective lighting, plenty of ice on fish etc. Since many customers may not be familiar with the large variety of seafood, face-to-face selling and product knowledge is very important. Customers may know only a few species of fish, one or two ways of preparing or cooking seafood, and may also lack knowledge on the storage and shelf life of seafood. Therefore, seafood retailers who talk to their customers about handling, quality, and meal suggestions are more likely to enhance their sales. Staff appearance and manners too have a marked effect on customer perceptions. Positive memories are what motivate customers to return.

Development of materials for sustainable fishing and aquaculture: Role of Garware

Food is a basic need of all human beings, and the global food industry faces the challenge of meeting the food needs of a growing population economically and sustainably. Technology plays an important role in filling the gap between supply and demand. In India, the Green Revolution launched in the mid-1960s converted the traditional, low-productivity agriculture system in many parts of the country into a high-productivity industrial system through the adoption of modern methods and technology. The White Revolution, also known as Operation Flood, launched in 1970s, made India self-dependent in milk production, and turned it into the world's largest milk producer.

After these epoch-making developments, India is now poised for a Blue Revolution in the fisheries sector. For the first time since independence, the Government of India has created an independent ministry for fisheries with a structured development plan under the Pradhan Mantri Matsya Sampada Yojana (PMMSY). An unprecedented investment of over Rs. 20,000 crores will fuel growth in India's fishing industry and enhance its position in the global market.

Half a century before these developments, a visionary businessman with a social commitment, the late Dr. Bhalchandra (Abasaheb) Digambar Garware, undertook the mission of transforming Indian fishing through technology solutions.

Realising the various hardships faced by the country's largely poor fishing community, he took the initiative of developing innovative nets that would increase fishermen's returns, at low initial investment. Remarkably, Abasaheb undertook this onerous task without any government support. Further, with supreme confidence, he decided to manufacture nets that would meet global standards and find acceptance even in developed nations.

Till Garware's intervention, fishing nets in India were made out of natural fibres (cotton, manila, sisal, hemp, etc.), which posed major disadvantages such as low breaking strength, vulnerability to damage by fungus, deterioration in water, and heavy maintenance cost incurred for boiling nets in preservative petroleum-tar. Using his knowledge of plastic materials, Abasaheb Garware took the daring step of replacing natural fibre with nylon as the basic building block for fish nets. Through the change of base material, several chronic problems faced by fishermen were overcome.

Dr. Bhalchandra Digambar Garware

(21.12.1903 – 2.11.1990) - Revolutionary industrialist, who, first time in India replaced traditional natural fibre fishing nets by synthetic Nylon and HDPE fishing nets, which are, not only many times durable and stronger but also lighter in weight to reduce drag and save fuel and energy. Owing to his legendary contribution to fisheries and other sectors, the Government of India bestowed Mr. B.D. Garware the 'Padma Bhushan' award in 1971.

The pioneering company founded by Abasaheb Garware, Garware Nylons Ltd., soon emerged as a major manufacturer of nylon, plastics and polyester yarns in India, for use in the fishing and aquaculture sectors, as also in textiles and other industrial applications. Through processing of nylon filaments, nylon yarn and twines were produced for the manufacture of fishing nets, which gave a great economic boost to the fishermen community.

After this success, BDG, as he was fondly called, undertook backward integration in 1967 by taking up manufacture of nylon chips, which were then being imported. This step enabled production of nylons that are especially suited for manufacture of nylon yarn for fisheries. Garware's backward integration also led to a reduction in the input cost, and ensured sustainability of the technology innovation.

Keeping a moderate margin on its products, Garware Nylons rapidly increased the user base for its fishing nets. The 210-denier multifilament nylon yarn and twines manufactured by the company remained the gold standard in the Indian fishing sector through the 1970s to the mid-1990s.

In 1977 the company set up a new plant at Ahmednagar, Maharashtra, for producing 'machine-made' nylon fishing nets—a first in India, Replacing handmade nets, the machine-made nets met high quality standards, with a uniform mesh and tight knots. These nets were ideal for gill netting and the purse seine method of fishing.

In 1981, a new plant was commissioned at Aurangabad, Maharashtra, for manufacture of PVC floats required for fishing. Subsequently, the company introduced nylon monofilament fishing nets for the first time in India. These nets revolutionised gill net fishing. As the nets were not visible, fish catch was high.

Working on a parallel track, BDG pioneered the manufacture of polypropylene ropes in 1977 through Garware Wall Ropes Ltd, a company set up in technical collaboration with M/s Wall Works, USA. Till then, India's entire requirement of shipping ropes was met through imports. Garware Wall Ropes broke this cycle of dependency by manufacturing inexpensive, lightweight, strong and shock- resistant PP ropes. The floating characteristic of the ropes was due to lower specific gravity of the raw material compared to High Density Poly Ethylene (HDPE). Garware Wall Ropes then introduced HDPE monofilament yarn and fishing twine, which are ideally suited for use in mechanised trawlers.

The path-breaking, nation-building work of B.D. Garware was recognised by the Government of India, which bestowed upon him the 'Padma Bhushan' award in 1971, in appreciation of his contribution to fisheries and his vast philanthropic work.

BDG's legacy was carried forward in the 1990s by the second generation, his son, Mr Ramesh Garware, till 2014. Blue HDPE fishing nets introduced in the early 1990s under his leadership have negligible maintenance cost, high impact resistance and are well suited for fishing round the year. Enabling trawler fishing deep into the sea, the nets opened a new geography for fishing, and led to increased investment on mechanized boats.

Mr Ramesh Garware expanded the HDPE nettings business to cover a wide range of needs in both the domestic and international markets. Under his strategic vision, a large manufacturing facility was set up at Wai, Maharashtra (India). Considering the heavy import duty on imported fish-net machines and the increasing domestic demand, Mr Ramesh Garware set up a facility for manufacturing machinery to produce ropes and fishing nets. He also expanded the product portfolio of Garware Wall Ropes Ltd to include fishing gear, namely fabricated net assemblies with all the required accessories. These products were well accepted in the demanding international markets of Europe and North America. Manufacturing was directed towards forward integration and value-added solutions, with a focus on shifting from handmade nets to machine-made HDPE nets for mechanized boats.

Since 1995, the Garware machine-made HDPE nets have found wide acceptance in mechanized fishing across the globe, due to value-adding characteristics such as high strength, good floating capacity, lower water absorption, and minimum shrinkage. In India, the Garware nets encouraged more and more fishermen to explore untapped fisheries resources.

With the entry of Mr. Vayu Garware in Garware Wall Ropes Ltd in 1995, the third generation of the Garware family joined in the endeavour of providing appropriate solutions for profitable and sustainable fishing. Under his leadership, Garware Wall Ropes has undergone a re-branding exercise and emerged as Garware Technical Fibres Ltd. (GTFL), a well-diversified company offering a range of value-adding solutions to several industry sectors.

Listening closely to the voice of customer, GTFL has launched major innovations such as the GARFIL range of nets for trawl fishing, Star and Sapphire Excel nets for eco-friendly fish farming, and X2 aqua mooring ropes and nets for protective cultivation. Other revolutionary products include the recently developed V2, an eco-friendly aqua net that reduces the costs related to bio-fouling by more than 50%; and X 12, a non-pharmaceutical shield against many harmful marine algae.

For the trawl industry, GTFL has introduced 'Sapphire Net', which is designed to provide better water flow and reduce drag. This results in fuel saving and higher fish catch due to its superior mouth-opening feature. GTFL's GARFIL STR nets weigh 25% less than other nets, can be easily handled, and are very cost effective. Another range of new nets is produced by the company from 'Dyneema' fibres, which are significantly lighter yet stronger than steel. This product range is in great use in power transmission, shipping, aquaculture, wild-catch fishing and other sectors of fishing.

GTFL (a member of Garware Group) is today a leading player in the global aquaculture market, with a significant market share in US, EU, UK, Canada, Norway, Chile and other developed countries. The company has a portfolio of application-focused cages such as V2 Star, which promotes environment-friendly fishing by avoiding usage of hazardous copper coating for preventing growth of marine organisms. Copper-based coating has a negative impact on farmed fishes as also the oceanic environment. Due to increasingly stringent environmental regulations, and for improved fish health / quality, aquaculture companies across the world are moving away from copper-coated cages. GTFL's Star cages meet their needs.

GTFL has provided innovative solutions to many other chronic challenges in fishing and aquaculture. Globally, the aquaculture industry suffers enormous losses due to attacks of seals and sea lions, which consume the fish grown for harvesting and drive away fishes. Conventionally used predator cages made of synthetic materials are only partially effective in preventing the attacks. GTFL has addressed the problem by developing cages made from Sapphire 'Ultra Core' and 'Sapphire Sealpro' nets. Compared to polyamide and polyester nets, the GTFL nets are stiffer and more resistant to predator bites. The nets are blended with stainless steel wires in the core to ensure that predators cannot bite through the net cords. A special coating increases the stiffness of the netting. Biological fouling, or the accumulation of microorganisms, plants, algae, or animals on wetted surfaces, is another serious challenge in marine aquaculture. Biofouling hampers efficient and sustainable production in three ways: (a) it restricts water exchange (b) increases risk of disease and high stress-level of fish due to lowered dissolved oxygen levels from poor water exchange and (c) causes cage deformation and structural fatigue due to extra weight imposed by fouling. To deal with these issues, GTFL has developed anti-fouling nets and in-situ cleaning systems, which have been appreciated globally.

Similarly, GTFL has provided an innovative solution to the problem of sea lice, which causes losses of millions of dollars to the industry every year. For decades, only medicines and pesticides were used to fight the lice. GTFL offers a non-pharmaceutical lice controlling method, through lice SKIRTS coated fabric, which blocks harmful algal cells and harmlessly resolves the issue at an attractive cost.

Recently, along with research institutes of the Indian Council of Agricultural Research (ICAR) and other state institutes, GTFL introduced marine aqua cages in India. With two fully equipped R&D centres, a team of innovators with rich expertise and experience, and a focus on value-added solutions for diverse segments of users, GTFL is excellently poised to make significant contributions to the growth of the fisheries sector in India. With over 50% of its business coming from international markets, and a strong presence in key overseas markets, GTFL will also provide a ready platform for efforts launched under PMMSY to match global standards.

As the Indian fisheries sector readies itself for a quantum leap, the Garware name will always be etched in the history of the Blue Revolution in India.

Solutions for global aquaculture industry

Production of farmed aquatic organisms in caged enclosures is a relatively recent development. The use of marine commercial cages was pioneered in Norway in the 1970s. With the rise of salmon farming, the cage aquaculture sector has grown rapidly, particularly during the last two decades. Cage aquaculture is presently undergoing rapid changes in response to pressures from globalization and growing demand for aquatic products in both developing and developed countries. To meet the challenges faced in the aquaculture industry, whether in offshore or deep-ocean locations, Garware Technical Fibres Ltd (GTFL) is offering innovative and application focused solutions targeted at the top sustainability-related challenges in the salmon aquaculture industry, namely:

- Fish escapes
- Bio Fouling
- Sea lice
- Carbon footprint

To deal with these issues, and considering the limitations of comparatively expensive nylon fibre, there was a need to develop a new fibre for the aquaculture industry. GTFL has modified High Density Poly Ethylene (HDPE) to offer technically competent and cost-effective solutions. Lower specific gravity resulting in flotation has been resolved in newly developed V2 nets with use of poly ethylene (PE) that has higher specific gravity than regular PE. Another challenge, due to the hydrophobic property of the material, was application of coating. This was

INNOVATIONS IN CAGE CULTURE

Technologies developed by Garware Technical Fibre Limited (GTFL)

2010: GTFL developed HDPE knotless cages for Indian aquaculture farms. Supplied to India's first commercial aquafarm for pangasius. Low cost, easy to handle, less maintenance and durable

2012: GTFL's first HDPE Sapphire cage trial conducted in Canada. GTFL provided solution for legal restrictions imposed on use of copper paint in traditional nylon cages

Ferro Iceland purse seiner, using GTFL purse seine net bought from M/s Vonin, reported a recorded catch of 2500 MT Mackerel at one go. The fishing vessel was not able to accommodate all the fish and had to call few more industrial vessels to take up the caught fish. Further, the fishing quota of the entire season of the fishing vessel was utilized in just one voyage.

GTFL supplies purse seines to Norway, Denmark, USA etc. Weight of each purse seine ranges from 03 MT to 60 MT. Highest catch captured yet is 2000 MT; seine is from GTFL. High strength, low shrinkage, tight knots, long lasting treatment makes it unique.

GTFL's first Sapphire excel plus cage installed in Norway. 200 m circular - 08 MT - for salmon fish farm. Change over to HDPE from traditional nylon. Less maintenance, durable, organic fish growth etc.

2019: Largest cage 47*47mDq*50mD - 12 MT supplied to Norway for its innovative project. GTFL's innovative V2 technology fibre and Plateena rope is used in this netting. High strength, algae resistant, low maintenance, better water flow.

also eliminated by developing "coatable" HDPE. Garware cage nets made with V2 fibre are the industry's first aquaculture cage nets with built-in eco-friendly, anti-fouling technology. Providing long lasting anti-fouling effect, the nets can reduce bio-fouling maintenance costs up to 50 per cent.

Garware X12 solves the problem of infestation by tiny marine ectoparasites faced in salmon farms. Each year millions of dollars are spent on pharmaceutical treatments to keep the ectoparasite infestation in control. The Garware X12 Lice Skirt is a non-pharmaceutical product that keeps the ectoparasite infestation in check by acting as a mechanical barrier that allows healthy water exchange.

Garware's new PE solutions in aquaculture minimise fish escape and predation. Predator nets are especially developed with a mix of various synthetic fibres that improve mesh and knot-breaking strength, provide high stiffness and have high cut-resistance. Garware's Sapphire PE predator nets have been quite effective in preventing predation and thus lead to increased salmon farm protection and reduction in seal culls.

Another innovative product, Sapphire Ultra Core, has been awarded a patent in Chile. As one customer reports, "The design has worked very well...So far we have not had seals or small sharks get through the net base to the dead fish." Salmon farms are also susceptible to harmful algal blooms that can cause mass mortality overnight. Garware's X12-AB blocks different harmful algal cells from entering into the cage but is water permeable, thereby reducing the use of external oxygenation.

Solutions for domestic fishing sector

Cage aquaculture was introduced in India in 2007 by ICAR-CMFRI with Garware cages, and the sector has tremendous potential as the country has a long sea-coast, untapped EEZ, extensive riverine and estuarine waters, and a wealth of fish fauna. The Pradhan Mantri Matsya Sampada Yojana (PMMSY) is focusing on this potential.

In the traditional fishing sector, GTFL offers solutions for both active fishing (trawling & purse-seine) and passive fishing (gill nets, hooks & lines, dol nets). In both cases, the solutions deal with the following requirements

- Better fish catch
- Fuel saving, for cost effective fishing
- Long life (higher breaking strength) of nets

After considering important aspects such as the fishing position (bottom / mid water / pelagic), specific design, targeted species and devices used, GTFL's R&D team has developed many innovative products. Some of these are as follows:

SNG NETS (Sapphire Next Generation): Developed with the right-size twine for easy drag, better drape in netting, and higher knot tightness, these nets are specially designed for the mouth portion of pelagic trawl. The net does touch the ground and has a better mouth opener. The net is very popular as it gives a better fish catch. As it is easy to drag, it leads to savings in fuel costs and higher speed of fishing boats.

BTR NETS (Bottom Trawl Nets): With better abrasion resistance, and higher knot tightness, these nets are specially designed for bottom trawls. Due to lower breakages and higher fish catch with lower fish losses, this product has been welcomed by the fishing sector.

GARFIL - Runner Right Sized: This product is designed for the belly portion of trawl, with special features such as better water flow, lower drag and better mouth opening.

GARFIL Nest – Readymade CODEND: Specially designed for codend portion of net with knotless netting and uniform mesh size, the product leads to lower fish damage, and better water flow. The light weight results in fuel efficiency.

X2 BRIDAL Rope: This rope has been designed for bridle operations with features like controlled elongation, kink-free operation, and better breaking strength. Due to these features, the rope leads to better catches, and balanced/ kink-free operations. Speed of boats is higher, and the rope is more durable than conventional options.

With its solution-based approach, GTFL works closely with fishing communities in India and abroad to develop products that provide significant cost savings and make the business sustainable in the long run. An example of such an output is an innovative product for shrimp fishing in the Gulf of Mexico, which helped the fishermen reap huge benefits compared to conventional nylon netting.

It is important to note that most of the developments and innovations in active fishing have taken place through data collection, vessel monitoring and use of information technology at sea. Very little development has happened in the area of fishing gears. This is where Garware is playing an important role, globally.

Conclusion

The total fish production of India increased from 7.52 lakh tonnes in 1950-51 to 125.9 lakh tonnes in 2017-18 recording significant growth in the fisheries sector. The percentage contribution of inland fish to the total landings showed significant increase from 29% to 71% while the share of marine fisheries reduced from 71%

to 29% in the last several decades. There is ample scope for harnessing the potential of inland capture fisheries in addition to enhancing production from culture fisheries in all ecosystems.

India was the fifth largest seafood exporter of the world in 2017-18 with an export figure of Rs. 45,106.89 crores. The contribution of Fisheries GVA to Agriculture GVA has grown from 4.53% in 2011-12 to 6.16% in 2017-18 where as it has contributed 0.93% to total GVA of India in 2017-18. The domestic market for seafood is growing in India with emergence of its sale through e-commerce besides traditional retail, wholesale and modern trade sector. The long-awaited demand of fisheries professionals for creating a separate Department of Fisheries at the country level has finally been met in 2019.

Retailing is multi-billion-dollar industry in India and seafood is one of the important sources of food protein. India has a long coast line as well as rightly developed farming techniques of culture for production of commercially important seafood species. Seafood forms an important category in any of the modern trade as well as in online sale format. The awareness of eating healthy and safe seafood is on the rise in India. The trend of increasing income among middle class, worldwide travel experience, convenience, easy access to technology are few of the factors that prompt buyers to buy seafood from online segment. Many of the modern trade companies follow the quality standards and adopt suggested measures thoroughly.

Introduction of synthetic fishing gear materials has revolutionized the fishing and aquaculture industry in India. Till Garware's intervention, fishing nets in India were made out of natural fibres which posed major disadvantages such as low breaking strength, vulnerability to damage by fungus, deterioration in water, and heavy maintenance cost incurred for boiling nets in preservative petroleum-tar. Chronic problems faced by fishermen were overcome through replacement of natural with synthetic materials for construction of fishing gears, accessories and cages.

Annexure 1 Causes and consequences of fish spoilage

Spoilage Consequences	Cause
Reduced shelf life	Due to the action of bacteria, enzymes and oxidation.
Off flavours and smells	From the breakdown of tissue through the action of bacteria and enzymes.
Taints	Off flavours that arise from contamination during handling and preparation.
Reduced quality	Deterioration of the visual, physical and chemical characteristics of seafood from the action of bacteria and enzymes.
Food poisoning	Predominantly from the contamination and growth of bacteria.

Annexure 2 Causes and mode of control of seafood spoilage

Cause	Effect	Controlling Factor	Mode of control
Bacteria	Live healthy fish can be covered in bacteria, while the flesh remains sterile. After death, incorrect handling can introduce bacteria to the flesh resulting in spoilage. Seafood handlers and the environment are other sources that may result in bacterial spoilage and the resulting 'off' odours and flavours	Temperature of seafood: below 5°C, or greater than 60°C	Prevents bacteria from growing or producing toxins Kills bacteria
		Rapid processing	Minimises the time when bacteria can grow
		Good storage practices: separation (raw & ready-to- eat)	Prevents cross contamination
		Covering food	Protects seafood from contamination.
		Good handling methods	Prevents contamination & cross contamination.
		Healthy & hygienic staff	Prevents contamination & cross contamination by Disease
		Clean & hygienic premises	Prevents contamination & cross contamination
		Using correct packaging	Prevents bacteria from getting on to the seafood
Enzymes	Enzymes are present in the flesh and various organs (e.g. gut) of all seafood. They function to break down food into energy. After death, enzymes will continue to function resulting in the breakdown of flesh to a soft texture, with	Temperature of seafood: between 2°C and -1°C	Slows the activity of enzymes.note: Between -1.5°C and -5°C enzyme activity is increased.
		Temperature greater than 75°C	Thorough cooking will halt enzyme activity.
		Thorough wrapping	Reduced exposure to air unpleasant odours and flavours slows enzyme activity

[Table Contd.

Contd. Table]

Cause	Effect	Controlling Factor	Mode of control
Oxidation	Oils in seafoods will react with oxygen in the air over a period of time causing rancidity (strong fishy odour and flavour). Oily fish will become rancid faster than lean fish. The bright red colour of muscle	Correct packaging reduces contact with oxygen, such as vacuum packing or thorough wrapping)	Physical barrier prevents contact with oxygen
		Rapid processing	Minimises length of time product is exposed to air
	will become dull and eventually turn brown. Oxidation is a major cause of spoilage during prolonged freezer storage. Dehydration will accelerate oxidation	Glazing, where a thin coating of ice is applied to frozen product	Physical barrier prevents contact with oxygen
		Holding frozen product at or below -30°C	Colder temperatures reduce level of oxidation
		MAP - modified atmosphere packaging	Evacuates oxygen from the atmosphere surrounding the product, replacing with another gas
Dehydration	The drying out of seafood causes a reduction of flavour, juices and loss of weight, Severe dehydration of frozen seafood is referred to as freezer burn. It may result in a dry 'woody' appearance or the buildup of icicles within pre-packaged goods	Using correct packaging	Physical barrier prevents dehydration
		Glazing, where a thin coating of ice is applied to frozen product	Physical barrier prevents dehydration
		Constant temperature of frozen product below-30°C	Temperature fluctuations result in the partial thawing and refreezing of product. This process results in the rupture of cell walls and the loss of fluids from within the cell

[Table Contd.

Contd. Table]

Cause	Effect	Controlling Factor	Mode of control
Physical damage	Torn, bruised, cracked or crushed products are all results of rough handling	Training staff in appropriate handling and storage techniques. Using appropriate containers for storage and handling	Staff realising the importance of correct handling procedures and following these instructions Containers that are suitable for protection of product
Rigor Mortis	Rigor mortis has three stages; pre-rigor, rigor and post rigor.Fish have very delicate connective tissue between muscle blocks, if a fish is handled roughly while in rigor mortis, it will damage this tissue resulting in a texture change and loss of moisture and flavour	Filleting should never occur before the commencement of rigor,	The muscle contraction without a supporting skeleton will result in a tough texture
	Fish going slowly and gently through the process of rigor mortis will extend the shelf life of product	While a fish is in rigor it should not be straightened out or bent. Chilling slows and extends the process of rigor	While in rigor the straightening or bending of a fish will result in the tearing of muscle blocks away from the connective tissues resulting in a 'gaping'or torn fillet While in rigor spoilage is minimal

CHAPTER 3

ISSUES IN INDIAN MARINE FISHERIES AND MEASURES FOR SUSTAINABLE DEVELOPMENT

Introduction

Fishing has grown into an industry with global relevance providing livelihood to 39.4 million fishers involved in capture fishery and contributing towards 16.7% of globally consumed animal protein (SOFIA, 2014). The capture fishery sector is dominated by marine fishery, accounting for 87.2% of the total capture production of 90.9 million tons. Marine fisheries around the world remain seriously threatened by overfishing, overcapacity and range of environmental problems.

Marine capture fisheries play a significant role in meeting the nutritional requirements of the population and achieving the Millennium Development Goals. Globally, marine fisheries resources are under enormous pressure with the estimated fishing effort which exceeds the optimum by a factor of three to four (Pauly *et al*., 2002). According to the World Bank, excess fishing effort resulted in economic losses estimated to be 50 billion US dollars annually (World Bank, 2009). As per FAO (2009), 52% of global fish stocks were fully exploited, 28% were overexploited or depleted, 20% were moderately exploited, and only 1% showed signs of recovery which were the consequences of the fishing effort expansion since 1970. As per more recent report (FAO,2016), 58% of global fish stocks are fully fished, 31% overfished and10% underfished.

According to the Living Blue Planet Report (WWF, 2015), the state of global marine fisheries is grave and worrying. The global fishing fleet is 2-3 times larger than what the oceans can sustainably support. As a result, 31% of global fish stocks are classified as overfished and a further 58% as fully exploited, with no ability to produce greater harvest (FAO, 2016).

The development of marine fisheries has undergone transformation from traditional fishing methods to mechanized fishing all along the Indian coast during the last four decades. Marine fisheries sector of the country is in its post-2000 modernisation phase which is characterised by declining fish catches, depleted fish stocks, increasing conflict over fish resources, and mounting investment needs.

India's marine fish production has increased more than seven times, from 0.53 million tonnes in 1950 to 3.83 million tonnes in 2017, even as exports of marine fish and fish products increased from Rs. 35 crores in 1970 to over Rs 45000 crores in 2017-18 with an increase of 19.11% compared to 2016-17 (DoF, 2019). About 80-88% of the estimated potential yield of 4.41 million tonnes (Anon, 2011) is under exploitation. Total marine fish landings of India for the year 2018 was estimated at 3.49 million tons showing a decline of about 3.47 lakh tons (9%) compared to 3.83 million tons in 2017 (FRAD, CMFRI, 2019).

Issues in marine fisheries

Overfishing

Overfishing and irresponsible fishing practices have long been recognized as leading causes that have reduced aquatic biodiversity, along with other causes such as pollution, habitat destruction and fragmentation, non-native species invasions and climate change (Boopendranath ,2012). World-wide, overfishing is one of the biggest threats to the health of seas and their inhabitants. The results not only affect the balance of life in the oceans, but also the social and economic well-being of the coastal communities who depend on fish for their way of life. Increasing fishing efforts over the last 50 years as well as unsustainable fishing practices are pushing many fish stocks to the point of collapse.

There are three recognized types of biological overfishing. Growth overfishing occurs when fish are harvested at an average size that is smaller than the size that would produce the maximum yield per recruit. Recruitment overfishing is when the mature adult population (spawning biomass) is depleted to a level where it no longer has the reproductive capacity to replenish itself—there are not enough adults to produce offspring. Ecosystem overfishing occurs when the balance of the ecosystem is altered by overfishing. With declines in the abundance of large predatory species, the abundance of small forage type increases causing a shift in the balance of the ecosystem towards smaller fish species.

Impacts of overfishing include reduction of the spawning biomass of a fishery below desired levels such as maximum sustainable or economic yields, change in species composition, modification or destruction of habitat, modification of food chain and the trophic relationships within the ecosystem.

Measures to be adopted include avoiding overfishing and not having too many boats chasing too few fish. Number of boats should be appropriate for the natural supply of fish. Data on distribution of juveniles based on experimental fishing should be generated using GIS that would help in resource management. Mapping of juvenile-abundant areas would help to suggest restrictions on fishing grounds and fishing seasons. Spatial and seasonal restrictions on fishing effort assist in avoiding biological overfishing that would be useful in the sustainable management of fishery resources. Maintenance of biomass and stock is crucial for sustainable harvest in maintaining sustainable yield and socio-economic benefits to fishers.

Destructive fishing

The term "destructive fishing" has been mostly used to refer to impact considered as severe or unacceptable on the broader environment of target populations and on the ecosystem. It is considered as an unsustainable fishing practice that fundamentally destroys the marine natural resource. Destructive fishing methods include Poisons, Explosives and Electrical fishing. Use of poisons is widespread and they stun or kill the fish indiscriminately. Use of poisons can have severe and harmful impact particularly on coral reefs. Fishing with dynamite, a dangerous method, leads to indiscriminate killing of fauna in addition to serious consequences on resources, environment and users. Electrical fishing, dangerous unless the crew is adequately trained, results in indiscriminate killing of fish; young ones as well as spawners.

Effects of destructive fishing can be direct or indirect. Methods like fishing with dynamite are highly unsustainable because they typically do not target particular fish species and often result in juveniles being killed in the process. Damage to the coral reef structure further reduces the productivity of the area, thus adversely affecting both the reef-dependent fish populations and also the livelihoods of fishers and nearby communities.

Destructive fishing occurs when fishing gear is used in the wrong habitat/ sensitive environment where there is sea grass, algal beds, coral reefs and sponges. Bulk fish catching gears such as trawl nets and purse seines are often destructive. Operation of such gears leads to depletion of several fish resources over a period of time either due to usage of small mesh sizes below the legally permissible limits or due to indiscriminate fishing over coral reefs damaging the ecosystem. Bottom trawling can result in high levels of bycatch besides causing damage to the sea floor by scraping the bottom.

These fishing practices have negative impact on the environment. The impacts due to the operation of destructive methods include changes in bottom structure, changes in benthic fauna, reduction in spawning biomass, catch below maximum sustainable yield, changes in species composition and biodiversity, significant by catch of juveniles & benthic organisms, marine mammals, birds and dumping of gears and plastic by vessels.

Plans and measures to combat destructive fishing can be categorised into gear or exploitation related or habitat related. Gear related measures include gear selectivity, banning specific gear practices, gear substitution, modifying or deploying a gear in a less harmful manner. Protecting vulnerable habitat and closed areas including MPAs comprise habitat related measures. Though fishing practices such as dynamiting and poisoning are banned as per the Marine Fishing Regulation Acts of coastal states of India, they are still practiced in many places. Stricter enforcement of the rules is the only possible solution.

Bycatch/discards/juveniles

Due to usage of trawl nets with small-mesh cod end coupled with the increase in trawl fishing, there has been considerable increase in the landing of 'bycatch'. Bycatch includes all non-target aquatic resources and non-living materials (debris) which are caught while fishing. Discards are that part of bycatch which are released or returned to the sea either dead or alive (FAO, 2007). Juveniles of finfishes and shellfishes form part of trawl bycatch. Catch of juveniles and spawners leads to biological (growth and recruitment) overfishing. If juveniles and brooders are removed from sea continuously, proper recruitment and production do not take place and then the stocks are likely to collapse. In order to maintain the sustainable production of fish stocks, it is necessary to reduce the bycatch of juveniles and spawners in the trawl catch.

FAO estimated a global discard level of 20 million tonnes (FAO, 1999). Average annual global discards have been re-estimated at 7.3 million tonnes, based on a weighted discard rate of 8%, during the 1992-2001 period (Kelleher, 2004). The decline in discards may be due to a number of reasons such as stock depletion, strict regulations in some fisheries in the form of improved fishing selectivity, anti-discard regulations and increased use of bycatch reduction devices. Globally, shrimp trawling contributed to the highest level of discard/catch ratios of any fisheries, ranging from about 3:1 to 15:1, and the amount of bycatch varied in relation to target species, seasons and areas.

Trawling remains a controversial method of fishing due to the perceived lack of selectivity of the trawl net and the resultant capture of a huge quantity and diversity of non-target species, including endangered species such as sea turtles, coupled with its effect on the marine ecosystem (Kumar and Deepthi, 2006). In India, the by-catch landed at fishing harbors are utilized mainly for the production of manure and animal feed.

Bycatch and discards are serious problems leading to the depletion of the resources and have negative impacts on biodiversity (Harrington *et al.,* 2005; Alverson & Hughes, 1996). In tropical countries like India, bycatch issue is more complex due to the multi-species and multi-gear nature of the fisheries. Among the different fishing gears, trawling accounts for a higher rate of bycatch, due to the comparatively low selectivity of the gear. The total discards in Indian fisheries were estimated at 57,917 tonnes, which formed 2.03% of the total landings (Kelleher *et al.*, 2005). The discards of bottom trawling pose a threat to marine biodiversity.

Tropical shrimp fisheries produce largest quantity of bycatch. Trawl bycatch in the tropics known to be constituted by a high proportion of juveniles, particularly of commercially important fisheries needs serious attention in development and adoption of bycatch reduction technologies (Dineshbabu *et al.,* 2010). Majority of fisheries in the world attempted to address the issue of catch and discards in trawls through physical modifications to trawls meant to improve selectivity. The types of modifications reflected the fishery-specific characteristics.

Exploitation of juvenile fish results in considerable economic loss and also causes serious damage to the fish stock in terms of long-term sustainability of the resources. A minimum legal size (MLS) is seen as a fisheries management tool with the ability to protect juvenile fish, maintain spawning stocks and control the sizes of fish caught. MLS could be used to protect immature fish ensuring that enough fish survive to grow and spawn, control the numbers and sizes of fish landed, maximize marketing and economic benefits and promote the aesthetic values of fish (Mohamed *et al.*, 2014). There is need for adopting policies and practices that reduce the level of by-catch, ecosystem-based management to ensure the long-term sustainability of oceanic resources, and the adoption of a precautionary approach with emphasis on reducing, and if possible avoiding discards.

Ghost fishing

Ghost fishing occurs when certain gear such as pots or gillnets have either been lost or abandoned at sea and continue to catch and kill fish until the gear falls apart or is retrieved. It is environmentally deleterious and the fish caught is wasted.

Ghost fishing normally occurs with passive fishing gears such as long lines, gill nets, entangling nets, trammel nets, traps and pots. Use of biodegradable materials for gear construction and collapsible traps can reduce the problem.

FAO Code of Conduct for Responsible Fisheries (CCRF)

Why CCRF?

Fisheries provide a vital source of food, employment, recreation, trade and economic well-being for people throughout the world. Fish being a perishable commodity, need to be harvested and utilized in a responsible manner. Recognising the need for conservation, management and development of fisheries, the UN Food and Agriculture Organization (FAO) conference adopted the Code of Conduct for Responsible Fisheries (CCRF or popularly called "the Code") on 31st October 1995 that was ratified by more than 170 nations (FAO, 1995). The code was the result of a series of intergovernmental meetings that sought to build international consensus on the basic principles required for responsible fisheries. It was formulated to address the issues which threatened sustainability of fisheries such as overexploitation of important fish stocks, damage to ecosystems, economic losses, and issues affecting the fish trade.

The principal goal of the code was to achieve sustainable benefits from fisheries for people the world over (Anon, 2000). The code was fundamentally a global response to the progressively failing state of many fisheries the world over (Hanchard, 2004). The code through its effective implementation aims at ensuring conservation, management and development of living aquatic resources, with due respect for the ecosystem and biodiversity while ensuring availability of sufficient quantity of fish both for present and future generations.

About the Code

The code is global consisting of a collection of principles, goals and elements for action. Though the code is voluntary, as it is partly based on the relevant rules of international law such as the United Nations Convention on the Law of the Sea (UNCLOS), the governments have the responsibility to implement the code in cooperation with stakeholders like representatives from industries, communities, non-governmental organizations, regulatory authorities etc. For effective implementation of the code, its principles and goals need to be incorporated into the national fisheries policies.

The code is now widespread and seen as a self-help guide for governments interested in taking their fisheries towards sustainability. It has been translated into several languages apart from English to promote and create greater understanding among the personnel at the grassroots level. It provides guidance to states on how to go about having a system and practice in place for effective fisheries administration. The code is one of the most important international instruments available for the conservation, management and development of aquatic resources across the globe. It sets out principles and standards for responsible practices in fisheries that cover all the activities in both inland and marine sectors and all the stakeholders involved.

The code has 12 articles and two annexes. Articles 1 to 5 cover, the nature and scope of the code; objectives; relationship with other international instruments; implementation, monitoring and updating; and the special requirements of developing countries respectively. There are 7 substantive articles (articles 6 to 12) each dealing with a separate aspect of fisheries conservation and management. The issues dealt with are: Fisheries Management (Article 7), Fishing Operations (Article 8), Aquaculture (Article 9), Coastal Area Management (Article 10), Post-Harvest Practices and Trade (Article 11) and Fisheries Research (Article 12).

Article-7 states that all those engaged in fisheries management should adopt measures for the long-term conservation and sustainable use of fisheries resources through appropriate policy, legal and institutional framework, (FAO, 1995). Article-8 provides guidelines to the States to ensure that only fishing operations allowed by them are conducted within waters under their jurisdiction and that these operations are carried out in a responsible manner. These include maintaining a record of all authorizations to fish, regularly updating statistical data and establishing systems for monitoring, control, and surveillance among others. Article 9 gives direction towards carrying out responsible aquaculture. Article-10 deals with integration of fisheries into coastal area management and stresses on institutional framework to be adopted to achieve sustainable and integrated use of resources. Article 11 provides guidelines for responsible post-harvest practices and trade. Article-12 recognizes that responsible fisheries require the availability of a sound scientific basis to assist fishery managers and other interested parties in making decisions.

CCRF in the Indian context

Under the Eleventh plan of the GOI, Ministry of Agriculture, Dept of Animal Husbandry, Dairying and Fisheries, CCRF was identified as one of the priority areas under the scheme, 'Management of Marine Fisheries'. As fisheries is vital to the economy of our country that provides livelihood to millions of fishers besides providing food, creating jobs and generating foreign exchange, it is very necessary to protect the aquatic resources from being over fished and to ensure that the resources are available for the future generations too by applying the essence of the code. The comprehensive marine fishing policy outlined by the Government of India (2017) recognized the need for incorporation of principles of the code into every component activity in India.

As the first step towards promoting implementation of code in India, the Bay of Bengal Programme (BOBP) in association with the Government of India organized a national workshop for coastal States and Union Territories at Chennai during 29-30 September 2000 (Yadava, 2000). The objectives of the workshop were to fully familiarize government functionaries with the elements of the Code and the technical guidelines, which were prepared by the FAO to assist member countries in implementing the Code. The code was translated into several Indian languages, and a number of awareness campaigns for fishers and fishery managers were organized by several agencies (Yadava, 2000; Ramachandran, 2002).

Shenoy and Biradar (2005) outlined the salient features of the Marine Fishing Regulation Acts (MFRAs) of coastal states of India. The code provides a suitable framework and guidelines for environmentally compatible, socially acceptable and economically viable development and management of the fisheries sector in the central Asian region (Kumar *et al.,* 2007). Guidance on how the code can be put into practice in India has been brought out as Indian Marine Fisheries Code (IMFC) which is expected to give an impetus to bring about a sea change in the manner in which marine fisheries is managed in the country. The IMFC explains in detail each sub-article of the FAO CCRF and provides information on how and by whom the article can be implemented (Mohamed *et al.,* 2017).

Though each State has formulated regulatory methods such as declaration of closed seasons, protection of endangered species, prohibition of destructive fishing methods, regulation of mesh size etc; enforcement of many of these measures is very limited and dismal due to several reasons. There is need to create awareness of the implications and strategies for implementation of the code through human capacity building programs.

In an effort to create awareness of the code, collaborative Regional Training Courses in CCRF were conducted by ICAR Central Institute of Fisheries Education (CIFE), Mumbai and the Bay of Bengal Program, Inter Governmental Organization (BOBP-IGO), Chennai. Training was imparted to the middle and Junior level fisheries officials from India and other countries neighbouring Bay of Bengal such as Srilanka, Maldives, Bangladesh and Myanmar. At the state level, collaborative capacity building programs in CCRF and sea safety were conducted by ICAR-CIFE, Mumbai in association with the Department of Fisheries, Maharashtra in an effort to take the code to the grassroots. (Shenoy,2011).

As part of the institutional efforts taken for facilitating implementation of the code, ICAR- Central Institute of Fisheries Education, Mumbai organized a national workshop on "Creation of awareness of the CCRF and capacity building for effective implementation in India" in collaboration with the Department of Animal Husbandry, Dairying and Fisheries, Ministry of Agriculture, Government of India during 1st- 2nd February, 2012 (Shenoy and Lakra, 2012).

Regularization of common property rights and introduction of the concept of responsible fishing posed some difficulties in India due to the open access nature and lack of preparedness to face stringent restrictive management measures (Bhat and Chembian, 2012).

Implementation

FAO has the responsibility to serve as a catalyst and facilitator in the implementation of the code. In this regard, FAO undertakes activities at different levels which includes identification of financial support; project identification and technical support; education and training; technology transfer; support to regional fishery bodies; preparation of resource materials (technical guidelines, manuals, etc) designed to provide practical direction for government officials and other stakeholders on how to implement different aspects of the code; and dissemination of information in the form of databases, bibliographic references, etc.

The implementation of the code at the national level will require extensive involvement of stakeholders. Creation of awareness of the code and capacity building are very crucial in the effective implementation. Collaboration with national agencies (government and nongovernmental organizations), regional and international organizations, institutions, individual scientists and technicians, the private sector, consumer advocacy groups and fishing communities will facilitate faster implementation. In order to effectively implement the code, it is essential to create awareness and better understanding of the code among all the stakeholders. It is necessary to adapt the contents of the code to local conditions prevalent in states and disseminate the same in local language to the grassroot level fishers. The FAO-CCRF has not been entirely put into practice in India, although there are different compliance levels. Indian Marine Fisheries Code aptly named IMFC provides information on how each article of the code can be implemented and by whom.

The Government of India had constituted a committee with three sub-groups in 1999 for effective implementation of CCRF in India. Government of India (2015/2018) imposed a uniform ban (61 days) on fishing by all fishing vessels in the Indian Exclusive Economic Zone (EEZ) beyond territorial waters on the east coast including Andaman & Nicobar Islands and west coast including Lakshadweep Islands for conservation and effective management of fishery resources and also for sea safety reasons. The ban is effective from 1st June to 31st July (both days inclusive) on the west coast and 15th April to 14th June (both days inclusive) on the east coast.

Compliance of CCRF in Indian marine fisheries

Problems like overfishing, resource depletion, habitat degradation and resource conflicts faced by marine fisheries sector at local as well as global level could be mainly attributed to the non-compliance of fisheries rules and regulations. It is important to examine the level of compliance, investigate the causes of non-compliance of the local fisheries with the code and review the existing policies so as to recommend necessary amendments for encouraging compliance of the code at the local level. Evaluation of the compliance of local fisheries with the relevant provisions of the FAO CCRF in marine fishery sector will facilitate the state regulatory authorities to suggest measures for conservation and sustainable management.

The FAO and other international agencies periodically conduct evaluation on the implementation status of the code in several countries. Fisheries Centre, University of British Columbia made a global assessment of compliance to the code (Pitcher *et al.*, 2006). Varkey *et al.* (2006) and Pitcher *et al.* (2006) reported score of 40% i.e. (failed grade) based on an estimation of compliance of Indian fisheries with Article 7of CCRF.

Reports based on few studies carried out in the Indian context are available which highlight how far the code has been implemented. A study focused on applications of the code at the grassroot level by local fisheries management authorities in marine fisheries of Kerala with reference to guidelines for fishing operations (Article 8 of FAO CCRF) showed compliance on many areas of Article 8 like documentation of catch and effort, registration and licensing of fishing vessels, safety of fishers and insurance coverage. However, mesh size regulations as per section 4 of Kerala Marine Fisheries Regulation Act (KMFRA), 1980 were not followed. Other areas suggested where improvement is required include Monitoring, Control and Surveillance (MCS), fishing gear selectivity and energy optimization.

Ail *et al.* (2014) reported an overall 54% score for the review of marine fisheries of Kerala with Article 8 of the code.

Evaluation of the compliance of the bag net fishery in Maharashtra, India with the relevant provisions of the Article7 of the CCRF revealed an overall compliance of 67.52%. Though the bag net fishery is the most common and traditional fishery of Maharashtra that supports livelihood of several fishers, there is no specific policy to regulate, develop and ensure sustainability of the bag net fish resources. This study highlights the need to revise the existing MFRA of Maharashtra by incorporating the requirements of the code for achieving sustainability of resources. In order to make the bag net fishery sustainable in the State, management measures suggested include modifications in the fishing gear systems to reduce bycatch/discards, amendments to the existing MFRA giving due consideration to the views and opinions of stakeholders along with scientific evidence and creation of awareness about CCRF amongst the stakeholders to effectively implement the code. The code could be taken as a reference document to facilitate responsible management of the bag net fishery taking the local conditions into account. (Kumawat et al., 2015)

The trawl net fishery of north and south Konkan region of Maharashtra coast was evaluated for its ccompliance with relevant provisions of FAO CCRF. Kharatmol (2018) reported overall compliance level of 52.91% for Article 7, 59.73% for Article 8, 64.00% for Article 10 and 68.15% for Article 12 with medium compliance for its adaptation and implementation indicating need for improvement of trawl net fishery in Maharashtra.

Major causes of non-compliance of the code

Limited livelihood opportunities as well as ignorance about the law are reported to be mainly responsible for the non-compliant behaviour of fishers. Ignorance of fishers regarding salient features of MFRA is a major deterrent in successful implementation of the code. This could be attributed to the weak institutional linkage of fishers with the Department of Fisheries (DoF) and other related government and non-government organizations responsible for managing the fisheries. The weak enforcement of regulations particularly relating to mesh size has led to more violations.

Mitigation measures

Following measures are suggested for enhanced implementation of the code:

1. Need to formulate and implement appropriate and updated specific policies for major fisheries.
2. Need to create alternate livelihood options to fishers particularly during fishing ban period.
3. Mitigate the communication gap between fishers and management authorities with regard to rules and regulations under MFRA.
4. Strict enforcement of existing rules and regulations of MFRA.
5. Need intensified efforts through conduct of regular seminars/ workshops/ meetings to spread the importance of responsible fishing.
6. The existing MFRA of states need revision to incorporate the requirements of the code for sustainability of resources. Fishers expect the State to bring amendments to the existing laws or draft new policy after giving due consideration to the views and opinions of stakeholders coupled with scientific evidence.
7. Restriction on the number of boats in operation to prevent overfishing
8. Diversify inshore to offshore trawl fishing (Deep sea fishing)
9. Make documentation of the catch and fishing information mandatory
10. Strengthen MCS functions by mandatory use of log books, colour coding of fishing vessels, automatic identification system (AIS) and issuance of biometric cards to fishers.
11. Need to prevent encroachment of large boats in the reserved area of traditional fishers as per MFRA.
12. Creation of awareness about CCRF amongst the stakeholders to effectively implement the code.

Conclusion

Marine capture fisheries resources are under tremendous pressure on account of overfishing, overcapacity, destructive fishing, degradation of habitats among others along the inshore and coastal areas. As per the recent report of FAO, 58% of global fish stocks are fully fished, 31% overfished and10% underfished. Management measures and technical interventions are required to mitigate impacts of biological overfishing. Gear related measures to combat destructive fishing include gear selectivity, banning specific gear practices, gear substitution, modifying or deploying a gear in a less harmful manner. Protecting vulnerable habitat and closed areas including MPAs comprise habitat related measures. Though fishing practices such as dynamiting and poisoning are banned as per the Marine Fishing Regulation

Acts of coastal states of India, they are still practiced in many places. Stricter enforcement of the rules is the only possible solution. Exploitation of juvenile fish results in considerable economic loss and also causes serious damage to the fish stock in terms of long-term sustainability of the resources. A minimum legal size (MLS) is seen as a fisheries management tool with the ability to protect juvenile fish, maintain spawning stocks and control the sizes of fish caught.

Under the Eleventh plan of the GOI, Ministry of Agriculture, Dept of Animal Husbandry, Dairying and Fisheries, the Code of Conduct for Responsible Fisheries (CCRF) was identified as one of the priority areas for conservation, sustainable management and development of marine fisheries of India. Problems like overfishing, resource depletion, habitat degradation and resource conflicts faced by marine fisheries sector could be mainly attributed to the non-compliance of fisheries rules and regulations. Limited livelihood opportunities as well as ignorance about the law are reported to be mainly responsible for the non-compliant behaviour of fishers. Ignorance of fishers regarding salient features of MFRA is a major deterrent in successful implementation of the code.

CHAPTER 4

AQUACULTURE

Introduction

Fish is the most important source of animal protein for the human population, it is rich in protein and essential amino acids. It is also a good source of calcium, vitamin A, B12 and omega-3 fatty acids. People irrespective of age who do not get enough nutrients from cereal-based diets, would be benefited from the inclusion of fish in the diet. Aquaculture not only supplies dietary essentials for human consumption, but provides excellent opportunities for employment and income generation, especially in the more economically backward rural areas. Sixty million people are directly engaged, part time or full time, in primary production of fish, either by fishing or in aquaculture, supporting the livelihoods of 10-12% of world population (FAO, 2016). Aquaculture currently accounts for over 50% of the global food fish consumption (FAO, 2016). Aquaculture in India has evolved as a viable farming practice over last three decades with considerable diversification in terms of species and systems, and has been showing an impressive annual growth rate of 6-7% (FAO 2019).

Aquaculture is currently one of the fastest growing food production systems in the world. Most of the global aquaculture output is produced in developing countries and significantly in low-income food-deficit countries. As defined by the United Nations Food and Agriculture Organization (FAO), aquaculture is the "farming of aquatic organisms including fish, mollusks, crustaceans and aquatic plants." With stagnating yields from many capture fisheries and increasing demand for fish and fishery products, expectations for aquaculture to increase its contribution to the world's production of aquatic food are very high, and there is also hope that aquaculture will continue to strengthen its role in contributing to food security and poverty alleviation in many developing countries. However, it is also recognized that aquaculture encompasses a very wide range of different aquatic farming practices with regard to species (including seaweeds, molluscs, crustaceans, fish and other aquatic species groups), environments and systems utilized, with very distinct resource use patterns involved, offering a wide range of options for diversification of avenues for enhanced food production and income generation in many rural and peri-urban areas.

Globally India stands second in culture fisheries production. China, with world's one fifth of population produces one-third of total fish harvested and two thirds of fish cultivated (FAO, 2016). While in India, the culture system is based on 3-6 species combination, Chinese have 10 or more species in a single pond thus maximizing productivity. Indian aquaculture has demonstrated a six and half fold growth over the last two decades, with freshwater aquaculture contributing over 95% of the total aquaculture production. India is bestowed with 3.15 million ha of reservoirs, 2.36 million ha of ponds and tanks as well as 0.19 million ha of rivers and canals. Freshwater aquaculture with a share of 34% in inland fisheries in mid 1980s has increased to about 80% in recent years (DADF, 2017). The technologies of induced carp breeding and polyculture in static ponds and tanks have brought about remarkable upward trend in aquaculture productivity and turned the sector into a fast-growing industry. The research and development programs of the Indian Council of Agricultural Research (ICAR), as well as the development support provided by the Government of India (GoI) through a network of Fish Farmers' Development Agencies (FFDA) have been the principal vehicles for this development. Additional support has been provided by several other organizations, state departments and financial institutions. So far, about 0.65 million ha of water area has been brought under fish farming covering 1.1 million beneficiaries. Currently the average annual yield is around 03 tonnes / ha. At the same time, training has been imparted to about 0.8 million fishers (DADF, 2017).

The main activity in the Indian aquaculture sector is the production of freshwater carp and brackish water shrimp. The total inland production of India put together capture and culture was 8.9 million tonnes in the year 2017-18 (FSI Handbook of Fisheries Statistics 2018). The three main Indian carps, namely rohu, catla and mrigal constitute the major part of production with more than 5.86 million tonnes; followed by silver carp, grass carp and common carp constituting a second important group with 0.58 million tonnes. Average national production at pond level increased from 0.6 tonnes / ha / year in 1974 to 3 tonnes / ha / year presently, with the existence of several aqua culturists with levels production as large as 8-12 tonnes / ha / year (Penman *et al.*, 2005). Technologies for induced carp reproduction and polyculture technologies in static ponds and reservoirs have revolutionized the freshwater aquaculture sector and transformed the sector into a rapidly growing industry. The sector has witnessed the increased interest in diversification with the integration of high value species, including medium and minor carp, catfish, etc. Carp and other fish are produced for the local market while a proportion of freshwater shrimp production is for export. In contrast, the development of aquaculture in brackish water has been limited majorly to a single species, the *Litopenaeus vannamei* and to some extent *Penaeus monodon*.

Origin and History of Aquaculture

Aquaculture refers to the long history of fish culture in Asia, ancient Egypt and central Europe. The Classic of Fish Culture, believed to have been written around 500 BC by Fan Lei, a Chinese politician-turned fish-culturist, is considered proof that commercial fish culture existed in China in his time, as he cited his fish ponds as the source of his wealth. Later writings of Chow Mit of the Sung Dynasty (Kwei Sin Chak Shik in 1243 AD) and of Heu (A Complete Book of Agriculture in 1639 AD) describe in some detail the collection of carp fry from rivers and, in the latter publication, methods of rearing them in ponds. Even though stews or storage ponds for eels and other fish existed in Roman times and later in monastic houses in the Middle Ages, and a 2500 BC bas-relief of fish in Egypt is believed to be of tilapia raised in a pond, the earliest form of fish culture appears to be of the common carp (*Cyprinus carpio*), a native of China. Indigenous systems of Indian carp culture seem to have existed in eastern parts of the Indian subcontinent in the 11th century AD. Fish culture was practised in Indo-China for many centuries and the early systems of pen and cage culture of catfish appear to have indeed originated in Cambodia. Probably starting as a means of holding fish alive before marketing, flow-through culture from fry to market size with artificial feeding developed in the course of time.

Dr Hiralal Chaudhuri (1921–2014) - Owing to his pioneering research in 'induced breeding of fish', Chaudhuri is regarded as the 'father of induced breeding' in the country. To honour this great achievement, which in fact was the harbinger of the 'first blue revolution' in the country, the Government of India has declared 10 July as the 'National Fish Farmers' Day' (commemorating the 'first induced breeding' of *Cirrhinus reba* in captivity on 10 July 1957 by Chaudhuri). Thus, Chaudhuri has been immortalized by his outstanding original research contribution

The history of aquaculture projects all over the world has led to the conclusion that the right selection of sites is probably the most important factor in determining the feasibility of viable operations. Although many of the factors will depend on the culture system to be adopted, there are some which affect all systems, such as agroclimatic conditions, access to markets, suitable communications, protection from natural disasters, availability of skilled and unskilled labour, public utilities security, etc.

With more than 8,000 kilometers of coastline, there is great potential for the development of mariculture which only started in recent years with the cultivation of mussels and oysters. Given the substantial contribution of aquaculture to socio-economic development in terms of income and creation of employment through the use of unused and little used resources in several regions of the country, ecological aquaculture has been considered as an engine of rural development, and food and nutritional security for the rural masses. It also has great potential as a source of foreign currency. Greater support for R&D with intense links between research and development agencies, increased investment in fish and shrimp hatcheries, establishment of aquaculture farms, crushers and auxiliary industries have all been identified as important areas to maintain the growth rate of the sector. Ecological aquaculture has been seen as a driver of rural development, and of food and nutritional security for the rural masses. It also has great potential as a source of foreign currency.

Aquaculture Practices

Based on Culture Practices

Extensive Aquaculture

Extensive aquaculture involves low degree of control over environment, nutrition, predators, competitors, and disease-causing agents. Plant and animal seed stock is obtained from nature. Cost, technology, stocking rates, and production levels are low.

Semi Intensive Aquaculture

Semi-intensive aquaculture involves a combination of some attributes of extensive and intensive aquaculture. It is usually done in man-made ponds and raceways. Cost, technology, stocking rate and production levels are all intermediate.

Intensive Aquaculture

Intensive aquaculture involves high degree of control over the systems. Seed stock is produced from domestic brood stock within the system. Cost, technology, stocking density, and production levels are all high.

Based on Species to be Cultured

Monoculture: Culture of one species in Extensive, intensive, or semi-intensive condition

Polyculture: Culture of multiple species usually under semi-intensive condition

Integrated farming: Integration occurs when outputs (usually by-products) of one production sub-system are used as inputs by another, within the farm unit (Semi-intensive system)

Based on Salinity

Freshwater Aquaculture

The freshwater aquaculture comprises of the culture of carps, catfishes, freshwater prawns, pangasius, and tilapia. With technological inputs (Induced breeding of carps and catfishes, hatcheries for mass-scale spawning, seed rearing and carp polyculture), entrepreneurial initiatives and financial investments, the pond productivity has gone up to 2000 kg/ha/yr, with several farmers and entrepreneurs achieving higher production levels of 6000-8000 kg/ha/yr (Ayyappan *et al.*, 2011).

Carp is the mainstay of culture practice in the country, which is supported by strong traditional knowledge base and scientific inputs in various aspects of management. Carps contribute 87% of the total aquaculture production. Although the country is enriched with large number of potential cultivable carp species, it is only the three Indian major carps; catla (*Catla catla*), rohu (*Labeo rohita*) and mrigal (*Cirrhinus mrigala*), that contribute a major share. Scientific interventions in the last five decades have led to the development of carp culture technologies with varied production potentials depending on the type and level of inputs. Further, other species like catfishes, freshwater prawns and molluscs for pearl culture have also been brought into the culture systems. In addition, a range of other non-conventional culture systems, like sewage-fed fish culture, integrated farming systems, cage and pen culture, running water fish culture have made freshwater aquaculture a growing activity across the country. Being mainly organic-based, the freshwater aquaculture practices are also able to utilize and treat a number of organic wastes including domestic sewage, enabling eco-restoration.

Culture and Seed production of major cultivable species

Seed being the basic input in any culture system, its production has been accorded highest priority in terms of broodstock management, establishment of hatcheries, refinement of induced breeding techniques, rearing and production of quality seed across the country.

Culture of Carps

The technology of induced breeding of carps under control condition has become a common practice of the farmers today. The induced breeding technology has made mass production of quality seed under control condition possible, thereby, reducing the dependence on natural seed collection. Besides Indian major carps, the technology of breeding of Chinese grass carp and silver carp has also been domesticated all over the country. Various carp species are domesticated to breed before and after the monsoon. The technology of multiple breeding of carps developed by Central Institute of Freshwater Aquaculture (CIFA) has been able to demonstrate 2-3-fold higher spawn recovery from a single female during season through 3-4 times breeding within an interval of about 45 days. The technological evolution of hatchery design and operation from initial earthen pits to double-walled hapa and subsequently to glass-jar and circular ecohatchery provided scope to produce and handle mass quantities of eggs during hatching. Carp hatcheries in the public sector have contributed to an increase in seed production from 6,321 million fry in 1985-86 to over 39,000 million fry at present (Source: NFDB). Even states like Assam and West Bengal are producing seeds much beyond their requirement, showing the prospects of export trade and its economic viability.

Raising of seed in the initial two stages is associated with high rates of mortality due to several management problems. Package of practices have been developed and standardized for raising fry and fingerlings with higher growth and survival levels. Higher survival levels of fry of over 40-60% through intensive rearing during nursery stage have been demonstrated at stocking densities of 5- 10 million/ ha in earthen ponds and up to 30 million in ferro-cement tanks (Ayyappan *et al.*, 2011). Further, at 2-3 lakh/ha stocking densities the technology of fingerlings rearing can result 60- 80% survival, with mean fingerlings size of 100 mm in a rearing period of 3 months (Ayyappan *et al.*, 2011).

Culture of Catfish

The pond culture of catfish involves mainly Philippine catfish 'magur' (*Clarias batrachus*) is currently propagated on a large scale along the north-eastern regions, mainly in the State of Assam. While the stinging catfish, 'singhi' (*Heteropneustes fossilis*) has potentials of mono- and polyculture (Clarias and Anabas). The induced breeding technology of induced breeding and seed production of these two important fishes has been standardized. The pond management measures were more or less similar to that of carp with stocking density of 30,000-50,000 fingerling/ha. Production level of 3-5tonnes/ha (Ayyappan *et al.*, 2011) is achieved in grow out culture of magur. Considering the high market demand for catfish and the availability of a huge potential resource in the form of swamps and derelict waters, commercial farming of these species is presently receiving important attention. Research with regard to development and standardization of induced breeding and grow-out technologies of several other non-air breathing catfishes like *Mystus* seenghala, *M. aor*, *Pungasius pungasius*, *Wallago attu*, *Ompak pabda* are also being envisaged in view of the high consumer preference for these in different parts of the country.

Culture of Giant River Prawn

The giant river prawn (*Macrobrachium rosenbergii*) is the largest and fastest growing species being farmed and possesses considerable demand both in domestic and international markets. It is cultured either as monoculture or as polyculture. The monoculture of giant river prawn is mostly confined to ponds with supplementary feeding and a production yield level of 1.0–1.5 tonnes/hectare in a 7–8-month production cycle. The polyculture of freshwater prawn with juvenile carps has also been demonstrated to be economically viable. There are about 71 scampi hatcheries in the country during the last decade including 43 in Andhra Pradesh with an installed capacity of about 8 billion Pl/yr (Ayyappan *et al.*, 2011).

Freshwater Pearl Culture

While marine pearl culture in India had its beginning in the early seventies, freshwater pearl culture remained an unexplored area till late eighties until the research programmes by the Central Institute of Freshwater Aquaculture, Bhubaneswar were initiated. The investigations by the institute in last one and half decades have not only led to development of the base technology of surgical implantation by using three commonly available freshwater mussel species, *viz.*, *Lamellidens marginalis*, *L. corrianus* and *Parreysia corrugata,* but also standardized different steps involved for the production of cultured pearls. Three different surgical procedures that is, mantle cavity insertion, mantle tissue implantation and gonadal implantation techniques have been standardized for obtaining different kinds of pearl products.

Integrated Fish Farming

It is the combination of two or more normally separate farming systems where byproduct i.e., waste from one subsystem is utilized for sustenance of other; for example fish-pig/poultry/duck farming. Though organized integrated farming systems are not very common in the country, use of organic manures in the form of cattle wastes and poultry droppings are common in most of the farms of the country, especially, in carp culture farms. Production levels 3-5 tonnes/ha/year have been demonstrated by the integration of fish with poultry/duck/pig, with waste derived from these farm animals as principal input and without provision of any supplementary feed.

Cage and Pen Culture

Commercial fish farming in cages is in the nascent stage in the country, even though the practice is widely accepted globally. The information on cage culture in the country is limited to a few experimental trials with major carps and catfishes, with a maximum-recorded production of 3.3 kg/m3/month during grow-out culture of grass carp (Karnatak and Kumar 2014). Few fisheries groups are doing cage culture of tilapia and pangasius in various reservoirs but its much limited considering over 3.15 million ha potential area which are available under reservoirs. The emphasis on cage culture is inevitable in coming years to meet the ever-increasing demand of fish. Pens are usually constructed in shallow margins of reservoirs, tanks and ox-bow lakes. They can effectively be utilized for raising fry and fingerlings, which has been demonstrated in several trials carried out all over the country.

Sewage-fed Fish Culture

The practice of recycling sewage through agriculture, horticulture and aquaculture is in vogue traditionally in several countries, including India. Sewage-fed fish culture in *bheries* of West Bengal is an age-old practice. Though the area of coverage is gradually reducing, about 5700 ha is still utilized for growing fish by intake of raw sewage into the system and as much as 7000 tonnes (Webster 2003) of fish, mainly contributed by carps, are produced annually from these water bodies. The Central Institute of Freshwater Aquaculture has evolved an aquaculture-based sewage treatment system incorporating duckweed and fish culture for treatment of domestic sewage.

Coldwater Fisheries

The country possesses significant aquatic resources in terms of upland rivers/ streams, high and low altitude natural lakes, manmade reservoirs, both in Himalayan region and western ghats, which hold large populations of both indigenous and exotic cultivable and noncultivable fish species. Important food fishes in the region are mahseers and schizothoracids among the indigenous species and trouts among the exotic varieties. Research efforts over the years have led to development of technology of seed production of important cultivable species like trout, mahseers and snow trout. High survival rates of hatchery seed in case of trout along with successes in production of mahseer seed under control conditions have led to possibilities of farming. Breeding of different species of snow-trout *viz.*, *Schizothoraichthys niger*, *S. esocinus*, *S. micropogon* and *S. planifrons* and *Schizothorax richardsonii* has also become possible and the technology has been perfected for mass production of the seed under controlled farm conditions (Sehgal 1999). In India total 25 trout hatcheries are installed with a capacity of 1.5-2.0 million trout seeds (Ayyappan *et al.*, 2011).

Brackishwater Aquaculture

Brackishwater aquaculture in India is an age-old practice in *bheries* of West Bengal and *pokkali* fields of Kerala. The modern and scientific farming in the country is only about a decade old. The country possesses huge brackishwater resources of over 1.2 million hectares suitable for farming (Abraham & Sasmal, 2009). However, the total area under cultivation is just over 13% of the potential water area available i.e. 157,400 ha in 2012-13 (Ayyappan *et al.*, 2011). Shrimp is the single commodity that contributes almost the total production of the sector. The production levels of shrimp recorded marked increase from 28,000 tonnes in 1988- 89 to 6,80,015 tonnes in 2017-18. The major contribution is from the *Litopenaeus vannamei* and rest from *P. monodon.*

Culture of crab species like *Scylla serrata* and *S. tranquebarica* has also been taken up by few entrepreneurs. There are several other finfish species like *Mugil cephalus*, *Liza parsia*, *L. macrolepis*, *L. tade*, *Chanos chanos*, *Lates calcarifer*, *Etroplus suratensis* and *Epinephelus tauvina* which possess great potential for farming, but commercial production of these species is yet to be taken up in the country. The studies on induced breeding of shrimps were initiated by the Central Marine Fisheries Research Institute in the early 70s'; an experimental hatchery was established by the Institute in 1975 at Narakkal, Kerala. MPEDA took the lead for establishment of two largescale hatcheries *viz.*, TASPARC (The

Andhra Pradesh Shrimp Seed Production Supply and Research Centre) and OSPARC (Orissa Shrimp Seed Production Supply and Research Centre) in 80's that gave a boost for the establishment of a number of commercial hatcheries in the private sector. The technology of hatchery production of shrimp seed involving broodstock development, induced maturation and spawning, larval-rearing and post-larval (nursery) rearing has been standardized. At present about 351 shrimp hatcheries are operational with a total production capacity of 14 billion PL20/year (Deo *et al.*, 2013). Though brackishwater farming in India is an age-old practice, the scientific and commercial aquaculture of the country at present is restricted to shrimp farming owing to the high export potential of the shrimps. Semi-intensive culture practices mainly with black tiger prawn have demonstrated production levels of 4-6 t/ha (FAO National Aquaculture Overview of India) in a crop of 4-5 months. The high return coupled with credit facilities from commercial banks and subsidies from MPEDA have helped in the development of shrimp farming in the country to a multi-billion-dollar industrial sector.

In India, a major shift in India's policy on shrimp took place with the introduction of an exotic species of shrimp, viz, *Litoenaeus vannamei*. The pilot-scale introduction of *L.vannamei* initiated in 2003 and after a risk analysis study large-scale introduction has been permitted in 2009. The introduction of vannamei in India occurred under controlled conditions with a clear procedure laid down by the government. Initially, two companies, Sarat Seafood and BMR hatcheries (FAO National Aquaculture Overview Country India), were given permission to import broodstock from approved countries and conduct trials in a restricted environment. The Central Institute of Brackishwater Aquaculture and National Bureau for Fish Genetic Resources conducted the risk analysis for the introduction of vannamei in India. Following the risk analysis studies, the government decided for a large-scale introduction of commercial use of vannamei in 2009. P. vannamei importation and cultivation guidelines were prepared by the Department of Animal Husbandry, Dairying and Fisheries. Coastal Aquaculture Authority (CAA), of the Government of India, Chennai is the agency for granting permission to import vannamei broodstock and for giving permissions for vannamei culture by farmers. To facilitate farmers in getting quality SPF vannamei seed, the Government of India set up a quarantine center at Chennai and all vannamei broodstock is allowed to enter India after the consignment is cleared at this quarantine center at Chennai. Currently, CAA has given permissions to farmers for farming vannamei in 22 715 hectares and allowed 135 hatcheries (Ayyapaan *et al.*, 2011) for importing vannamei broodstock for production and supply of quality SPF vannamei seed to farmers.

Inland saline aquaculture

Inland saline aquaculture is a flagship program of CIFE carried out at its Rohtak centre. It has the unique distinction of being the only central research centre in India dedicated to research on use of inland saline soils and ground water for fish and shrimp culture (Inland saline aquaculture).Some of the technologies developed using inland saline ground water include commercial farming of Pacific white shrimp (*Litopenaeus vannamei*) ; commercial farming of Tiger shrimp (*Penaeus monodon*) in saline affected soils using inland saline water; Giant freshwater Prawn (*Macrobrachium rosenbergii*) seed production and culture technology ; Grow out culture technologies for Indian Major carps (IMC) & Minor carps in low saline water areas; and Pangasius (*Pangasianodon hypophthalmus)* culture technology for low inland saline water / soil areas.

Mariculture

Intensive researches during last two decades by the Central Marine Fisheries Research Institute have led to the development of several viable technologies with regard to seed production and culture of important marine crustaceans, molluscs and seaweeds. Several programmes on sea ranching of exploited stocks such as pearl oyster, *Xancus pyrum*, *Trochus* sp., *Turbo* sp. and giant clam have been taken up in the country.

Culture of important marine molluscs and seaweeds

Mussel Culture

Green mussel, *Perna viridis* and brown mussel, *Perna indica* are the two important mussel species available in the country. The culture technology of these has been standardized. Mussel farming is carried out either in rafts or by long line methods. While long line system is very flexible and can withstand turbulent sea, raft system is more rigid and suited for more calm seas. Mussels attain harvestable size of 70-80 mm in 6-7 months of culture period (Gopakumar 2015); production levels of 12-14 kg mussels/meter (Gopakumar 2010) of rope have been reported. Economic analysis of the mussel farming made based on pilot scale studies on raft culture by CMFRI showed over 40% profit margins on investment of about Rs. 24,000 per raft of 8mx8m during a culture period of 6-7 months (Asokan 2013).

Edible Oyster Culture

The culture of edible oyster in India was initiated as early as the beginning of this century. However, intensive researches on various aspects of the culture were taken up only during seventies. The technique of oyster farming consists of two items, collection of spat and growing the spat to adult stage. *Crassostrea madrasensis* is the only species that is found to be important for commercial farming. This species reach harvestable size (80 mm) in a culture period of 7-8 months and production levels of 8-10 tonnes (Gopakumar 2010) of shell on oysters/ha are obtained.

Seaweed Culture

Seaweed forms an important component of the marine living resources, available largely in shallow seas, wherever, suitable substratum is available. Agar agar and algin are two principal industrial products of seaweeds. Seaweed is also used as food, fodder, fertilizers and in several other industrial and pharmaceutical products. The seaweed resources of the country are mainly confined to the coasts of Tamil Nadu and Gujarat. Since 1972, CMFRI is involved in experimental culture of different seaweed species and developed technologies for important agarophytes like *Gracilaria edulis, G. corticata* and *Gelidiella acerosa*. Both net and rope culture technologies have been standardized. In case of *G. acerosa* both coral stone method and net culture method have been standardized. Culture practices of several other species are on experimental scale.

As far as marine fish farming is concerned, culture of Lethrinus, Epinephelus, *Mugil cephalus*, milk fish, cobia, pompano and pearl spot has been tried, either in monoculture or in the integrated systems. Success has been achieved in the broodstock development and spawning of greasy grouper, seabass, cobia, pompano and *M.cephalus*.

During 2006–2009, the Central Marine Fisheries Research Institute (CMFRI), Kochi, has made commendable progress in marine cage culture. In 2007, a cage of 3m diameter and 4m depth was floated in Vizhinjam bay and stocked with juveniles of *Caranx sexfasciatus* of average size 81.7 mm and 7.8 g weight. Within four months, they grew to an average size of 210 mm (186g). Sturdy open-sea cage culture protected by outer predator nets and special shock absorber to withstand and absorb pressure was moored at a depth of 11 m about 300 m from beach at Visakhapatnam in Andhra Pradesh State. About 1350 barramundi, *Lates calcarifer* of 14.5 g seeds were stocked in the cage and reared by feeding low-value trash fish. The barramundi cage culture is getting impetus due to availability of its seed and necessary funding provided by the National Fisheries Development Board (NFDB), in Hyderabad (FAO National Aquaculture Oveview, Country India).

Institutional Framework of Aquaculture Development in India

Department of Fisheries under Ministry of Fisheries, Animal Husbandry and Dairying was created in 2019. This agency is responsible for planning, monitoring and the funding of several centrally sponsored developmental schemes related to fisheries and aquaculture in all of the Indian States. Centrally sponsored schemes like the 422 FFDAs cover almost all districts in the country and the 39 BFDAs in the maritime districts have also contributed to aquaculture development.

The Indian Council of Agricultural Research is an autonomous body responsible for co-ordinating agricultural education and research in India. The Fisheries division of ICAR undertakes the R&D on aquaculture and fisheries through a number of research institutes. There are about 400 Krishi Vigyan Kendras (Farm Science Centres) in the country, operated through State Agricultural Universities, ICAR Research Institutes and NGOs, Central Agricultural Universities, most of which also undertake aquaculture development within their scope of activities. The MPEDA functioning under the Ministry of Commerce, besides its role in the export of aquatic products also contributes towards the promotion of coastal aquaculture. Many other organizations and agencies also support R&D and include the Departments of Science and Technology; Biotechnology departments, University Scholarship Commissions, NGOs and private industry. Presently there are 30 professional fisheries colleges in India (Kumar, *et al.*, 2018).

Major Institutes involved in the Development of Aquaculture Sector of India

Rajiv Gandhi Centre for Aquaculture (RGCA)

RGCA is the Research & Development arm of Marine Products Export Development Authority (MPEDA) and is dedicated to augment the Indian seafood exports through sustainable culture technologies.

Mission of RGCA

1) To promote sustainable aquaculture with long term vision
2) To establish Technology Development Centres in Aquaculture in various locations in India for developing and disseminating appropriate technologies for scientific aquaculture

3) To develop and introduce world class sustainable technologies in aquaculture
4) To transfer technical know-how, plans, designs and other relevant information for establishing aquaculture units in various states of India
5) To give consultancy and technical services to the entrepreneurs and farmers for establishing aquaculture units
6) To impart training in various aquaculture technologies developed at its centres
7) To conduct pilot scale operations and to set up demonstration farms to popularize the technologies developed/acquired
8) To scale up the technologies developed in research institutes by joining hands with the concerned scientists and disseminate the same through extension, education and demonstration programmes
9) To assist National institutes, agencies both in public as well as in private sectors for developing innovative technologies which are having scientific application
10) To undertake execution of Aquaculture projects entrusted by Government agencies/ departments like Department of Bio-Technology (DBT), Department of Ocean Development (DOD), Ministry of Agriculture, Ministry of Commerce, Ministry of Food Processing etc
11) To take up such activities as to re-seed and replenish the over exploited stock of the sea and other large inland water bodies through ranching with hatchery reared young ones for sustainable development of fisheries
12) To introduce proven aquaculture technology of the selected species which are commercially successful elsewhere in the world but not yet introduced in India. The centre will buy the technology from national or international source, blend the same under Indian condition with local technology if available and sell the same to Indian entrepreneurs after assuring the commercial viability

Coastal Aquaculture Authority (CAA)

The Coastal Aquaculture Authority (CAA) was established under the Coastal Aquaculture Authority Act, 2005 and notified vide Gazette Notification dated 22nd December, 2005. The main objective of the authority is to regulate coastal aquaculture activities in coastal areas in order to ensure sustainable development without causing damage to the coastal environment. The authority is empowered to make regulations for the construction and operation of aquaculture farms in coastal areas, inspection of farms to ascertain their environmental impact, registration of aquaculture farms, fixing standards for inputs and effluents, removal or demolition of coastal aquaculture farms, which cause pollution etc.

National Centre for Sustainable Aquaculture (NaCSA)

National Centre for Sustainable Aquaculture (NaCSA) was established by MPEDA in the year 2007 as an outreach organization for uplifting the livelihood of small-scale shrimp farmers. The long term objective of NaCSA is to enable aquaculture farmers to adopt sustainable and environment friendly farming practices to produce quality and safe aquatic products such as shrimps, scampi and fish for export and domestic markets. NaCSA will facilitate links between aquaculture stakeholders and strengthen farmer societies, and farmers to facilitate formulation of common policies, strategies and voluntary guidelines to benefit farming community as a whole in the country aiming at a 'gross root level' approach. The main objectives of NaCSA are summarised as follows,

1) To develop social contacts and spirit of fellow feeling among its members and to maintain a library with good books, periodicals and newspapers.
2) To strive for the eradication of illiteracy
3) To develop the social and cultural activities for the welfare of the society
4) To function as the prime mover of the extension activities among aquaculture farmers
5) To provide updated information to technical staff of aquaculture societies
6) To act as a federation of all the aquaculture societies and facilitate formulation of common policies, strategies, etc
7) To provide common infrastructure facilities like testing laboratories etc
8) To act as a central agency for standardization of inputs in aquaculture

National Fisheries Development Board (NFDB)

The National Fisheries Development Board (NFDB) was established in 2006 as an autonomous organization under the administrative control of the Department of Fisheries, Ministry of Fisheries, Animal Husbandry and Dairying, Government of India to enhance fish production and productivity in the country and to coordinate fishery development in an integrated and holistic manner. One of its main objectives is to provide focused attention to fisheries and aquaculture (Production, Processing, Storage, Transport and Marketing).

NFDB works closely with all the Central Fisheries Organizations and Institutes, Universities, States' Department of Fisheries, besides various stakeholders thus contributing to the development and management of fisheries sector. NFDB has published "Aquaculture Technologies Implemented by NFDB" in November 2019 which covers different aquaculture techniques related to wetland fisheries development, cage culture in inland open water bodies, brackishwater cage culture, farming of silver pompano in brackishwater ponds, sea cage farming, seaweed cultivation, backyard recirculatory aquaculture system, aquaponics system, integrated paddy cum fish culture, aqua one centre – an ICT enabled aquaculture support service etc.

Central Institute of Freshwater Aquaculture (CIFA)

The Central Institute of Freshwater Aquaculture (CIFA) is a premier research Institute on freshwater aquaculture in the country under the aegis of the Indian Council of Agricultural Research (ICAR), New Delhi. The institute has had its beginnings in the Pond Culture Division of Central Inland Fisheries Research Institute (CIFRI) established at Cuttack, Orissa in 1949 with a view to face challenges in the field of fish culture in ponds, tanks and other small aquatic body. Subsequently, CIFRI, in a major effort to give emphasis to freshwater aquaculture research, initiated steps to establish the Freshwater Aquaculture Research and Training Centre (FARTC) over 147 ha campus at Kausalyaganga, Bhubaneswar, Orissa. The centre gradually developed into its full capacity and became an independent institute during 1987 as Central Institute of Freshwater Aquaculture (CIFA). The institute is also the Lead Centre on 'Carp Farming in India' under Network of Aquaculture Centres in Asia-Pacific (NACA) operative under Food and Agriculture Organisation of United Nation (FAO).

The vision of CIFA is to make an Indian freshwater aquaculture globally competitive through eco-friendly and economically viable fish production systems for livelihood and nutritional security. Their mandate involves doing basic and strategic research for the development of sustainable culture systems for freshwater finfish and shellfish, species and systems diversification in freshwater aquaculture & human resource development through training, education and extension.

Central Institute of Brackishshwater Aquaculture (CIBA)

Central Institute of Brackishwater Aquaculture (CIBA) was established under Indian Council of Agricultural Research, New Delhi, the Ministry of Agriculture, Government of India on 1-4-1987. The institute is based in Chennai city of Tamil Nadu.

The main mandate of CIBA is to do basic and strategic research for sustainable brackishwater culture system, species and systems diversification in brackishwater aquaculture & human resource development through training, education and extension. CIBA acts as repository of information on brackishwater fishery resources with a systematic database and also do the work of human resource development, capacity building and skill development through training, education and extension.

Central Marine Fisheries Research Institute (CMFRI)

The Central Marine Fisheries Research Institute was established by Government of India on February 3rd 1947 under the Ministry of Agriculture and Farmers Welfare and later it joined the ICAR family in 1967. During the course of over 65 years the Institute has emerged as a leading tropical marine fisheries research institute in the world.

Mariculture is one of the important divisions of CMFRI and thrust research areas of this division include 1. Captive broodstock development of marine finfishes and induced breeding techniques 2. Development of high density larviculture techniques 3. Evolving protocols for diversified farming practices such as sea cage farming and coastal mariculture 4. Demonstration of mariculture technologies and participatory farming programmes 5. Development of growout feeds and health management protocols for mariculture.

Central Inland Fisheries Research Institute (CIFRI)

Recognizing the role of inland fisheries, Government of India established a Central Inland Fisheries Research Station at Calcutta on 17 March 1947, under the then Ministry of Food and Agriculture. Later, in 1959, the Central Inland Fisheries Research Station was elevated to a full-fledged research institute, christened as "Central Inland Fisheries Research Institute". In 1967, it came under administration of Indian Council of Agricultural Research, Ministry of Agriculture and Farmers Welfare, New Delhi. Vision of CIFRI is to develop sustainable fisheries from inland open waters for environmental integrity, livelihood and nutritional security.

Directorate of Coldwater Fisheries Research (DCFR)

Coldwater fisheries has a great potential in generating rural income and providing food security in Indian Uplands. To utilize the available resources and opportunities in the coldwater fisheries sector, the involvement of Indian Council of Agricultural Research in this sector started in late sixties which subsequently culminated in the creation of National Research Centre on Coldwater Fisheries (NRCCWF) as an independent research center on 24 September 1987 during the VII five-year plan. NRCCWF is the only national facility in the country to take up the research investigations on capture and culture aspects with a focus on exotic and indigenous coldwater fish species. NRCCWF has been converted to Directorate of Coldwater Fisheries Research (DCFR) in XI five-year plan. DCFR is located at Bhimtal.

Main mandate of the DCFR is to do basic, strategic and applied research in Coldwater fishes and aquaculture, act as a repository of hill fisheries resources and human resource development through training, education and extension.

Central Institute of Fisheries Education (CIFE)

ICAR-Central Institute of Fisheries Education (CIFE) is a leading Fisheries University having a distinguished heritage and has nurtured many illustrious scholars and leaders over the years. In over 50 years of existence, CIFE has emerged as a Centre of excellence for higher education in fisheries and allied disciplines. The institute was established on 6th June 1961, under the Ministry of Agriculture, Govt. of India with assistance from FAO/UNDP. It came under the administrative control of Indian Council of Agricultural Research (ICAR) on 16th April 1979 and subsequently, the scope and mandate have been widened to include education, research and extension. Recognizing the pivotal role played by CIFE in human resources development in fisheries, the institute was conferred the status of Deemed-to-be-University on 29th March 1989.

Vision of CIFE is to be a world-class organization providing leadership in fisheries education and research and a mission to achieve academic and research excellence by creating state-of-the-art infrastructure and attracting globally competent faculty and students. Mandate of the institute are (1) to conduct post-graduate academic programmes in core and emerging disciplines of fisheries science (2) To conduct basic and strategic research in frontier areas of fisheries (3)to conduct demand-driven training and educational programmes for different stakeholders in fisheries sector and (4) to provide technical support, inputs for policy development and consultancy services.

CIFE has well established brackish-water farm at Kakinada, Andhra Pradesh, freshwater fish farm at Powerkheda, MP and inland saline soil farm at Rohtak, Haryana.

National Bureau of Fish Genetic Resources (NBFGR)

National Bureau of Fish Genetic Resources (NBFGR) was established in December 1983 at Allahabad under the aegis of Indian Council of Agricultural Research (ICAR) to undertake research related to the conservation of fish germplasm resources of the country. The institute is presently located at Lucknow, U.P. The institute's vision is assessment and conservation of fish genetic resources for intellectual property protection, sustainable utilization & posterity.

Central Institute of Fisheries Technology (CIFT)

Central Institute of Fisheries Technology (CIFT) set up in 1957 and located at Kochi, Kerala is the only national center in the country where research in all disciplines relating to fishing & fish processing is undertaken. The institute has mandate i) To do basic and strategic research in fishing & processing ii) Design & develop energy efficient fishing systems for responsible fishing and sustainable management iii) Development of implements & machinery for fishing and fish processing iv) Human resource development through training, education & extension.

Tamil Nadu Dr J Jayalalithaa Fisheries University (TNJFU)

The Department of Fisheries was started in Tamil Nadu Agricultural University (TNAU) during 1973. The Fisheries College, Thoothukudi was established in 1977 and it was upgraded as Fisheries College and Research Institute (FCRI) during the year 1989. The Fisheries College and Research Institute (FCRI), Thoothukudi became a constituent Institute of Tamil Nadu Veterinary and Animal Sciences University (TANUVAS) with effect from 20.09.1989. Tamil Nadu Fisheries University (TNJFU) was established on June 19, 2012 with headquarters at Nagapattinam.

The University has aquaculture quality testing labs at various places in Tamil Nadu. Mobile aquaculture testing lab facility is also made available by this university. It has developed exclusive aquaculture research farm facilities. The university has established Directorate of Sustainable Aquaculture (DSA) & Directorate of Incubation and Vocational Training in Aquaculture (DIVA) under which there are various centers operational for carrying out the activities in sustainable aquaculture.

Kerala University of Fisheries and Ocean Studies (KUFOS)

The Kerala University of Fisheries and Ocean Studies (KUFOS) is an autonomous public funded institution established on 20th November 2010 and governed by the Kerala University of Fisheries and Ocean Studies Act, 2010. As per the Notification No.19540/Leg.I1 /2010/Law dated 28.01.2011, the Government of Kerala has promulgated Act 5 of 2011 forming the Kerala University of Fisheries and Ocean Studies (KUFOS). KUFOS is the first Fisheries University formed in India, and comes under the Kerala State Ministry of Fisheries.

Central Agricultural University (CAU) College of Fisheries (COF)

Government of India, established the Central Agricultural University (CAU) with its headquarters at Imphal, Manipur through the Central Agricultural University Act, 1992, No 40 dated 26th December 1992. Through the ordinance No 2 of 1996 promulgated by the Ministry of Agriculture, Department of Agricultural Research & Education (DARE), Government of India – the College of Fisheries started functioning from 03rd October 1998. The main mission of the college is to achieve the excellence in teaching, research & extension education in the field of fisheries and allied sciences.

New Approaches to Fish Farming

Good Aquaculture Practices

To tackle larger issues such as habitat destruction, climate change and disease pandemics, a novel approach to farm management is required. The Good Aquaculture Practices (GAqP) typically address environmental impacts, food safety, animal welfare and social aspects of aquaculture operations. It emphasises planning new farms through ecosystem and social assessments of the proposed sites and cooperation with other stakeholders. This holistic approach to farm management embraces the interconnectedness of farms and carrying capacities of the surrounding environment. GAqP considers the potential effects of climate change and community conflicts on new and existing farms, to ensure farms are more resilient to environmental and social issues (Bone *et al.,* 2018).

An example of effective, coordinated disease management in inland systems is the control of Spring Viremia in Carp (SVC) in the UK. The UK adopted a coordinated control and eradication programme in 2005 and was free of the pathogen in 2010. This was possible due to legislation that allowed for a passive surveillance programme. In addition to passive surveillance, the authorities educated stakeholders and conducted active surveillance with government and industry bodies, which focused on testing aquaculture sites, traders and imports at particular risk. On suspicion of the disease, the authority would place temporary movement limitations on a site until testing. Once the disease was confirmed, the movement of animals on or off site was prohibited until the stock was culled and the site disinfected. When culling and disinfection was not possible, sites had to introduce strict biosecurity measures.

Farming Alternative Species

Fish and livestock farming operations release high quantities of nitrogen-based pollutants into marine ecosystems. In addition to innovations in farm practices, infrastructure and technology, shifting production to non-carnivorous species could help increase production and improve environmental performance of the sector. Currently, carnivorous fish, such as salmon, shrimp and tuna, are most popular with Western consumers. Farming these carnivorous fish requires more energy and feed resources compared to farming herbivorous fish. This trend results in pressure on wild fish stocks to supply feeds and overall these operations generate more waste.

By shifting to different species, global seafood production can be increased without requiring more resources. A recent academic report led by Professor Jennifer Jacquet, New York University, notes that recent moves to industrialise octopus farming are "counterproductive from a perspective of environmental sustainability as it would increase, not alleviate, pressure on wild aquatic animals". Demand for octopus in several Asian and Mediterranean countries is driving this trend, despite octopus having an FCR of 3:1 (Jacquet *et al.*, 2019). Considering these issues, farmers and legislators should rethink the strong focus on the farming of predatory fish to increase seafood production while improving sustainability impacts.

Farming filter feeders, such as bivalves, can have a positive impact on marine pollution. Some research has shown that filter feeders such as bivalves remove nitrogen more efficiently and at a lower cost than sewage treatment facilities (Chopin, 2013). Researchers estimated the value of eutrophication reduction services from shellfish and seaweed farms in the EU to be •11 to •17 billion in 2008 (Ferreira *et al.*, 2009). In 2016, bivalve farms in the EU were responsible for the removal of 46,800 tonnes of nitrogen (Ferreira and Bricker, 2016). Bivalve production is increasing. It has already surpassed the value of salmon farming in the US, with oysters ($173 million in 2015) and clams ($112 million in 2015) being the most popular (Maloney, 2018).

In addition, farming of certain marine species may play a role in sequestering carbon, therefore mitigating climate change. Marine algae and seaweeds also account for over half of all global carbon fixation and 70% of all carbon stored in the world's oceans (Chopin, 2013). Bivalves are thought to sequester carbon through shell production. Carbon makes up approximately 12% of bivalve shells.

However, as carbon is also emitted during bivalve shell production, researchers have not concluded whether bivalve production could serve as a carbon sink (Olivier *et al.*, 2016).

Carbon trading credits and nutrient trading credit systems have been proposed by academics in recent years to incentivise farmers to produce species with a positive impact on pollution (Chopin, 2013; Lindahl *et al.*, 2005; Olivier *et al.*, 2016). While still largely exploratory, building momentum with legislators and the private sector around these credit systems could support more sustainable farming practices.

Integrated Multi-Trophic Aquaculture (IMTA)

Several measures can reduce effluent pollution, such as reducing excess feed through standardising feed inputs and removing and treating excess nutrients in ponds. The FAO also proposes integrated multitrophic aquaculture systems that allow the waste of one species to serve as food for another as a cost-effective way of minimising water pollution.

As recently highlighted by FAO, farming multiple complementary species within the same production area allows farm operators to follow a more ecologically sound approach with better resilience to risks like disease outbreaks and climate change. Moving away from monoculture and adapting more natural polyculture farming systems is a fundamental part of Integrated Multi-Trophic Aquaculture (IMTA). This is the farming of selected species in a way that allows one species' uneaten feed, waste, nutrients and by-products to be recaptured and converted into feed and energy for other crops and/or marine animals farmed in the system. Farmers can combine production of carnivorous species, such as salmon, with seaweeds or filter feeders that consume excess nutrients from fed aquaculture for their growth (Chopin, 2013). This means the culture of fish in combination with filtering species, like seaweeds and bivalves, has a net-positive effect on water quality. Bivalves improve water quality for cultured fish, but research from Canada shows that seaweeds and mussels also benefit from being cultured in the vicinity of salmon farms, with up to 50% higher growth rates compared to monocultures.

In China, the government encouraged businesses to experiment with upscaling sustainable culture systems; one company received permission to upscale an IMTA system across a 40,000-hectare coastal area. In 2005, the company produced 28,000 tonnes of assorted seafood valued at over $60 million with a net profit of $18 million (Troell *et al.*, 2009). In 2017, the US, following a similar approach, awarded the first aquaculture permit in federal waters to Catalina Sea Ranch, which adopts IMTA to farm mussels, seaweeds and other sustainable marine crops. The company became profitable within two years of operation and in January 2019 applied for permission to expand its operations (Bartholomew, 2018).

Implementing IMTA systems is not currently a viable option for all aquaculture businesses as the technology has not been scaled in many areas and would require an overhaul of current monoculture production systems. However, open cage or pond systems cannot handle high stocking densities without compromising other valuable ecosystem services and without endangering the productivity and viability of their own operations over the long term (Goldberg and Naylor, 2005).

Offshore Aquaculture

Currently most marine farming takes place in coastal areas. Offshore aquaculture systems are situated further out at sea and may reduce pollution and eutrophication relative to coastal systems. Open sea areas cover less valuable habitats and are less fragile than coastal zones, resulting in greater carrying capacities for large-scale aquaculture operations. Higher flush rates also prevent farms from polluting and weakening their own fish stocks through farm effluents. This minimises disease outbreaks and limits the spread of diseases between farms.

Finally, as offshore facilities are situated far from local communities, the potential for social backlash is reduced (Buck *et al.*, 2018). While there are several benefits, areas further from the coast generally experience more adverse weather, which incurs higher equipment costs and may bring increased physical risk. It is also important to note that effluents are still discharged into the sea: however, it is expected that they will have fewer negative impacts in the open sea relative to shallower, more valuable coastal environments.

Offshore aquaculture is still in early stage of development and faces several sustainability and safety risks. Offshore areas typically experience stronger winds and bigger waves. Cage structures must be larger and stronger, resulting in higher start-up costs. Secondly, there are still many regulatory uncertainties, especially in Europe and North America. China was in the process of constructing 178 offshore pilot farms (China Dialogue Ocean, 2018) in 2018. For example, De Maas SMC, a large gas and oil contractor, was awarded a $150 million contract from the Chinese government to construct a pilot offshore farm (Harkell, 2018:B). Another Chinese consortium has reserved $955 million to develop three pilot farms. Norway is also making headway, with the first offshore trials for salmon farming approved in December 2018 (Aadland, 2018).

Bag nets

When businesses are unable to implement IMTA or offshore aquaculture, cage farming operations can significantly reduce effluent discharge and disease risk by employing bag nets instead of open nets. These bags are semi-closed containment systems that collect farm waste like faeces and uneaten feeds, so they do not pollute the environment (NOFIMA, 2014). They also create barriers against diseased wild fish and parasites in order to reduce infections in cultured stocks. Improvements are achieved with a small amount of energy compared to land-based farming systems. For example, in a study on salmon farming systems in Canada, farms with bag nets had the smallest overall environment footprint (Ayer and Tyedmers, 2009). Implementation is a challenge as bag nets present an added expense for producers.

Aquamimicry

Aquamimicry is the most advanced technology to the shrimp farming industry that provide natural live diets "Copepod" for post larvae prior to stocking, pond water stability, enhances good survival rate, fastest growth rate, high profitability, totally sustainable and without any destruction to earth. It is the intersection of aquatic biology and technology (synbiotics) mimicking the nature of aquatic ecosystems to create living organisms for the well-being development of aquatic animals.

Synbiotics refer to nutritional supplements combining probiotics and prebiotics in a form of synergism, hence synbiotics. The synbiotic concept was first introduced as "mixtures of probiotics and prebiotics that beneficially affect the host by improving the survival and implantation of live microbial dietary supplements in the gastrointestinal tract, by selectively stimulating the growth and/or by activating the metabolism of one or a limited number of health-promoting bacteria, thus improving host welfare". Aquamimicry helps in Efficiency Microbial Systems (EMS): Creates naturally synbiotics environment that balances both heterotrophication and autotrophication under biodynamics management. Early Maturity Shrimps (EMS): Enhances fastest growth rate by creating natural biocolloids, invertebrates, copepod and naturally live microdiets for the post larvae prior to stocking along with fermented plan protein throughout the culture cycle. Expanding Monetary Systems (EMS): Expands large number of organic ponds with excellent harvest under natural biosecurity, high profitability and totally environmentally friendly.

Aquamimicry is relatively new technique and spreading fast especially in Asia. Aquaculture entrepreneurs in India have also started adopting this technique for shrimp aquaculture however it is in very initial stages and efforts in right direction will help setting the trend.

Preventing Early Mortality Syndrome (EMS) in shrimp through Bio-Floc Technology

Water quality is key to preventing the spread of infections. Research finds that high acidity and salinity in ponds affects the risk of EMS infection (Sajali *et al.*, 2019). To reduce the risk of infection, farmers can line ponds with nets to prevent pathogens from entering and raise post larvae in separate tanks or net pens for ten to 20 days, to ensure they are strong enough and disease free before transferring them to freshwater ponds (The Fish Site, 2016).

Biofloc technology is an emerging method of preventing EMS infection. These systems convert toxic chemicals such as nitrate and ammonia into useful products such as fish feed. Studies show that shrimp infected with the EMS causing virus are more likely to survive in biofloc systems than clear seawater systems. These systems also reduce the risk of bacterial infection, which could decrease antibiotic use on shrimp farms. This is particularly important in the context of EMS, as several studies show that the bacteria that causes EMS (AHPND) is increasingly resistant to antibiotics (Sajali *et al.*, 2019).

While the technology is promising, there are barriers to its adoption. Biofloc systems are energy intensive as the water must be consistently aerated, they are also sensitive to light conditions and local water chemistry, making it difficult for farmers to implement (Thong, 2014).

Recirculatory Aquaculture System

Recirculatory Aquaculture Systems (RAS) is essentially a technology for farming fish or other aquatic organisms by reusing the water in the production. The technology is based on the use of mechanical and biological filters, and the method can in principle be used for any species grown in aquaculture such as fish, shrimps, clams, etc. Recirculation technology is however primarily used in fish farming. It is a technology that uses minimal resources to produce maximum output of fish. Development of RAS will modernise and diversify traditional aquaculture. This system rears fish in indoor tanks under control conditions.

Enhancing Information Technology / IoT (Internet of Things) use in Aquaculture

Digitalisation can boost production, curb diseases and enhance profitability. Feeds account for 50-70% of aquaculture costs (Partos, 2010), but up to 30% of feeds are left uneaten and pollute the local environment (Cubitt, 2008). One technology company, Cage Eye, created a hydro-acoustic system that detects noise levels among fish. As salmon make less noise when they stop eating, operators know when to stop feeding. Lingalaks CEO Erlend Haugarvoll says the technology could save his firm between $900,000 to $1.3 million in overheads, with savings estimated at $114 million yearly if the system is fully adopted in Norway (Baraniuk, 2018).

Facial Recognition Technology

In 2018, Cargill partnered with Cainthus to develop facial recognition technology for both terrestrial livestock and farmed seafood. The feed system uses predictive imaging to identify individual animals and track key data such as food intake, heat detection and behaviour patterns. This generates predictive analytics to help operators make better decisions. Cargill's recently developed iQuatic software also gives operators real-time access to dissolved oxygen levels, weather shifts, animal health and other factors which use machine learning to optimise advice over time. Another new company, Umitron, combines solar-powered sensors and machine learning to optimise feeding.

Automated Video Surveillance

It is being used to determine when and how much to feed caged salmon, while looking out for potential pests like sea lice. Self-guided laser systems that kill sea lice without harming fish have also been developed. These laser systems are placed in the centre of fish cages. When a camera detects sea lice, the laser targets the lice without harming the salmon. As a relatively new and patented solution, there is little documentation assessing its efficacy.

Blockchain Technology

Blockchain technology is being used to improve the traceability and transparency of seafood supply chains. The technology is used by companies ranging from start-ups such as Provenance, to multinationals like IBM, to record key data about the journey of seafood from farm to consumer.

Using blockchain to record data about where fish has come from and how it was produced helps to eliminate cases of food fraud. In addition, blockchain data systems are also being considered as potential solutions to overcome challenges associated with the trustworthiness of existing certification schemes.

Examples from India for Using Information Technology/IoT in Aquaculture

Aqua One Center (AOC) of National Fisheries Development Board (NFDB)

National Fisheries Development Board (NFDB) is one of the lead organizations in India in fisheries and aquaculture development sector. NFDB is facilitating development of the fisheries sector in line with the technological advancement and adoption by way of promoting new technologies and practices to suit local resources and conditions on a continued basis. Strengthening support systems, institutional arrangements and networking at different levels as supportive and complementary systems are some of the initiatives taken to build a new landscape for significant growth in the sector. The Aqua One Center (AOC), an Information & Communication Technology (ICT) enabled aquaculture support service, will disseminate proven technologies and innovations and facilitate their wider adoption by registered fish farmers thereby facilitating the sector's overall growth.

AOC provides technologies for pond culture, cage culture in reservoirs, culture-based-capture fisheries in wetlands, Recirculation Aquaculture System (RAS), integrated farming etc., Better Management Practices (BMPs) including inputs management, data collection and management, setup water quality and disease diagnostic laboratory, advisory services with respect to life-cycle of species cultured, water quality, growth, health management, disease diagnosis, surveillance, etc, establish an *e*-traceability system.

Eruvaka Technologies

Eruvaka Technologies firm based at Vijayawada in Andhra Pradesh of India develops on-farm diagnostic equipment for aquaculture farmers to reduce their risk and increase productivity. They integrate sensors, mobile connectivity and decision tools for affordable aquaculture monitoring and automation. Their mission is to accelerate the use of technology in aquaculture. It helps farmers to monitor their ponds in a better way and reduce their investments to make it sustainable.

Feed Innovation

The feed sector is actively developing alternative protein sources that are more efficient and environmentally friendly. Feed innovation presents an exciting opportunity for investors, as new products could be used beyond the farmed seafood sector and adopted by the pet and consumer food markets as well. Most innovation in feed is occurring in the salmon sector, which is under intense scrutiny from investors and consumers to shift to more sustainable practices.

For salmon feeds, plant-based ingredients have replaced a larger percentage of the fishmeal and fish oils traditionally required. However, some innovations also come with adverse sustainability and health drawbacks. Many terrestrial plant-based ingredients come with equally high environmental impacts because of intensive production techniques. The low digestibility of some of these ingredients also generates more waste through an increase in fish faeces. As a result, alternative feed ingredients that provide comparable nutrition to fish meal or soy but at lower environmental and financial costs are being developed. The biggest gamechangers include feed ingredients made from bacteria, algae and insects.

Skretting, a global producer of aquaculture feeds, acknowledges that balancing and prioritising different environmental impacts is a challenge. For example, to protect marine biodiversity, a feed producer may wish to replace fishmeal with soy – however this increases exposure to deforestation and GHG emissions. Skretting aims to utilise life-cycle analysis tools to ascertain how to measure and manage trade-offs between GHG emissions, land use, eutrophication, marine biodiversity and other environmental impacts. The company is prioritising impact against two significant environmental challenges: biodiversity loss and climate change. It is therefore exploring raw materials that do not require further use of agricultural land. Among the most fundamental game-changers are feed ingredients made from bacteria, algae and insects.

Regardless of what innovation companies decide to pursue, decoupling the continuing dependence of the aquaculture sector on capture fisheries should be a main priority for companies, legislators and investors.

Repurposing Waste as Feed

Waste and by-products from agriculture, aquaculture and livestock production, particularly poultry, are increasingly used as feed ingredients. World Bank research indicates that the use of fish processing by-products in aquafeed may increase up to 2030, reducing pressure on wild stocks. Its projections show that if all producing countries integrated fish processing waste into fishmeal production, this would increase global fishmeal production by 11.8% compared to the baseline scenario. This would be accompanied by a corresponding 14.1% drop in fishmeal price and 1.9% growth in global aquaculture production. Vertically integrated producers may be at an advantage in this scenario, as they will have better access to fish by-products through other parts of their businesses.

Fish Oil Replacement from Algae

Inroads are also being made in the use of algae, particularly as a replacement for fish oil and high-grade hatchery feeds. Fish oil is an excellent source of omega-3 fatty acids or DHA, but with rising prices of forage fish, the feed industry is seeking alternate sources of omega-3 fatty acids. Companies like Corbion, which acquired TerraVia for $20 million in 2017, are pioneering ways to produce omega-3 fatty acids from heterotrophic algae, which require no light and can be produced efficiently and in greater quantities than light-dependent algae. The company distributed 40,000 tonnes of its AlgaPrime salmon feeds from September 2016 to March 2017 alone. Algal applications for other species are in the pipeline. (Source: Shallow returns: ESG Risks and Opportunities in Aquaculture, 2019)

Seaweed from Polyculture

As part of IMTA, the by-products and waste of fish or shrimp can be used to grow seaweed. This process increases the efficiency and output of ponds and outdoor operations. A 2014 trial led by Cooke Aquaculture in Canada produced organically grown kelp and blue mussels around salmon pens, minimising pollution by converting fish waste to food. When used as additives or fed directly to fish, kelp can grow up to 30 times faster than land plants, with the added benefits of providing habitats for marine life, absorbing CO, and fighting ocean acidification. Overall, seaweed production for feeds, cosmetics and human food is rapidly expanding. The annual global market for seaweed is about 12 million tonnes, valued up to $5.6 billion with seaweed grown for human consumption accounting for $5 billion.

Single Cell Proteins

While still in early stages of development, several companies are innovating to produce single cell proteins for aquafeed. Single cell proteins are proteins from the cells of microorganisms such as yeast, fungi, bacteria and algae.

New Product Developments

In addition to the above technologies in the aquaculture sector, new plant-based products are being developed that imitate seafood. As in the meat and dairy sectors, alternatives to seafood are being launched and cell-cultured seafood is in development.

These innovations could disrupt the protein industry by providing attractive, viable products through significantly less resource-intensive production methods. On average, annual global sales of plant-based meat alternatives have grown by 8% a year since 2010 and are projected to achieve a compound annual growth rate (CAGR) of 5.8% between 2018 and 2026, signalling growth potential for seafood alternatives. These products are some way behind meat alternatives as the flaky texture of fish is harder to replicate with plant-based ingredients.

In the livestock sector, leading companies are increasing exposure to these alternative proteins, largely through acquisitions, venture investments and new product development. If successful, these products could meet growing future demand for seafood while bypassing the sustainability challenges faced by capture and farm production methods today.

Snapshot: Plant-based seafood alternative products

Sophie's Kitchen, a plant-based seafood brand, participates in an accelerator programme overseen by Chipotle Mexican Grill. The quick-service restaurant chain runs an accelerator programme for food-focused startups through the Chipotle Cultivate Foundation. The programme aims to drive change in the food industry by assisting impact-focused growth-stage ventures to scale up and expand their positive impact. Sophie's Kitchen reported a 72% jump in sales from Q1 2017 compared with Q1 2018 (Fox 2018), demonstrating early consumer and corporate appetite for its products.

Good Catch Foods, an emerging vegan food company, has expanded international distribution of its plant-based tuna product in 2019 following a US release the previous year. In August 2018, Good Catch Foods closed an $8.7 million Series A funding round including investment from PHW Group, a European poultry processor. (Food Business News, 2018).

Quorn Foods, an established player in the meat alternative market, launched a new range of plant-based 'Fishless Fillets' in March 2019. The frozen fish segment is an £850 million category in the UK and as the first company to produce a product that imitates the most popular breaded and battered fish products, the company envisions that this new product will be popular.

Conclusion

Aquaculture is one of the important sectors globally in providing food supply, security and nutrition to the people. Indian Major Carps are the main fishes in the domestic aquaculture sector followed by Tilapia and Pangasius. *Litopenaeus vannamei* is the major cultivable shrimp species in India which is widely cultured with *Penaeus monodon* at certain parts of the country. There are many other potential fish species for culture like cobia, Asian seabass, pearlspot, milk fish, silver pompano, groupers etc. Various aquaculture technologies have been developed by lead institutions like Central Institute of Brackishwater Aquaculture (CIBA), Central Institute of Freshwater Aquaculture (CIFA), Central Marine Fisheries Research Institute (CMFRI) etc. Directorate of Coldwater Fisheries Research (DCFR) is the institute which is mandated for cold water fisheries development in the country. There is need to adopt new approaches of fish farming. With newly formed Department of Fisheries in 2019 & launch of PMMSY schemes, fisheries and aquaculture sector of India is poised for remarkable growth in the coming years.

CHAPTER 5

REGULATIONS FOR FISHERIES AND AQUACULTURE IN INDIA

Introduction

India being the one of the largest producers of fish in the world is blessed with a vast coastline of 8,118 km and 2.02 million sq. km. of Exclusive Economic Zone (EEZ) including 0.5 million sq. km. of continental shelf. As per Marine census carried out in 2010 by Govt. of India; altogether 52,982 traditional crafts, 73,410 motorized crafts and 72,749 mechanized crafts are in operation along the coast of India which contributes to the total marine landing of the country. The marine capture fisheries is characterized by almost stagnant or at times declining fish catches, overfishing, overcapacity, increasing landing of bycatch especially low value bycatch, increasing discards in multiday trawls, increasing conflict over fish resources, mounting investment needs, and export market fluctuations. Since the last decade most of the major commercially exploited stocks are showing signs of over exploitation. On the contrary, demand for fish and fishery product is increasing considerably, both at the domestic and export front.

The policies by the central Government and the State governments have a direct influence on marine fisheries of India. Marine fishing regulation and sustainable management of fisheries within the territorial waters come under the jurisdiction of concerned state governments/Union territories and therefore there is no uniform pattern followed by all states. Fisheries development and management of the off-shore and deep-sea waters come under the purview of the Union Government. All the maritime states and Union Territories along the east and west coast have enacted Marine Fishing Regulation Act (MFRA) as per the directives of the Model Bill of the Union Government.

Salient features of Indian marine fisheries regulations

The Indian Fisheries Act, 1897

The Indian Fisheries Act, 1897 offers protection to fisheries against destructive fishing methods like explosives or dynamites. Provincial governments were empowered to make rules with regard to mesh size regulations, seasonal ban in selected waters.

The Merchant Shipping Act, 1958

The Merchant Shipping Act, 1958 has been formulated to ensure efficient maintenance of Indian Mercantile Marine in a manner best suited to serve national interest. The manning requirements are given in section 76(4) of the Act, Prevention of collision at sea in 285(1), lifesaving appliances in 288(1), rules relating to fire appliances in 289, inspection of life-saving and fire-fighting equipment in 290(1), information about ship's stability in 298(1) and seaworthiness of vessel in 335 (1).

Indian Wild life Protection Act, 1972

It offers protection to marine biota. Creates conditions favorable for in situ conservation of fauna and flora. Amended in 1991 to prohibit fishing within the sanctuary area of Gahirmatha, annual mass nesting place for Olive Ridley turtle, and endangered species, accorded the status of marine sanctuary in 1997. Amended in 2001 to include several species of fish, corals, sea cucumbers and sea shells in Schedule I and III.

The territorial waters, Continental shelf, EEZ and other Maritime zone act, 1976

The act describes limits and rights of various zones such as territorial waters, contiguous zone, EEZ, Continental shelf etc.

The Coast Guard Act, 1978 (Safeguarding life and property at sea)

The Coast Guard Act, 1978 offers protection to maritime zones of India in the national interest. It ensures safety and protection of artificial islands, offshore terminals, installations and other structures and devices in any maritime zone; protection to fishermen including assistance to them at sea while in distress; preserve and protect the maritime environment and to prevent and control marine pollution;

assisting the Customs and other authorities in anti-smuggling operations; and other matters, including measures for the safety of life and property at sea and collection of scientific data, as may be prescribed.

Marine Fishing Regulation Model Bill, 1979

A model Act that provides guidelines to the maritime states to enact laws for providing protection to marine fisheries by regulating fishing in the territorial waters. The measures include: regulation of mesh size and gear, reservation of zones for various fishing sectors and also declaration of closed seasons. Laws are framed and amended from time to time by different maritime states.

Marine Fishing Regulation Acts (MFRA)

The Model Marine Fisheries Regulations Bill was provided by the Government of India in 1979 to all the maritime states of India/UTs to enact suitable legislation for respective states. With a view to conserving the marine fisheries resources against over-exploitation and for regulating fisheries in the state, each state formulated legislation to provide for regulation, protection, conservation and development of fisheries in its territorial waters. Kerala and Goa were the first to enact the Marine Fisheries Regulation Act in 1980 followed by other maritime states/UTs.

The Acts cover issues of registration and licensing of fishing vessels, zonation, regulation of mesh size, area of operation, declaration of closed season, protection of endangered species, prohibition of destructive fishing methods and regulating fleet strength of the fishing craft under different categories to ensure sustainable yield from fisheries.

Salient features of Regulations / Restrictions / Prohibitions in the Maritime States of India

State	Particulars
Andhra Pradesh	• No vessel to be engaged in fishing using nets with mesh size below 15 mm • No mechanized fishing vessel to operate in water upto 4 nautical miles from the shore. • Mechanized vessel below 15 m length to operate from 4 nautical miles from the coast. • Shrimp trawlers engaged in fishing without TED shall be liable for confiscation of entire catch and imposition of fine.

[Table Contd.

Contd. Table]

State	Particulars
Goa	• Nets with mesh size less than 24 mm for catching fish and 20 mm for catching prawns are prohibited. • Fishing of juveniles of fishes like Sardines, Mackerels etc. is banned throughout the year. • Mechanized fishing by means of trawl nets and purse seines banned during monsoon period from 1st June to 31st July every year. • No mechanized vessel to operate upto 3 nautical miles.
Gujarat	• In case of trawl net, square mesh of minimum 40 mm size at cod end needs to be used. • No gill net with mesh size less than 150 mm can be operated • Electric fishing is prohibited. • Use of wounding gears like spears, arrows, harpoons etc; is prohibited • No fishing can be carried out in the territorial water from 1st June to 31st July every year • Catching of whales, sharks and turtles is prohibited under Wild Life Protection Act, 1972 • Non-mechanized vessels can fish within 5 nautical miles and mechanized vessels beyond 5 nautical miles.
Karnataka	• The government regulates, restricts or prohibits fishing in any specified area by such class of fishing vessels or use of such fishing gear or catching of such species of fish as may be prescribed in the notification issued from time to time. • Area up to 3 nautical miles from shore is reserved for traditional craft. • Mechanized boats up to 16 M length are allowed to operate beyond 3 nautical miles.
Kerala	• Use of bottom trawl nets having less than 35 mm cod end mesh size is prohibited. • Use of purse seine, ring seine, pelagic and mildwater trawls for fishing in the specified area mentioned in the Govt. notification is prohibited. • Prohibition of use of bottom trawl nets from sunset to sunrise in specified areas. • Trawl ban for 61 days from 1st June to 31st July every year. • Area up to 5.5 nautical miles is reserved for traditional craft.

[Table Contd.

Contd. Table]

State	Particulars
Maharashtra	• Fishing in any specified area by such class of fishing vessel as specified is prohibited. • Number of fishing vessels which may be used for fishing in any specified area is regulated. • Catching in any specified area of such species of fish and for such period as specified is restricted. • Use of such fishing gear in any specified area as may be prescribed is prohibited.
Orissa	• No mechanized vessel to operate up to 3 nautical miles. • Mechanized vessel below 15 m length to operate from 3 nautical miles onwards. • mechanized vessel above 15 m length to operate only beyond 5.5 nautical miles from the coast.
Tamil Nadu	• No gill net of mesh size less than 25 mm shall be used • No shrimp trawl net with mesh size less than 37 mm at cod end shall be used. • No fish trawl net with mesh size less than 40 mm at cod end shall be used. • No bottom trawling operation within 3 nautical miles from the coast line. • Ban on fishing operation for 61 days from 15th April to 14th June every year. • Ban on pair trawling and purse seining in the territorial waters.
West Bengal	• No gill net with mesh size less than 25 mm shall be used. • Trawl net of standard mesh size fitted with turtle excluder devices to be used. • No shore seine / drag net with mesh size below 25 mm to be used. • Mechanized vessel above 15 m to operate beyond 27 nautical miles. • Ban on fishing operation for 61 days from 15th April to 14th June every year.
Andaman and Nicobar Islands	• Only Gill nets with mesh size above 25 mm are allowed to operate. • Only shore seine / drag net of mesh size above 25 allowed to operate. • Trawl nets of standard mesh size fitted with turtle excluder device alone are permitted. • Vessels up to 30 HP are allowed to operate beyond 5.5 nautical miles

[Table Contd.

Contd. Table]

State	Particulars
	• Only vessels above 30HP are allowed to operate beyond 5.5 nautical miles. • 15th April to 14th June shall be closed season for bottom trawlers and vessels engaged in shark fishing. • 1st May – 30th September every year is closed season for fishing sea shells. • Fishing nets below 20 mm mesh size are prohibited.
Lakshadweep	• Use of purse seine, ring seine, pelagic, mid water and bottom trawl of 20 mm mesh size is prohibited except live bait net. • Use of drift gill net of less than 50 mm mesh size and shore seine of less than 20 mm mesh size is prohibited.

The maritime Zones of India (Regulations of fishing by foreign vessels Act, 1981)

This act was introduced to control activities of foreign fishing vessels within Indian Maritime Zone. The Act provides basis for joint ventures and chartered vessels and also for bilateral / multilateral fishing access agreements. The Act's main provisions deal with the grant of licenses, prohibition of Indian citizens using foreign vessels, procedures for granting of permits or licenses, and the responsibility of permit holder for compliance.

The Environment (Protection) Act,1986

It authorizes the Central Government to protect and improve environmental quality, control and reduce pollution from all sources and prohibit or restrict the setting and or operation of any industrial facility on environmental grounds. It also makes it mandatory to conduct Environmental Impact Assessment (EIA) for specified developmental activities.

Biological Diversity Act, 2002

Main objective of the Act is to protect biological diversity of India. The Act provides for the conservation of biological diversity, the sustainable use of its components and the fair and equitable sharing of the benefits arising out of the use of biological resources, knowledge and related matters. There is a provision for setting up of National and State Biodiversity Boards. The Act encourages conservation and has a provision to declare a fish stock threatened if it is over exploited.

Deep-Sea Fishing Policy

The first deep-sea policy was announced by Government in 1977, providing for chartering arrangements with foreign operators, which was followed by a Charter Policy in 1981 for introduction of sophisticated foreign fishing vessels for promotion of deep sea fishery that was followed by 1991 Deep Sea Fishing Policy which envisaged joint venture, test fishing and leasing besides allowing the vessels chartered under 1986 policy to continue till the validity of their permit period. The Murari Committee Constituted by the Central Government in1996 came up with recommendation to sustain the resources along with protecting the interest of the fishermen and utilisation of the deep-sea resources in the EEZ. In late 2002, a new set of Guidelines for deep-sea fishing was announced by the Government.

The policy allows foreign fishing vessels into Indian waters beyond 12 nautical miles. After protests from local fishermen Charter and leasing operations of foreign trawlers was suspended in 1997. No granting of new licenses to joint venture companies operating in the EEZ. Deep Sea Fishing Policy, 1991 was practically scrapped in 1997.

In 2014, Government of India has issued the guidelines for fishing operations in Indian exclusive economic zone. Further, National Fisheries Development Board has created "Guidelines for Deep Sea Fishing and Tuna Processing". Central Government encourages developmental activities of Deep Sea Fisheries and in regard of the same, there was addendum announced in 2017 to the guidelines on central plan scheme on "Blue Revolution Integrated Development and Management of Fisheries" regarding "Assistance for Deep Sea Fishing". Under PMMSY, there is assistance available for deep sea fishing vessel of size 18-24m OAL for traditional fishermen and their societies in coastal states/UTs (financial assistance on actual cost, with a ceiling of Rs 120 lakhs; 50% subsidy; subject to a maximum of Rs 48 lakhs per fishing vessel).

Draft National Marine Fisheries Bill 2019

The Marine Fisheries (Regulation and Management) bill 2019 was drafted after four rounds of consultative meeting of the committee for drafting Marine Fisheries Regulation and Management bill in consultation with experts and stakeholders. The bill provides for regulation and management of fisheries in the Exclusive Economic Zone of India and the high seas and for conservation and sustainable use of marine fisheries resources; maintenance of law and order in the maritime zones of India (for fishing and fishing related activities); supporting the social security, livelihoods and safety at sea of fishers and fish-workers, in particular the traditional and small-scale fishers. This Act may be called the Marine Fisheries (Regulation and Management) Act, 2019. It specifies regulation of fisheries in the exclusive economic zone and high seas that states that no Indian fishing vessel shall engage in any fishing or fishing related activity within the exclusive economic zone of India or the high seas, except with a permit issued by the Central Government or any authority notified under this Act for fishing, and shall be subject such conditions and restrictions as prescribed. It also specifies offences and penalties for violations.

Major highlights of Marine Fisheries (Regulation and Management) Act, 2019

- The Central Government shall notify a system of monitoring, control and surveillance towards safety and security of fishermen and fishing vessels and for the compliance of fishery management measures in the maritime zones of India.
- The fishing vessel licensed under relevant section of this Act shall comply with the manning, insurance, sea safety and seaworthiness norms of fishing vessels as provided under the Merchant Shipping Act, 1958 and as prescribed.
- The Central Government may, through a special license to be issued in writing, allow a vessel to carry out any scientific research, survey or investigation related with fisheries or for any experimental or recreational fishing or any other purpose in accordance with such terms and conditions as may be prescribed.
- Fishing shall be prohibited in marine protected areas, ecologically and biologically significant areas and vulnerable marine ecosystems, to ensure protection of endangered and threatened species and for the maintenance of the ecological balance of the marine environment and its biodiversity.
- In order to ensure sustainable utilization of resources, and for safety of fishermen, during the breeding season, the spatial and temporal closures shall be implemented for conservation of stocks, as prescribed.
- The Central Government may, from time to time, notify one or more plans for the management of one or more fish species and fishing related activities consistent with the basic principles underlying the United Nation's Code of Conduct for Responsible Fisheries and ecosystem approach to fisheries management, in such area(s) of the maritime zones of India as may be prescribed.

- All licenses granted under this Act, shall be subject to fisheries management plans as may be notified by the Central Government and in the event of any inconsistency between a license so granted and a fisheries management plan, the said plan shall take precedence and its provisions shall be deemed to be applicable to the said license.
- Destructive fishing methods including use of dynamite or any other explosive substance, poison or noxious chemicals, light or other destructive materials or any other methods to catch or destroy the fish in the maritime zone of India, as prescribed, is prohibited.
- The Central Government may regulate, restrict, or prohibit fishing in any area of EEZ by such class or classes of fishing vessels, the use of such fishing gear, the number of fishing vessels which may be used for fishing in any area within the EEZ, and the capture of such species of fish, of such size of fish and for such periods as may be prescribed.
- A 'National Marine Fisheries Authority' shall be established for development, management and regulation of marine fisheries in the maritime zones of India.

Challenges in implementation of Regulations

Despite formulation of MFRA in all the coastal states and the UTs, level of enforcement of the same is dismal. Low level of enforcement could be mainly attributed to limited designated manpower for enforcement in the maritime states particularly in the context of a vast Indian coastline.

The proposed policy on marine fisheries envisages a deviation from the open access system to a limited access system and also incorporation of principles of the Code of Conduct for Responsible Fisheries. Shift from open access to limited access is a big challenge considering dependence of fishers on coastal fishing for their livelihood.

Difficulty in enforcement of the fishing zone regulations due to paucity of requisite navigational equipment having mechanism of accessing distance covered by fishing craft as well as insufficient number of patrol boats is a hindrance in the proper implementation of MFRA.

Uneconomical fishing operations on most occasions mainly due to reduction in the commercial catch have prompted fishers to use nets with mesh sizes much smaller than prescribed in MFRA. This has led to widespread violations of mesh size regulations resulting in increased landing of Low Value Bycatch (LVB) and catch of juveniles of commercially important fishes.

Ignorance of marine fishers about the consequences of unsustainable fishing practices such as destructive fishing and overfishing coupled with lack of awareness about CCRF poses difficulties in adopting responsible fishing practices.

Suggesting common management strategies for fish resources and areas within the Indian EEZ poses problems as both resources as well as areas are in different stages of exploitation.

Intersectoral conflicts arise at times due to lack of harmony between state MFRA and national legislation for Indian owned fishing vessels that operate beyond territorial waters in the Indian EEZ.

Aquaculture Legislation in India

India is a federal republic, subdivided into 28 states and nine union territories. According to the Constitution, state legislatures have the power to make laws and regulations with regard to a number of water-related sectors (water supply, irrigation, canals, drainage and embankments, water storage and water power), land (rights on or above land, land tenure, transfer and disposition of agricultural land) and fishing, as well as conservation, protection and improvement of stocks and prevention of animal diseases. Although there are many laws and regulations appropriate to aquaculture, adopted at the state level, at the central level, several key laws and regulations govern aquaculture. The Indian Fishing Act (1897) , has a century of existence and it penalizes the massacre of fish by stocking of water or the use of explosives. The Environment Protection Act (1986) is an umbrella act containing provisions for all aspects related to the environment. It also includes the Water Act (Preventing and Combating Pollution) (1974) and the Life Protection Law (1972). These legislations should be read at the same time to get a full picture of the rules that are applicable to aquaculture.

On December 11, 1996, the Indian Supreme Court presented a landmark decision with major implications for the aquaculture sector regarding the establishment of shrimp farms in the coastal region. The Supreme Court - among others - has prohibited the construction or installation of shrimp culture ponds in the Coastal Regulatory Zone and less than 1,000 meters from Lake Chilka and Lake Pulikat with the exception of ponds of traditional types and improved traditional types. It is also stipulated that an authority should be established to protect coastal and ecologically fragile areas, such as shores, water fronts and other coastal areas and particularly to deal with the situation. To perform its functions indicated by the Supreme Court, Notification SO 88 (E) (1997) established the Aquaculture Authority in accordance with the Environment (Protection) Act. The Authority, which has specific responsibilities for Aquaculture, comes under the administrative control of the Ministry of Agriculture.

National Inland Fisheries and Aquaculture Policy, 2019

Recognising the social, economic and nutritional importance of inland fisheries and aquaculture, GoI constituted a committee that prepared a draft National Inland Fisheries and Aquaculture Policy (NIFAP). The draft was circulated among the State Government/ UTs and stakeholders for comments. The revised draft policy after incorporation of comments from various sources including State Fisheries Departments/ICAR Institute and Organizations was approved in 2019. The Department is in the process of initiating the next steps for notification of the National Inland Fisheries & Aquaculture Policy.

Major highlights of the policy

Aquaculture Development

1) States/UTs have to develop Action Plans for introduction and expansion of shrimp/prawn farming in inland saline/alkaline and freshwater areas as per the recommendations of scientific organizations like ICAR Institutes, Universities, etc. and the Department of Fisheries, Government of India with due consideration to sustainability and ecosystem health. States/UTs are encouraged to prepare Integrated Coastal Aquaculture Development Plan for each of the coastal blocks and districts.
2) Aquaculture zoning along with area specific plans has to be formulated using modern scientific tools for scientific and planned development of aquaculture and its regulation.
3) Development of aquaculture may be promoted in low-lying areas, lands with saline and alkaline soils, and lands not suitable for crop cultivation and large land blocks of such areas may be earmarked for development as aquaculture zones.
4) In order to utilize vast unutilized resources including government owned land for aquaculture, the scope of land use categories at state level would require redefinition and enlarging to specifically include fisheries and aquaculture as integral components of agriculture.
5) It is essential to distinguish the needs and aspirations of farmers with small aquaculture holdings from those having large holdings. Union in consultation with States/UTs need to define, document and handhold farmers with small

aquaculture holdings through schemes and programmes. Fish Farmers Producer Organizations (FFPOs), Common Interest Groups (CIG), Self-help Groups (SHG), etc. need to be encouraged to cater the needs of these farmers.

6) Mandatory registration of aquaculture farms, simplification of legal and environmental requirements for farm registration and leasing has to be ensured.
7) Best Management Practices (BMPs) / Good Aquaculture Practices (GAPs) are to be promoted to minimize disease incidence and other ecological externalities thereby ensuring sustainability. The practice of screening of seed for causative organisms before stocking would be strictly implemented.
8) Foolproof mechanisms have to be put in place to ensure traceability of the aquaculture produce along with requisite regulatory framework and infrastructure.
9) Spurt in growth of cage culture in open water resources and shrimp/prawn farming in inland saline and freshwater areas warrant a regulatory framework for inland aquaculture by State Authorities on the lines of Coastal Aquaculture Authority (CAA).
10) Aquaculture development efforts will be in alignment with relevant national and global instruments, guidelines and good practices including Sustainable Development Goals (SDGs), Code of Conduct for Responsible Fisheries (CCRF) Guidelines and Voluntary Guidelines on Sustainable Small Scale Fisheries (VG-SSF).
11) In order to ensure that the fish produced from waste water aquaculture is safe for consumption, appropriate regulatory, management and precautionary measures need to be put in place in coordination with relevant agencies.
12) Trout farming is the main stay of fish production in high altitude areas of the country. Trout seed and feed production has to be promoted in suitable areas following a cluster approach and increased private sector participation.

Seed, Feed and Other Aquaculture Inputs

1) Farmers and Private sector will be encouraged for setting up of hatcheries and seed rearing farms across the country, especially in seed deficient areas.
2) Adequate broodbanks especially at the national and state level are to be established to cater the requirements of hatcheries.
3) Registration and accreditation of hatcheries is to be made mandatory by States/UTs. Use of seeds produced from certified broodstock, hatcheries and seed production units will be encouraged.
4) Private sector will be encouraged to establish fish feed mills and use of locally available ingredients in feed formulations.

5) Registration of all aquaculture inputs needs to be made mandatory. The specifications of aquaculture inputs, type and quality of ingredients used therein has to be notified, accredited and inspected for compliance. Requisite regulatory framework and infrastructure has to be put in place.
6) The use of drugs and chemicals including antibiotics and pesticides in aquaculture needs to be regulated through suitable mechanism.

Introduction and Regulation of Exotic Species

Entry of any exotic species meant for aquaculture including broodstock, seed and Specific Pathogen Free (SPF) stocks etc. needs to be regulated as per the existing National Laws/ Rules for import, breeding and farming.

Disease Surveillance

The present disease surveillance and reporting system will be further strengthened with inbuilt provisions to identify and contain any emerging diseases with the involvement of States/UTs.

Diversification of Aquaculture

1) Major thrust need to be given on diversification of species in both freshwater and brackish water systems by establishing hatcheries, brood stock multiplication centres and nuclear breeding centres.
2) R&D programs require focus on developing breeding, hatching, seed rearing and grow-out technology for identified alternative species suitable for pond aquaculture.
3) Thrust will be given bring in/adopt advanced technologies and practices to enhance production and productivity.

Ornamental Fish Culture

Collection and trade of native ornamental fish species from natural waters require regulation by the concerned States/UTs. The database on native ornamental fish species need to be further strengthened. Institutional support and efforts for breeding, rearing and promotion of trade of indigenous ornamental fishes need to be intensified to facilitate growth in this segment.

National Mariculture Policy, 2019 (Revised Draft)

As the marine fish production in India over the years has been almost stagnant in the capture sector, there is scope for increasing production through mariculture . The National Mariculture Policy (NMP), 2019 aims to ensure sustainable farmed seafood production for the benefit of food and nutritional security of the nation and to provide additional livelihood and entrepreneurial opportunities to the coastal communities for better living. The overall strategy of NMP is to increase seafood production in sustainable and responsible manner, ensure socio-economic development, enhance food, health and nutritional security and safeguard gender, social equity and environment.

The vision and mission of NMP have been drafted for increasing farmed seafood production in the country-based on following.

1) Recognizing that the demand for seafood is increasing year after year.
2) Realising that additional seafood requirement of the country in future years cannot be met by capture fisheries and inland aquaculture alone.
3) Recognising that to enhance the living conditions of coastal fishermen, additional livelihood options are needed.
4) Noting further that sea farming sector is still in its infancy in the country.
5) Acknowledging that there is immense potential for sea farming in the country.
6) Recalling that there are many mariculture technologies developed in the country which can be commercialized.
7) Viewing with appreciation that mariculture has already contributed to substantial seafood production in many countries and is growing.

A sustainable and responsible mariculture sector that contributes to the food and nutritional security of the country and enhances the quality of life of the stakeholders is the vision of NMP. It is a policy framework leading to widespread adoption of mariculture technologies to meet the additional seafood demand while ensuring the environmental sustainability, socio-economic upliftment of stakeholders and facilitating the responsible development, co-ordination and management of mariculture production in the country. Objectives of NMP are given below.

1) To enhance Mariculture production in the country and increase income, employment and entrepreneurship opportunities in a sustainable and responsible manner.
2) To promote co-operative partnership in mariculture by encouraging the infrastructural, technical and financial inputs.

3) To adopt an environmentally sustainable approach for development of mariculture.
4) To provide an enabling environment for sustainable development of mariculture in India by providing the required policy and legal framework and support to entrepreneurs venturing to the area of mariculture.

Conclusion

The marine capture fisheries are characterized by almost stagnant or at times declining fish catches, overfishing, overcapacity, increasing landing of bycatch especially low value bycatch, increasing discards in multiday trawls, increasing conflict over fish resources, mounting investment needs and export market fluctuations. Though the demand for fish and fishery product is on the increase both at the domestic and export front, most of the commercially exploited fish stocks show signs of overexploitation. The policies by the Central government and the State governments have a direct influence on the fisheries and aquaculture of India. Despite formulation of MFRA in all the coastal states and the UTs, level of enforcement of the same is dismal. Low level of enforcement could be attributed to among others, limited designated manpower for enforcement in the maritime states particularly in the context of a vast Indian coastline. The Marine Fisheries (Regulation and Management) bill 2019 was drafted after four rounds of consultative meeting of the committee for drafting Marine Fisheries Regulation and Management bill in consultation with experts and stakeholders. The proposed policy on marine fisheries envisages a deviation from the open access system to a limited access system that is a big challenge considering dependence of fishers on coastal fishing for their livelihood.

National Inland Fisheries and Aquaculture Policy, 2019 has been drafted and the government is in the process of initiating the next steps for notification of the National Inland Fisheries & Aquaculture Policy (NIFAP). Major highlights with regard to Aquaculture in the policy includes guideline for aquaculture development; seed, feed and other aquaculture inputs; introduction and regulation of exotic species; disease surveillance; diversification of aquaculture; and ornamental fish culture. Realising that additional seafood requirement of the country in future years cannot be met by capture fisheries and inland aquaculture alone, National Mariculture Policy (NMP), 2019 has been drafted and the goal is to ensure sustainable farmed seafood production for the benefit of food and nutritional security of the nation and to provide additional livelihood and entrepreneurial opportunities to the coastal communities for better living.

CHAPTER 6

FAO STRUCTURE AND ITS CONTRIBUTIONS TO INDIA'S FISHERIES DEVELOPMENT PRIORITIES

FAO in India

The Food and Agriculture Organization of the United Nations (FAO) started its operations in India in 1948. Since then, it continues to play a catalytic role in India's progress in the areas of crops, livestock, fisheries, food security, and management of natural resources. FAO has significantly contributed in terms of technical inputs besides investment in agricultural development in India. In the words of M.S Swaminathan, world renowned agricultural scientist, "FAO has played a catalytic role in India's progress in the areas of crop and animal production and food security". With its global experience, FAO has provided key policy and technical inputs in wide range of areas involving the food and farm sectors. Management of water and natural resources, crops, livestock, food security information systems and fisheries comprise the major part of the work under FAO partnership. The policy goals of the government of India have been strengthened by the FAO activities in fisheries sector by addressing sustainability and inclusiveness needs of the sector. Technological interventions by FAO have helped in improving the livelihoods of millions of fisher folk. Participatory approach adopted by FAO and focus on gender empowerment has been mainstreamed into fisheries sector of India.

The Bay of Bengal program (BOBP) is the regional fisheries flagship program of FAO based in Chennai, India. Highlights of BOBP include revolutionizing boat building techniques using glass reinforced plastic in hull construction and widespread introduction of small motorized and large mechanised boats. Besides BOBP, direct FAO support to fisheries development in India increased through a wide range of projects over the years. Another major involvement of FAO was in two community managed projects after the tsunami in 2004. The interventions centred around recovery and rehabilitation in line with the Government of India policies. Similarly, project on capacity building in support of cleaner harbours addressed issues of producing management models that reduce wastage, improve standards of hygiene and apply principles of hazard analysis and critical control points (HACCP). Support to safety at sea for small scale fishermen project helped to create system for participation of fishermen.

There is a huge potential in marine, inland and brackish water segments of India. Development of water resources for irrigation has underpinned crop production. However, water scarcity and falling water-tables have been a key concern in recent years. Fortunately, the importance of judicious use of water is being increasingly recognised. Over the years, the work of FAO in India has gone well beyond the realm of food production, covering issues like access to food and nutrition, livelihoods, rural development and sustainable agriculture. With the looming impact of climate change and outbreaks of new strains of pests the work of FAO has become even more complex, making it an important knowledge partner to assist the country with informed decision-making.

Government's development priorities

The NITI Aayog is the government's think tank and policy planning agency that has identified critical areas and recommendations necessary to sustain and accelerate agricultural growth in the country through its action plans. The priorities set in the NITI Aayog's seven-year National Development Agenda and the medium-term Three-Year Action Agenda as well as the Union Budget represent the key overarching framework for the agriculture sector. The main objective of the government is to double farmers' income 2022-23 by solving the twin problems of maximising efficiency and ensuring equity in a sustainable manner. Achieving this goal would require significantly faster growth in nearly all variables that positively impact farmers' incomes.

The first priority area is of transforming the agricultural produce marketing policies and marketing interventions ensuring farmers receive remunerative prices. Agricultural marketing suffers from policy distortions, fragmentation resulting from large number of intermediaries, poor infrastructure and lack of vertical integration in the value chain. The government intends to reform acts that regulate the marketing of agricultural produce that fosters competition and improved price realisation for farmers. The government intends to encourage contract and group farming that improves access to technology, inputs and price realisation for the farmer. The government in April 2016, launched the electronic National Agricultural Market (e-NAM). The next stage aims to unify markets across the nation into a single market through electronic trading whereby a buyer anywhere in India can place an order anywhere in India.

The second priority for the government is boosting productivity of agriculture in India. The government has a fourfold agenda for achieving increased productivity: irrigation, seeds and fertiliser, technology and diversification to high-value agriculture, animal husbandry and fisheries.

The Skill India mission and employment training through DAY – NRLM are seen as important schemes that can be used to transfer technology through one to one exchange of information. The government also sees the diversification into production of high-value commodities such as horticulture, animal husbandry, forestry and fishery to meet the changing food demands of the nation and to ensure dietary diversification as a means of achieving food security in the country. The government has a key focus on developing marketing infrastructure and supply chain that would promote agribusiness in these sectors.

Several initiatives by the Government of India, through respective Missions, address the need for sustainable and 'climate-smart' agriculture. The National Initiative on Climate-resilient Agriculture (NICRA) aims to enhance the resilience of agricultural production to climate variability in vulnerable regions. These efforts are augmented by agro-ecological initiatives, implemented by a range of agencies across the country, that support small-scale farmers.

FAO's Country Programming Framework (CPF) for India

Contribution of FAO in terms of CPF is based on the following.

Sustainable development of agriculture, including natural resource management: (i) applying approaches and tools that have been pioneered by FAO in India to restore degraded lands and match water demands to supplies; and (ii) piloting tailormade strategies with farming communities so that they can adapt to climate change.

Food and nutrition security, focusing on: (i) improving nutritional status through appropriate analyses, strategy formulation and nutrition education; and (ii) diversifying livelihood sources in rain-fed areas, and building small-scale farmers resilience through risk management.

Transboundary cooperation and enhancement of India's contribution to global public goods, including facilitating India's sharing of expertise with other developing countries in the areas of agriculture and rural development.

Jointly prepared with the Government and other partners, the CPF reflects relevant priorities in key national development policies and is fully aligned with the UN Development Assistance Framework (UNDAF) for India 2013-2017. Further, successive CPF was aligned with the United Nations Sustainable Development Framework (UNSDF) cycle, which began in 2018 extending till 2022. FAO-India will play a critical role in the achievement of country strategic priorities laid out in UNSDF.

All CPF priority areas identified clearly support smallholders in developing productivity and competitiveness and in improving livelihood and reducing rural poverty for disadvantaged groups. Wherever relevant synergies will be created between the priority areas and the activities being implemented under each priority area. The five strategic objectives through their alignment into Regional Initiatives and Regional Priorities will govern FAO's support, in addition to the Government of India's priorities and the priorities and outcomes laid out in the UNSDF. Detailed list of outputs and targets are given in the resource requirements matrix and main implementing partners of CPF outputs (Annexure 3).

PRIORITY AREAS

Priority Area No 1. Sustainable and improved agricultural productivity and increased farm incomes

Under this priority area, FAO will facilitate adaptation of Farmers Water School (FWS) in Uttar Pradesh on groundwater management to surface irrigation practices to increase crop productivity and improve water-use efficiency. It will replicate learnings from Andhra Pradesh Farmer Managed Groundwater Systems (APFAMGS) and Groundwater Governance Pilot in AP to scale up in the International Fund for Agricultural Development (IFAD) funded AP Drought Mitigation Project to strengthen the adaptive capacity and productivity of agriculture in the rain fed areas across the country. Furthermore, FAO will provide technical assistance to the states of Nagaland and Mizoram for the IFAD funded FOCUS project in Nagaland and Mizoram for implementation support aimed at productivity enhancement through sustainable farming practices.

FAO will also implement the grant project of IFAD in Odisha and North-eastern states for diversification of livelihoods into sustainable forest based agro-enterprises. Through the Global Environment Facility (GEF) funded Green Agriculture project, FAO will work on developing farmer capacities for promotion of value chains of low-input alternative crops linked to adoption of sustainable agricultural and natural resources practices. Furthermore, it will provide technical guidance to National Rural Livelihood Mission (NRLM) to establish efficient and effective institutional platforms of the rural poor that enable them to increase household income through sustainable livelihood enhancements. In addition, FAO will pilot producer prices incentives monitoring and analysis mechanism in six states of the country to improve the evidence basis for agricultural and food policies with a particular focus on smallholder farms. FAO is also conducting a study on improving income of farmers by enhancing and sustaining pulse production in the country.

Priority Area No 2. Stronger food and nutrition security systems

Under this priority area, FAO's technical assistance will focus on providing technical assistance that drive the "Zero Hunger" initiative of FAO. FAO will collaborate with IFAD and World Food Program (WFP) in Odisha for pilot projects that promote nutrition-sensitive agricultural practices and positive nutritional behaviours including hygiene and sanitation practices targeted at tribal populations through low cost technology options. With NRLM FAO will work on reduction of absolute poverty by supporting initiatives that link with other government programs to improve health and nutrition situations of the marginalised population and break the cycle of poverty and malnutrition especially amongst women and children. It will also focus on capacity building in selected regional and national nutrition training institutions/universities to improve capacities to effectively design, implement and monitor nutrition education for behaviour change with a focus on healthy diets.

Priority Area No 3. Effective natural resource management, community development and assistance in transboundary cooperation

Under this priority area, FAO will implement the GEF funded Green Agriculture project that will provide models for successful landscape approaches to address the interface of biodiversity conservation in and around key protected areas. It will also provide technical assistance to the states of Nagaland and Mizoram for the IFAD funded FOCUS project to assist them in developing smallholder farmers' adaptive capacity to climate change by making jhum cultivation, the predominant mode of production in the two states more sustainable and gender inclusive. FAO is also providing technical assistance for pilot projects for strengthening Agriculture and Allied Sector Contributions to India's National Biodiversity Action Plan (NBAP), 2008 and the National Biodiversity Targets (NBT). Furthermore, under the regional BOBLME project, FAO plans to promote enhanced sustainable livelihoods and diversification for selected coastal communities. FAO through this project intends to promote better coordination, monitoring, awareness, innovative technology on marine pollution control, enhanced resilience and reduced vulnerability to natural hazards, climate variability and change of selected coastal communities.

FAO will provide technical assistance for sustained advocacy on combating anti-microbial resistance and implementation of the National Action Plan for Anti-microbial Resistance (AMR) that has been submitted by the government to the World Health Assembly of WHO and FAO. FAO will also promote innovative pilot projects on biomass-based energy generation for better utilisation of farm-based assets and agricultural products.

Priority Area No 4. Enhanced social inclusion, improved skills and employment opportunity in the agriculture sector

Under this priority, FAO will focus on the building capacities and skills of the poor for gainful and sustainable livelihoods through employment-generating agribusiness and enterprise clusters and other projects that are being supported under the DAY-NRLM and on grazing-based livestock production that is crucial to the livelihood security of the landless and the socially marginalised. It will provide assistance and build capacity to strengthen agro-ecological systems and farmer field school approaches that are currently being practised in various parts of the country with the objective of supporting employment-generating agribusiness and enterprise clusters. FAO will also help in highlighting and generating evidence on the importance of small ruminants and backyard poultry in enhancing overall returns from agriculture and making dryland and highland farming systems more climate resilient and thereby reducing the vulnerability of small and marginal farmers.

The Bay of Bengal Large Marine Ecosystem (BOBLME) project will provide regional mechanisms for planning, coordination, monitoring, knowledge management, programme coordination, partnership arrangements and inter-sectoral coordination for fisheries. Another output under this priority area will be increased awareness of animal health emergencies, antimicrobial resistance, capacity building in surveillance, promoting rational use of antimicrobials, support to research to elucidate AMR epidemiology in animal health sector with impact on human health and improved response of animal health sector to minimize impact of AMR on human health. FAO will work in close coordination with Indian Council of Agricultural Research (ICAR), Indian Council of Medical Research (ICMR), National Cooperative Development Corporation (NCDC), World Health Organization (WHO) and USAID to achieve the objectives.

Few recent projects of FAO in India

Priority area - Make agriculture, forestry and fisheries more productive and sustainable

Strengthening Agriculture and Allied Sector Contributions to India's National Biodiversity Action Plan (NBAP) 2008 and the National Biodiversity Targets (NBTs) (01 Feb 2017 - 31 Dec 2019)

India is home to four of the 34 "global biodiversity hotspots". This rich biodiversity is of immense ecological, economic, socio and cultural value. Threats and constraints to biodiversity conservation are mainly due to increasing population and urbanization. The main and important mechanism for conservation of biodiversity is the enactment of Biological Diversity Act in 2002. This progressive legislation has the potential to address lacunae in several aspects of biodiversity conservation and its allied sectors.

The project on Strengthening agriculture and allied sector contributions to India's National Biodiversity Action Plan (NBAP) 2008 and NBTs was taken up to support the efforts of GOI with increased capacity at national, state and local levels through policy development on mainstreaming biodiversity conservation into agriculture and allied sectors. The expected outcome includes increasing the capacity of national, state and local level actors in strengthening agriculture and allied sectors contributions in biodiversity conservation.

Support to the implementation of Indian Ocean Tuna Commission Regional Observer Scheme (January 2018- June 2020)

Priority area - Increase and improve provision of goods and services from agriculture, forestry and fisheries in a sustainable manner

Fisheries observer data is important for fisheries management, providing an independent source of detailed, high quality information on fishing activities and catches at a sufficient level of resolution to be used for analyses such as the standardisation of catch rates and analysis of bycatch mitigation measures. A large number of observer programmes have now been established for industrial fishing fleets across the Indian Ocean and these are used to collect scientific fisheries data by on-board observers, according to specific research requirements specified by each of the coordinating organisations.

This project aims to build on minimum standards by implementing these new tools and materials in three priority CPCs identified by the Commission for the establishment of observer schemes. The project expects to provide intensive and sustained support to CPCs to establish their programmes by first training observer managers and trainers, establishing databases, data management, quality control and reporting procedures and also directly supporting observer training.

Outcome includes improved scientific information and management advice on the fisheries for tuna and tuna-like species operating in the IOTC area of competence besides improved capacity (knowledge, understanding, tools, skills, systems and good practices) of individual observers and national bodies to collect information as required by the IOTC.

Theory of Change (ToC) for the Country Programming Framework (CPF) of India

The ToC summarises how concrete actions that FAO takes at the country level contribute to the impact and outcomes across each of the four priority areas and the factors enabling FAO's work in obtaining results.

In accordance with the national priorities of the government of India, the impact statement for the CPF 2018-22 is "Improved incomes of farmers, especially women and the most vulnerable and marginalised, through secure and resilient livelihood options in agriculture, improved food and nutrition security, improved resilience to climate change and climatic variability and improved access to social security measures". The envisaged impact will be contributed through the four priority areas. Additionally, to capture its contributions to the Sustainable Development Goals (SDGs), FAO has framed its work through outputs contributing to four outcomes (priority areas) that will be measured through output indicators, baselines and targets. This approach to developing a ToC seeks to explain the link between change strategies and the delivery of outputs that contribute to higher-level results, including the SDGs.

Structure of Food and Agriculture Organization of UN

The United Nations is an international organization founded in 1945. It is currently made up of 193 Member States. The mission and work of the United Nations are guided by the purposes and principles contained in its founding Charter. General Assembly, Security Council, Economic and Social Council, Trusteeship Council, International Court of Justice and a Secretariat comprise the established principal organs of the United Nations. Due to the powers vested in its Charter and its unique international character, the United Nations can take action on the issues confronting humanity in the 21st century, such as peace and security, climate change, sustainable development, human rights, disarmament, terrorism, humanitarian and health emergencies, gender equality, governance, food production, and more. The UN also provides a forum for its members to express their views in the General Assembly, the Security Council, the Economic and Social Council, and other bodies and committees. By enabling dialogue between its members, and by hosting negotiations, the organization has become a mechanism for governments to find areas of agreement and solve problems together. The UN's Chief Administrative Officer is the Secretary-General. The year 2020 marks the 75th anniversary of the United Nations.

India has been a founding member of the United Nations since signing the UN Charter in San Francisco in 1945. It is a major participant in the formulation and implementation of the United Nations sustainable development agenda. Implementing the Millennium Development Goals (MDGs), India was a leading voice in the negotiations on the Sustainable Development Goals (SDGs) and the 2030 Agenda for Sustainable Development. India also played a lead role in arriving at the Addis Ababa Action Agenda for financing development and the international community's collective success in adopting the Paris Agreement under the UN Framework Convention on Climate Change. India has all the necessary credentials to be included as a permanent member of the reformed United Nations Security Council and is at the forefront of efforts on UN reform process underway.

Over the decades, India has urged the UN to play a more central and effective role in pursuing a more equitable international order and an economic environment that is conducive to rapid economic growth and development in developing countries. India remains committed to seeking equity, policy space and real voice and participation for developing countries in global governance.

It has been a major partner of the UN specialized agencies and funds and programmes on development issues and has been one of the bigger contributors to their core funding from developing countries. India has played an active role in support of the special needs of the Least Developed Countries (LDCs), Landlocked Developing Countries (LLDCs) and Small Island Developing States (SIDS), both at the UN as well as bilaterally as part of the support extended to these countries under South-South Cooperation.

India is fully committed to the promotion and pursuit of sustainable development, with balanced emphasis on the economic, social and environmental pillars. It believes that as the greatest global challenge, poverty eradication is an indispensable requirement for sustainable development and that environmental concerns cannot be viewed in isolation from developmental imperatives of developing countries. India also believes that the Rio principle of common but differentiated responsibilities remains the cornerstone of international cooperation on sustainable development, as reaffirmed by Heads of States and Governments at the Rio+20 Conference and by several other recent UN Conferences and events. For developing countries like India, the issue of climate change goes beyond environmental sustainability and directly impacts their overriding priorities of development and poverty eradication. India believes that the global efforts to address climate change must be in full accordance with the principles and provisions of the Convention, in particular the principles of 'equity' and 'common but differentiated responsibilities'.

The Food and Agriculture Organization (FAO) is a specialized agency of the United Nations that leads international efforts to defeat hunger. The specialized agencies are independent international organizations funded by both voluntary and assessed contributions. Goal of FAO is to achieve food security for all and make sure that people have regular access to enough high-quality food to lead active, healthy lives. With over 194 member states, FAO works in over 130 countries worldwide and believes that everyone can play a part in ending hunger. The principal organs of FAO are the FAO Conference, the FAO Council, and the Secretariat, headed by a Director General.

The FAO comprises several departments namely Agriculture and Consumer Protection; Climate, Biodiversity, Land and Water Department; Corporate Services; Economic and Social Development; Fisheries and Aquaculture; Forestry; Technical Cooperation and Programme Management.

Committee on Fisheries (COFI)

The Committee on Fisheries (COFI), a subsidiary body of the FAO Council, was established by the FAO Conference at its Thirteenth Session in 1965. The Committee presently constitutes the only global inter-governmental forum where major international fisheries and aquaculture problems and issues are examined and recommendations addressed to governments, regional fishery bodies, NGOs, fish workers, FAO and international community, periodically on a world-wide basis. COFI has also been used as a forum in which global agreements and non-binding instruments are negotiated.

COFI membership is open to any FAO Member and non-Member eligible to be an observer of the Organization. Representatives of the UN, UN bodies and specialized agencies, regional fishery bodies, international and international non-governmental organizations participate in the debate, but without the right to vote. COFI may establish sub-committees on certain specific issues such as the COFI Sub-Committee on Trade and COFI Sub-Committee on Aquaculture. Such subsidiary bodies meet in the intersessional period of the parent Committee.

Functions of COFI

The two main functions of COFI are to review the programmes of work of FAO in the field of fisheries and aquaculture and their implementation, and to conduct periodic general reviews of fishery and aquaculture problems of an international character and appraise such problems and their possible solutions with a view to concerted action by nations, by FAO, inter-governmental bodies and the civil society. The Committee also reviews specific matters relating to fisheries and aquaculture referred to it by the Council or the Director-General of FAO, or placed by the Committee on its agenda at the request of Members, or the United Nations General Assembly. In its work, the Committee supplements rather than supplants other organizations working in the field of fisheries and aquaculture.

COFI Sub-Committee on Trade was established by the Committee on Fisheries (COFI) at its Sixteenth Session (1985). Membership is open to all Member Nations of the Organization. Non-Member states of the Organization that are members of the United Nations, or any of its Specialized Agencies or the International Atomic Energy Agency, may be admitted by the Council of the Organization to membership in the Sub-Committee. The terms of reference of the Sub-Committee as decided by COFI are the following: The Sub-Committee shall provide a forum for consultations on technical and economic aspects of international trade in fish and fishery products including pertinent aspects of production and consumption. In particular, the work of the Sub-Committee will include:

1. periodic reviews on the situation and outlook of principal fishery commodity markets covering all factors influencing them;
2. on the basis of special studies, discussion of specific fish trade problems and possible solutions;
3. discussion of suitable measures to promote international trade in fish and fishery products and formulation of recommendations to improve the participation of developing countries in this trade, including trade-related services;
4. in conjunction with the FAO/WHO Codex Alimentarius Commission, formulation of recommendations for the promotion of international quality standards and the harmonization of quality control and inspection procedures and regulations;
5. consultation and formulation of recommendations for economically viable fishery commodity development, including processing methods, the upgrading of products and production of final products in developing countries.

COFI Sub-Committee on Aquaculture was established by the Committee on Fisheries (COFI) at its Twenty-fourth Session in 2001. It provides a forum for consultation and discussion on aquaculture. It also advises the Committee on Fisheries (COFI) on technical and policy matters related to aquaculture and the work to be performed by the Organization. Membership is open to all Member Nations of the Organization. Non-Member states of the Organization that are members of the United Nations, or any of its Specialized Agencies or the International Atomic Energy Agency, may be admitted by the Council of the Organization to membership in the Sub-Committee. The Terms of Reference of the Sub-Committee on Aquaculture, based on the recommendations of the Expert Consultation, would be as follows: The Sub-Committee shall provide a forum for consultation and discussion on aquaculture and advise COFI on technical and policy matters related to aquaculture and on the work to be performed by the Organization in the subject matter field of aquaculture. In particular the Sub-Committee shall:

1. identify and discuss major issues and trends in global aquaculture development;
2. determine those issues and trends of international importance requiring action to increase the sustainable contribution of aquaculture to food security, economic development and poverty alleviation;
3. recommend international action to address aquaculture development needs and, in this regard:
   i. to advise on mechanisms to prepare, facilitate and implement action programmes identified, as well as on the expected contribution of partners;
   ii. to advise on the liaison with other relevant groups and organizations with a view to promoting harmonization and endorsing policies and actions, as appropriate;
   iii. to advise on the strengthening of international collaboration to assist developing countries in the implementation of the Code of Conduct for Responsible Fisheries.
4. advise on the preparation of technical reviews and of issues and trends of international significance;
5. address any specific matters relating to aquaculture referred to it by its Members, the Committee on Fisheries or the Director-General of FAO.

FAO Fisheries and Aquaculture Department

Fish be it from a marine or freshwater origin, has a key role to play in the fight against hunger as it reduces poverty by generating income and malnutrition as it provides valuable animal protein and essential micronutrients to vulnerable populations. The fisheries and aquaculture department support all efforts to promote Blue Growth – with its emphasis on reconciling social and economic development with environmental performance – to all fisheries and aquaculture policies. The department leads efforts to promote and support implementation of the Code of the Conduct for Responsible Fisheries and its related instruments, in addition to providing scientific advice, strategic planning and training materials. Moreover, it serves as a neutral forum in bringing together relevant actors to discuss issues related to international co-operation and multi-stakeholder approaches to fisheries and aquaculture management.

Mandate

- Develop methodology, assess and monitor the state of wild resources and elaborate resources management advice.
- Monitor and advise on the development and management of aquaculture.
- Collect, analyse and disseminate information on the sector (capture and aquaculture production, trade, consumption, prices, fleet, employment).
- Provide socio-economic analysis of fisheries and aquaculture and assist in the elaboration of development and management policies and strategies and institutions.
- Monitor and advise on technology development, fish processing, food safety and trade.
- Ensure skilled resourcing and effective delivery of FAO's Strategic Objectives in the field of marine and inland capture fisheries, aquaculture and food systems, and provide leadership to the FAO Blue Growth Initiative.
- Support and assist a network of regional fishery commissions and promote aquaculture networks.

Functions and Structure of FAO Fisheries and Aquaculture Department

The Department comprises one division with six branches employing more than 200 staff and consultants, and manages an overall total delivery of USD 45 million per year through approximately 280 fisheries/aquaculture related projects, in close collaboration with the colleagues based in the FAO decentralized offices.

The division is headed by ADG with Director and two Deputy Directors and respective branch heads along with staff.

Organogram of FAO's Fisheries and Aquaculture Department

The Fishery **Policy, Economics and Institutions Branch** oversees economic, social, institutional, governance, policy and management aspects of fisheries and aquaculture with particular emphasis on improving human well-being, food security and poverty reduction.

The **Aquaculture Branch** is responsible for programmes and activities related to development and management of marine, coastal and inland aquaculture, with a special emphasis on technical, socio-economic and environmental aspects.

The **Product, Trade and Marketing Branch** focuses on the improved post-harvest utilization of fisheries and aquaculture resources and the reduction of food losses along the entire value chain. It develops codes of practice, guidelines and standards related to the safety of products, utilization, marketing and responsible trade.

The **Statistics and Information Branch** is responsible for the collection, compilation, validation, analysis and dissemination of reliable and up-to-date information on all aspects of world fisheries and aquaculture.

The **Fishing Operations and Technology Branch** has primary responsibility to transfer and promote the use of appropriate technologies, infrastructure, equipment and practices dealing with the operation of marine and inland fisheries.

The **Marine and Inland Fisheries Branch** focuses on the management and conservation of fishery resources, including mainstreaming biodiversity and ecosystem concerns in fisheries management with special attention paid to threatened species and vulnerable habitats.

The **FISHCODE** serves as the principal instrument for project cycle management and resource mobilization in Fisheries and Aquaculture Department of FAO.

Role of FAO in Fisheries

Fishing is the capture of aquatic organisms in marine, coastal and inland areas. Marine and inland fisheries, together with aquaculture provide food, nutrition and a source of income to around 820 million people around the world from harvesting, processing, marketing and distribution. For many, it also forms part of their traditional cultural identity. One of the greatest threats to the sustainability of global fishery resources is illegal, unreported and unregulated fishing.

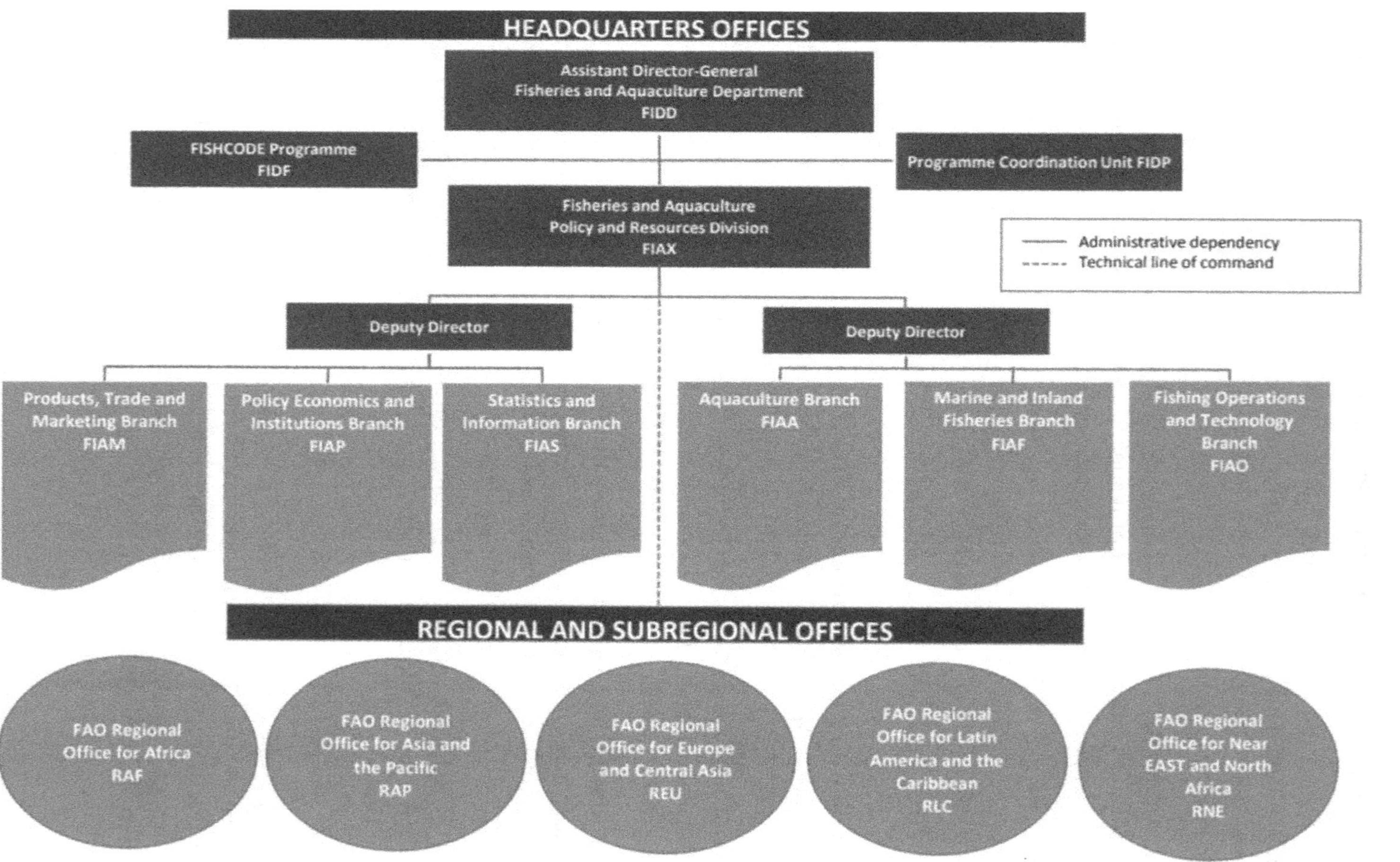

Source: http://www.fao.org/fileadmin/user_upload/COFI/docs/orgnigrammes/Flowcharts1_EN_5June.pdf

FAO recognizes the importance of fish and its many associated products for:

- Food security and nutrition.
- Economic growth through fish production and trade.
- Poverty alleviation and the creation of employment opportunities in rural areas.

FAO plays a leading role in international fisheries policy, including through the Committee on Fisheries (COFI) and related sub-committees on Fish Trade and Aquaculture. FAO works with a wide range of partners, including governments, regional fisheries bodies, cooperatives, fishing communities and others on:

- Implementing the Code of Conduct for Responsible Fisheries and the Ecosystem Approach to Fisheries (EAF).
- Compiling and publishing the global capture production database, including fleet, fishers and trade-related data.
- Reducing the negative impacts of fishing on the environment through technological and community-based management solutions.
- Implementing the Port State and Flag State Measures Agreements to prevent, deter and eliminate illegal, unreported and unregulated fishing.
- Assisting members countries in disaster preparedness as well as providing assistance to fishing communities affected by emergencies and natural disasters.
- Supporting member countries in developing and implementing international guidelines relating to fisheries operations including bycatch management and reduction of discards; eco-labelling and traceability; reduction of fish loss and waste; and supply chain efficiency.
- Improving understanding of the socio-economics of fisheries taking into account value chain dynamics and market access, the status of fisheries resources; access and user right; issues related to decent work conditions and social protection; equitable revenue distribution and profitability and value-addition.
- In close collaboration with intergovernmental organizations (e.g. CITES, CMS, IUCN and NGOs), implementing the International Plans of Action (IPOA) for: Reducing Incidental Catch of Seabirds in Longline Fisheries; Conservation and Management of Sharks; Management of Fishing Capacity; and Prevent, Deter, and Eliminate Illegal, Unreported and Unregulated Fishing.
- Raising the profile of inland fisheries due to its importance for food security and poverty alleviation.

- Providing assistance in disaster preparedness planning and in dealing with the impacts of climate change at the national, regional and international levels as well as assisting fishing communities affected by natural disasters and prolonged emergencies.
- Recognizing small-scale fisheries as a fundamental contributor to poverty alleviation and food security, FAO supports the development of the sector, including through the development of a dedicated instrument the voluntary Guidelines for Securing Sustainable Small-Scale Fisheries in the Context of Food Security and Poverty Eradication.

Major FAO outputs in fisheries

- Compilations of global catch, fleet and employment statistics by country and information about stock status and bio-ecological characteristics of commercially exploited aquatic species.
- Compilation and dissemination of worldwide fish price reports, market studies and trend analysis.
- Publications on fisheries, providing up-to-date information that is valuable to both developed and developing countries.
- Publication of global stock status reviews and bio-ecological information on aquatic species.
- Guidelines for Members countries about the use of local ecological knowledge and participatory approaches in fisheries management.
- Global fisheries statistical and data standards to improve data exchange and integration through partnerships with regional and national institutions.

Role of FAO in Aquaculture

Aquaculture is the farming of aquatic organisms in both coastal and inland areas involving interventions in the rearing process to enhance production. It is probably the fastest growing food producing sector and now accounts for 50% of the world's fish that is used for food.

Aquaculture development

FAO provides a wealth of information and tools on aquaculture development, issues and opportunities worldwide. FAO recognizes the fast-growing contribution aquaculture is making to food security, providing technical assistance through the implementation of the **Code of Conduct for Responsible Fisheries**, which:

- promotes sustainable aquaculture development, especially in developing countries, through better environmental performance of the sector, through health management and biosecurity
- provides regular analysis and reporting of aquaculture development status and trends at global and regional levels, sharing knowledge and information
- develops and implements efficient policies and legal frameworks which promote sustainable and equitable aquaculture development with improved socio-economic benefits

About 580 aquatic species are currently farmed all over the world, representing a wealth of genetic diversity both within and among species. Aquaculture is practiced by both some of the poorest farmers in developing countries and by multinational companies. Eating fish is part of the cultural tradition of many people and in terms of health benefits, it has an excellent nutritional profile. It is a good source of protein, fatty acids, vitamins, minerals and essential micronutrients. Aquatic plants such as seaweed are also an important resource for aquaculture as they provide nutrition, livelihood and other important industrial uses. Eighty percent of current aquaculture production is derived from animals low in the food chain such as herbivorous, omnivorous fish and molluscs. Based on its dynamic performance over the last 30 years, and with fairly stable catches from capture fisheries, it is likely that the future growth of the fisheries sector will come mainly from aquaculture. A sustainable aquaculture strategy needs

- a recognition of the fact that farmers earn a fair reward from farming
- to ensure that benefits and costs are shared equitably
- to promote wealth and job creation
- to make sure that enough food is accessible to all
- to manage the environment for the benefit of future generations
- to ensure that aquaculture development is orderly, with both authorities and industry well organized
- The ultimate aspiration is for aquaculture to develop its full potential so that:
    - communities prosper and people are healthier
    - there are more opportunities for improved livelihoods, with an increased income and better nutrition
    - farmers and women are empowered.

Conclusion

Food and Agriculture Organization (FAO) aims to achieve food security for all and make sure that people have regular access to enough high-quality food to lead active, healthy lives. FAO assistance in India is shaped by the 2019-2022 FAO Country Programming Framework (CPF) which is centred on priority areas such as sustainable natural resource management, stronger food and nutrition security system, increase in resilience of rural livelihoods, increase in farmers' income through improved skills; market linkage and value addition in agriculture and allied sectors. Country Programming Framework (CPF) reflects relevant priorities in key national development policies and is aligned with the UN Sustainable Development Framework (UNSDF) for India 2018-2022.

Annexure 3

Resource requirements matrix and main implementing partners of CPF outputs

Outcome 3: Promotion of sustainable management of natural resources and environmentally friendly agricultural practices for building resilience to climate change and climate variabilityUNSDF Focus Area:By 2022, environmental and natural resource management (NRM) is strengthened and communities have increased access to clean energy and are more resilient to climate change and disaster risks. By 2022, India's voice and participation in multilateral is dramatically increased on matters of peace, security, human rights, development and humanitarian assistance.

Output and Indicator	CPF Indicator target and year of achievement	Indicative Resource Requirements (USD)			Implementing partners (Government actors and other)
		Total estimated resources required	Available / Secured funding	Remaining Resources to be mobilized	
3.1 Improved capacity of national, state and local level actors in strengthening agriculture and allied sectors contributions for gender sensitive and equitable sustainable environment management	a. By 2018, institutional frameworks, mechanisms and capacities to support decision- making and stakeholder participation in Green Landscape planning and management are in place for the nine identified districts	3,828,2679	3,828,267	–	Ministry of Agriculture & Famers Welfare (MoA&FW) Ministry of Environment, Forest and Climate Change (MoEFCC), National Biodiversity Authority (NBA), State Government of Kerala, Punjab & Mizoram, Rajasthan, Odisha, Uttarakhand, Madhya Pradesh
SO Output 2.1.1., 2.1.2, 2,2,1,2.2.2, 2.3.2	b. By 2018, review policy and regulatory frameworks for agricultural and allied sectors for national agrobiodiversity commitments	150,000	150,000	153	Ministry of Agriculture and Farmers Welfare (MoA&FW) Ministry of Environment Forests and Climate Change (MoEF&CC) Ministry of Rural Development (MoRD)
SDG Target 2.4	c. By 2018, identify gaps in strengthening contributions of agriculture and allied sectors in effectively implementing NBAP and achieving NBTs at state level identified				Associated departments Selected State Governments and UTs

[Table Contd.

Contd. Table]

Output and Indicator	CPF Indicator target and year of achievement	Indicative Resource Requirements (USD)			Implementing partners (Government actors and other)
		Total estimated resources required	Available / Secured funding	Remaining Resources to be mobilized	
3.2 Contribute to sustainable management of fisheries, marine living resources and their habitats in the Bay of Bengal region for the benefit of coastal states and communities including women and the poor and disadvantaged communities SO Output 2.2.1, 2.3.2 SDG Target 14.2	a. Two Fisheries management units identified for developing gender and poverty sensitive EAFM plans b. Two MMA sites identified for capacity development programme for promoting gender and poverty sensitive best practices in management and evaluation c. One hotspot identified on river/ coastal/marine waters for addressing pollution from discharge of untreated sewage and wastewater, solid waste and marine litter, and nutrient loading andpromotion of cleaner fishing portsand addressing abandoned fishing gears applying ICM approaches	30,00010	–	30,000	Ministry of Agriculture and Farmers Welfare (MoA&FW) Ministry of Environment Forests and Climate Change (MoEF&CC)
3.3 Increased institutional capacity and enhanced policy focus for gender and poverty sensitive sustainable mountain development, with focus on mountain agriculture and associated value chains and women economic empowerment SO Output 2.1.2, 2.3.2 SDG Target 15.4	a. Conduct 11 Workshops and training (1 in each state) of state forum personnel and development of focus groups in mountain agriculture and allied sectors. b. By 2018, develop specific gender and poverty sensitive policy advocacy related to mountain agriculture and livelihoods of mountain communities, stems migration of mountain people, reduces risk to key ecosystem services andbiodiversity and promotes women's economic empowerment	20,000	20,000	–	Integrated Mountain Initiative (IMI)

[Table Contd.

Contd. Table]

Output and Indicator	CPF Indicator target and year of achievement	Indicative Resource Requirements (USD)			Implementing partners (Government actors and other)
		Total estimated resources required	Available / Secured funding	Remaining Resources to be mobilized	
3.4 Improved advocacy on combating anti-microbial resistance (AMR) and responses of animal health sector to minimise impact of AMR on human health SO Output 5.1.1 and 5.2.2 SDG Target 3d	a. Strengthen capacity of veterinary laboratories of ICAR with regard to AMR surveillance in the country b. Coordinate with ICMR for developing capacity of veterinary micro biologists for selected ICAR veterinary laboratories c. Initiate sharing of pertinent data among veterinary, fisheries and aquaculture laboratories for AMR surveillance under the Indian Network for Fisheries and Animal Antimicrobial Resistance(INFAAR)	127,000	127,000	–	Indian Council of Agricultural Research (ICAR) Indian Council of Medical Research (ICMR) National Centre for Disease Control (NCDC) World Health Organisation (WHO) United States Agency For International Development (USAID)
Total resource requirements for FAO's contribution to government priority 3		4,155,267	4,125,267	30,000	
Additional considerations on FAO'sassistance					

Source: FAO, Country Programming Framework for India, 2018

CHAPTER 7

SDG 14: CONSERVATION AND SUSTAINABLE USE OF THE OCEANS, SEAS AND MARINE RESOURCES (LIFE BELOW WATER)

Introduction

Oceans cover more than two thirds of the Earth's surface and provide billions of people with food and livelihoods. They produce about half the oxygen we breathe, and act as a climate regulator, absorbing atmospheric heat and more than one quarter of man-made CO_2. However, decades of increasing carbon emissions have led to a build-up of heat in the oceans and changes in their chemical composition. The resulting adverse effects of ocean acidification, climate change (including sea-level rise), extreme weather events and coastal erosion exacerbate ongoing threats to marine and coastal resources from overfishing, pollution and habitat degradation. Protected areas and policies and treaties that encourage responsible extraction of ocean resources are critical to confronting these threats.

The expansion of protected areas for marine biodiversity and existing policies and treaties that encourage responsible use of ocean resources are still insufficient to combat the adverse effects of overfishing, growing ocean acidification due to climate change and worsening coastal eutrophication. As billions of people depend on oceans for their livelihood and food source and on the transboundary nature of oceans, increased efforts and interventions are needed to conserve and sustainably use ocean resources at all levels. Protected areas play a critical role in sustainable development if they are both effectively managed and located in areas important for biodiversity. As of December 2018, 17 per cent of waters under national jurisdiction were covered by protected areas. This is a significant increase from 12 per cent in 2015 and more than double the coverage level in 2010. The mean percentage of marine key biodiversity areas (KBAs) covered by protected areas also increased from 31.2 per cent in 2000 to 45.7 per cent in 2018 (Sachs *et al.*, 2019).

CONSERVE AND SUSTAINABLY USE THE OCEANS, SEA AND MARINE RESOURCES FOR SUSTAINABLE DEVELOPMENT

CO_2 CO_2 CO_2

OCEAN ACIDITY HAS INCREASED BY

26% SINCE PRE-INDUSTRIAL TIMES

IT IS EXPECTED TO RAPIDLY INCREASE

BY 100–150% BY 2100

THE INCREASE IN OCEAN ACIDITY IS A NEGATIVE PHENOMENON. IT IMPACTS THE ABILITY OF THE OCEAN TO ABSORB CO_2 AND ENDANGERS MARINE LIFE.

104 OUT OF 220

COASTAL REGIONS IMPROVED THEIR COASTAL WATER QUALITY (2012–2018)

THE PROPORTION OF FISH STOCKS WITHIN BIOLOGICALLY SUSTAINABLE LEVELS DECLINED FROM

90% (1974)

TO

67% (2015)

87 COUNTRIES

SIGNED THE AGREEMENT ON PORT STATE MEASURES, THE FIRST BINDING INTERNATIONAL AGREEMENT ON ILLEGAL, UNREPORTED AND UNREGULATED FISHING

17% OF WATERS UNDER NATIONAL JURISDICTION ARE COVERED BY PROTECTED AREAS

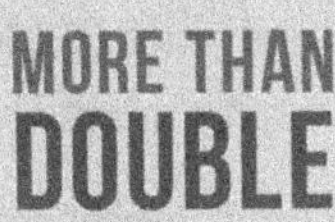

MORE THAN DOUBLE THE 2010 COVERAGE LEVEL

Source: The Sustainable Development Goals Report 2019

Coastal areas worldwide are affected by land-based pollutants, including sewage and nutrient runoff, leading to coastal eutrophication, degraded water quality and the impairment of coastal marine ecosystems. Analysis of the clean water indicator, a measurement of the degree of ocean pollution, shows that water quality challenges are widespread, but are most acute in some equatorial zones, especially in parts of Asia, Africa and Central America. Nearly all countries have room to improve their coastal water quality. Analysis of trends from 2012 to 2018 shows that positive change is indeed possible: 104 of 220 coastal regions improved their coastal water quality over that period. Such improvements require policy commitments at the country level to expand access to wastewater treatment and to reduce chemical and nutrient runoff from agricultural sources, along with global commitments to reduce plastic debris.

Ocean acidification is caused by the uptake of atmospheric CO_2 by the ocean, which changes the chemical composition of the seawater. Long-term observations of ocean acidification over the past 30 years have shown an average increase of acidity of 26 per cent since pre-industrial times, and at this rate, an increase of 100 to 150 per cent is predicted by the end of the century, with serious consequences for marine life. Ocean acidification threatens organisms as well as ecosystem services, including food security, by endangering fisheries and aquaculture. It also impacts coastal protection (by weakening coral reefs, which shield the coastline), transportation and tourism. As the acidity of the ocean rises, its capacity to absorb CO_2 from the atmosphere decreases, hampering the ocean's role in moderating climate change.

To achieve sustainable development of fisheries and to preserve the health and productivity of fisheries, fish stocks must be maintained at a biologically sustainable level. Overfishing not only reduces food production, but also impairs the functioning of ecosystems and reduces biodiversity, with negative repercussions for the economy and society. Analyses reveal that the fraction of world marine fish stocks that are within biologically sustainable levels declined from 90 per cent in 1974 to 67 per cent in 2015. However, this decreasing trend appears to have stabilized since 2008. More focused efforts are needed to rebuild overfished stocks, particularly in severely depleted regions.

Illegal, unreported and unregulated fishing remains one of the greatest threats to sustainable fisheries, the livelihoods of those who depend upon them and marine ecosystems. A framework of international instruments has been developed that addresses different aspects of fisheries management. Most countries have taken measures to combat such fishing and have adopted an increasing number of fisheries management instruments in the past decade. For example, the Agreement on Port

State Measures to Prevent, Deter and Eliminate Illegal, Unreported and Unregulated Fishing, the first international binding agreement to combat such fishing, entered into force in June 2016. The number of parties to the Agreement has rapidly increased and stood at 58 as of February 2019.

Small-scale fisheries are present in almost all countries, accounting for more than half of total production on average, in terms of both quantity and value. To promote small-scale fishers' access to productive resources, services and markets, most countries have developed targeted regulatory and institutional frameworks. However, more than 20 per cent of countries have a low to medium level of implementation of such frameworks, particularly in Oceania and Central and South Asia.

Sustainable Development Goals

The 2030 Agenda for sustainable development recognizes that eradicating poverty in all its forms and dimensions, including extreme poverty, is the greatest global challenge and an indispensable requirement for sustainable development. All countries and all stakeholders, acting in collaborative partnership, will implement this plan. The 17 Sustainable Development Goals and 169 targets which were set to demonstrate the scale and ambition of new universal Agenda seek to build on the Millennium Development Goals and complete what they did not achieve. They seek to realize the human rights of all and to achieve gender equality and the empowerment of all women and girls. They are integrated and indivisible and balance the three dimensions of sustainable development: the economic, social and environmental. The Goals and targets will stimulate action over the next few years in areas of critical importance for humanity.

The Sustainable Development Goals and targets are integrated and indivisible, global in nature and universally applicable, taking into account different national realities, capacities and levels of development and respecting national policies and priorities. Targets are defined as aspirational and global, with each government setting its own national targets guided by the global level of ambition but taking into account national circumstances. Each government will also decide how these aspirational and global targets should be incorporated into national planning processes, policies and strategies. It is important to recognize the link between sustainable development and other relevant ongoing processes in the economic, social and environmental fields.

Goal No	Goal Description
Goal 1	End poverty in all its forms everywhere
Goal 2	End hunger, achieve food security and improved nutrition and promote sustainable agriculture
Goal 3	Ensure healthy lives and promote well-being for all at all ages
Goal 4	Ensure inclusive and equitable quality education and promote lifelong learning opportunities for all
Goal 5	Achieve gender equality and empower all women and girls
Goal 6	Ensure availability and sustainable management of water and sanitation for all
Goal 7	Ensure access to affordable, reliable, sustainable and modern energy for all
Goal 8	Promote sustained, inclusive and sustainable economic growth, full and productive employment and decent work for all
Goal 9	Build resilient infrastructure, promote inclusive and sustainable industrialization and foster innovation
Goal 10	Reduce inequality within and among countries
Goal 11	Make cities and human settlements inclusive, safe, resilient and sustainable
Goal 12	Ensure sustainable consumption and production patterns
Goal 13	Take urgent action to combat climate change and its impacts
Goal 14	Conserve and sustainably use the oceans, seas and marine resources for sustainable development
Goal 15	Protect, restore and promote sustainable use of terrestrial ecosystems, sustainably manage forests, combat desertification, and halt and reverse land degradation and halt biodiversity loss
Goal 16	Promote peaceful and inclusive societies for sustainable development, provide access to justice for all and build effective, accountable and inclusive institutions at all levels
Goal 17	Strengthen the means of implementation and revitalize the Global Partnership for Sustainable Development

Goal 14. Conserve and sustainably use the oceans, seas and marine resources for sustainable development

Main aim under goal 14 is to prevent and significantly reduce marine pollution of all kinds, in particular from land-based activities, including marine debris and nutrient pollution by 2025. It is decided mutually to sustainably manage and protect marine and coastal ecosystems to avoid significant adverse impacts, including by strengthening their resilience, and take action for their restoration in order to achieve

Source: Sustainable Development Report, 2019

healthy and productive oceans by 2020. It is important to minimize and address the impacts of ocean acidification, including through enhanced scientific cooperation at all levels & task set to effectively regulate harvesting and end overfishing, illegal, unreported and unregulated fishing and destructive fishing practices and implement science-based management plans, in order to restore fish stocks in the shortest time feasible, at least to levels that can produce maximum sustainable yield as determined by their biological characteristics by 2020.

Conserving at least 10 per cent of coastal and marine areas, consistent with national and international law and based on the best available scientific information is one of the focus to be accomplished by 2020. Prohibiting certain forms of fisheries subsidies which contribute to overcapacity and overfishing, eliminate subsidies that contribute to illegal, unreported and unregulated fishing and refrain from introducing new such subsidies, recognizing that appropriate and effective special and differential treatment for developing and least developed countries should be an integral part of the World Trade Organization fisheries subsidies negotiation.

By 2030, economic benefits to be increased in small island developing States and least developed countries from the sustainable use of marine resources, including through sustainable management of fisheries, aquaculture and tourism. It is essential to increase scientific knowledge, develop research capacity and transfer marine technology, taking into account the Intergovernmental Oceanographic Commission Criteria and Guidelines on the Transfer of Marine Technology, in order to improve ocean health and to enhance the contribution of marine biodiversity to the development of developing countries, in particular small island developing States and least developed countries. Also, small-scale artisanal fishers should have access to marine resources and markets. Enhancing the conservation and sustainable use of oceans and their resources by implementing international law as reflected in the United Nations Convention on the Law of the Sea (UNCLOS) will provide the legal framework for the conservation and sustainable use of oceans and their resources.

Global efforts towards achieving SDG 14 : Role of FAO

Code of Conduct for Responsible Fisheries (CCRF)

FAO member countries drafted, negotiated and adopted a forward-looking instrument that seized upon the growing global interest in sustainable development popularly called the "Code" in 1995. The Code, which consists of a collection of principles, goals and elements for action, took more than two years to elaborate, involving representatives from members of FAO, inter-governmental organizations, the fishing industry and non-governmental organizations. The Code represents a global consensus on a wide range of fisheries and aquaculture issues. It lays forth principles of sustainable fisheries and aquaculture management. The Code's principles have given rise to various instruments that seek to improve the conservation, management and development of the fisheries and aquaculture sector. Member countries and all those involved in fisheries and aquaculture have been working to implement the Code through their policies. FAO is responsible for monitoring implementation and supporting countries in their efforts to implement the Code, providing capacity support when necessary.

FAO Port State Measure Agreement to prevent, deter and eliminate illegal, unreported and unregulated (IUU) fishing

It is a negotiated international treaty for preventing illegally caught fish from entering international markets through ports was adopted by FAO member countries in 2009 that officially entered into force as an international treaty on 5 June 2016. Illegal, unreported, and unregulated (IUU) fishing is believed to represent 20 percent of total catches per year. Estimates place the cost of illegal fishing between USD 10–23 billion annually. The Agreement itself recognizes the special requirements of developing states and includes provisions to establish funding mechanisms for implementation to countries that have become Party to the Agreement. These mechanisms are intended to be directed towards developing and enhancing capacity for monitoring, control and surveillance and compliance activities relevant to port state measures, as well as training for port managers, inspectors and enforcement and legal personnel. The Agreement promotes collaboration between fishers, port authorities, coast guards and navies to strengthen inspections and control procedures at ports and on vessels.

Global Record of Fishing Vessels, Refrigerated Transport Vessels and Supply Vessels

It's a country-certified repository of vessels involved in fishing. It provides a tool for eliminating illegal fishing. The Global Record of Fishing Vessels, Refrigerated Transport Vessels and Supply Vessels (Global Record) emerged from a process begun in 2005 with the adoption of the Rome Declaration on illegal, unreported and unregulated (IUU) fishing. The Global Record is a single tool in which state authorities and regional fisheries management organizations compile information about all vessels authorized for fishing operations in their countries or regions.

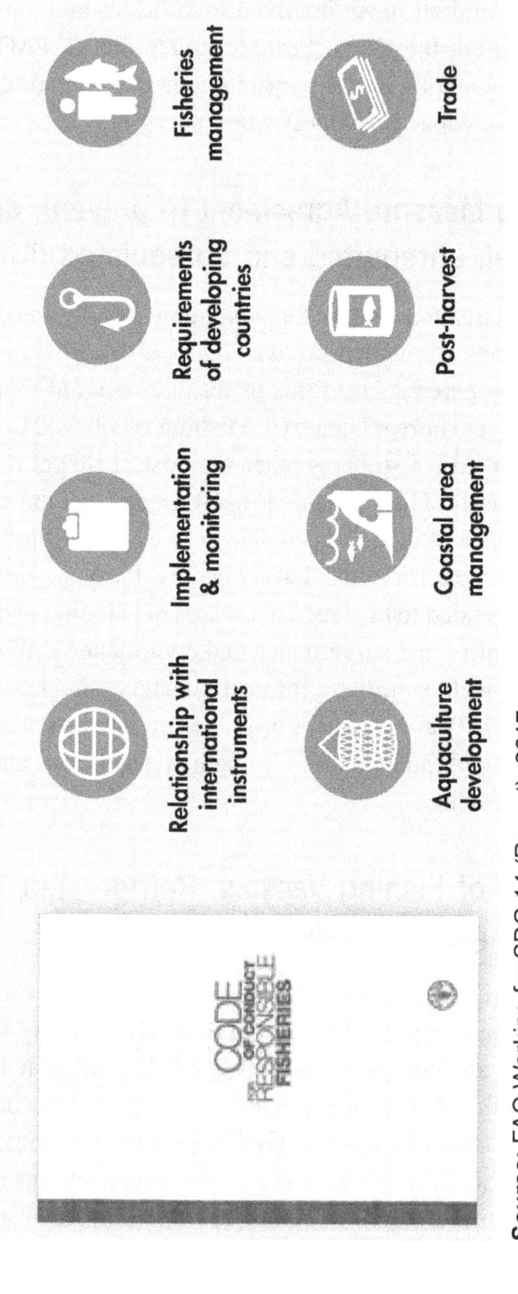

Source: FAO Working for SDG 14 (Report), 2017

Each vessel is registered into the database with a unique vessel identifier (UVI), which remains with a vessel throughout its lifespan, regardless of change of the vessel's name, ownership or flag. The database is crucial for the work of inspectors, port state authorities and flag state authorities, making it easier to identify vessels not regularly identified and registered by the proper national and regional authorities.

The Port State Measures Agreement is a cost-effective and efficient manner to combat IUU fishing by preventing vessels engaged in IUU fishing from using ports.

Two foreign vessels request entry to a port in a country that adheres to the Port State Measures Agreement:

REVIEW AND VERIFY

Fishing authorizations and gear

Purpose of visit

Transshipment information

Vessel identification

Catch on board and documentation

Vessel flag

Vessel marking

Compliance with fisheries regulations

Authorize use of port for all port services. Fish can be landed and transshipped.

Deny use of port, prompt notification to flag state, costal states and regional fisheries management organization and take other measures / prosecute.

Source: FAO Working for SDG 14 (Report), 2017

The Voluntary Guidelines for Catch Documentation Schemes

These are negotiated voluntary guidelines. It provides better and more harmonized traceability of fish along the value chain. The Voluntary Guidelines for Catch Documentation Schemes are aimed at combating, illegal, unregulated and unreported

(IUU) fishing. Catch documentation schemes are tracking and tracing systems that monitor the fish from the point of catch through the whole supply chain to its final destination, thereby documenting the legality of the seafood catch. A five-year negotiation process led by FAO successfully carried out the task set out for it in the Fisheries Resolution adopted by the United Nations General Assembly in December 2013 and the guidelines were unanimously approved in 2017 by a member country-driven FAO technical committee. These guidelines were presented for endorsement by the FAO Conference in July 2017.

Voluntary Guidelines for Securing Sustainable Small-Scale Fisheries in the context of Food Security and Poverty Eradication

These are negotiated voluntary guidelines. It recognizes and supports the important role of small-scale fishing communities. In 2014, FAO's Committee on Fisheries adopted Voluntary Guidelines for Securing Sustainable Small-Scale Fisheries in the Context of Food security and Poverty Eradication. The Guidelines comprise the first internationally agreed instrument for the small-scale fisheries sector. This ground-breaking instrument recognizes the key role small-scale fishing communities (comprising more than 90 percent of the world's capture fishers and fish workers) play in contributing to poverty alleviation and food security. The Guidelines support investing in health, literacy, and education, eradicating forced labour, promoting social security protection, mandating gender mainstreaming, and building fisheries' resistance to climate change and extreme weather events. FAO supports countries as they work towards implementing the voluntary guidelines in their national policies and programmes.

The Nansen Research Vessel

It is a marine research programme conducting marine research in developing countries for their benefit. Over four decades ago, the Government of Norway and FAO began a collaboration, to create a marine research partnership that was far ahead of its time. In the 1970s and 1980s, before environmental awareness was widespread, scientists on the *Nansen* embarked on survey voyages around the globe, measuring the health of oceans. The only marine research vessel to fly the UN flag, the R/V *Dr. Fridtjof Nansen* has carried out its research primarily in Africa but also in Asia, in some of the least observed waters on the planet.

Nansen surveys provide a platform for many developing countries that lack the proper infrastructure to conduct such marine research independently. All of the collected data are input into a dedicated database and made available to the countries and regions. This unique partnership allows many developing countries to achieve their efforts of managing sustainable fisheries and to obtain critical information key to their reporting on SDG 14 achievements. The newest Nansen vessel, the third since the start of this programme, was launched in Oslo's harbour on 24 March 2017. The new *Nansen* is the most advanced marine research vessel of its kind and new laboratories facilitate research in climate change and the study of marine plastics in addition to its fisheries management research activities.

Guidelines on Ecolabeling in Fisheries and Aquaculture

The Guidelines for the Ecolabelling of Fish and Fishery Products from Inland Capture Fisheries are of a voluntary nature and published in the year 2011 whereas the Guidelines for the Ecolabelling of Fish and Fishery Products from Marine Capture Fisheries were published originally in the year 2005 followed by revision 1 in the year 2009. They are applicable to ecolabelling schemes that are designed to certify and promote labels for products from well-managed capture fisheries and focus on issues related to the sustainable use of fisheries resources. The Technical Guidelines on Aquaculture Certification were developed by FAO upon the request of its Members attending the 3rd Session of the Committee on Fisheries (COFI) Sub-Committee on Aquaculture, held in India from 4-8 September 2006 and approved by the 29th Session of COFI, held in Rome from 31 January to 4 February 2011. The guidelines provide advice on developing, organizing and implementing credible aquaculture certification schemes.

Indian Scenario of Goal 14: Conserve and sustainably use the oceans, seas and marine resources

India has taken various steps to protect and enhance the coastal and marine ecosystem. The first Maritime Summit was organized in the country in April 2016. More than 4,500 delegates from across 40 countries participated in the Summit. A clear agenda has been formulated for promoting the 'Blue Revolution'. For tracking the levels of marine pollution along the coastline, the country has developed the Coastal Ocean Monitoring and Prediction System. Additionally, an oil spill management system has been put in place for responding to emergencies arising out of oil spills. Further, the Integrated National Fisheries Action Plan, 2016 is being implemented to promote the livelihoods of fishing communities as well as the ecological integrity of the marine environment. Giving new impetus to port-led development, the "Sagarmala" programme is improving port connectivity, port-linked industrialization and coastal community development.

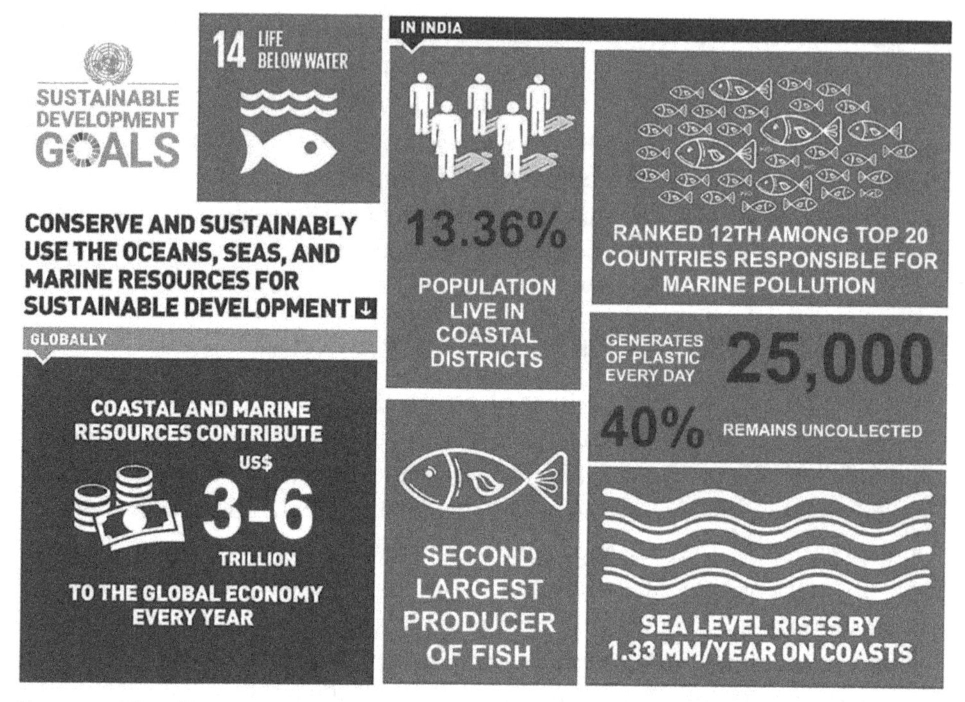

Source: https://in.one.un.org/page/sustainable-development-goals/sdg-14/

Mangroves and Coral Reefs

India has a long history of mangrove forest management. The Sundarbans mangroves, located in the Bay of Bengal, were the first in the world to be put under scientific management. Government of India supports research and development activities with an emphasis on mangrove biodiversity. There has been a net increase of 112 square km in the mangrove cover of the country as compared to the previous assessment. In fact, more than 15,000 ha. of mangroves has been planted in the state of Gujarat alone through active participation of local communities under the Integrated Coastal Zone Management project. Further, India is a part of the regional initiative 'Mangroves for the Future', being coordinated by the United Nations Development Programme and the International Union for Conservation of Nature. Four major coral reefs have also been identified in the country for intensive conservation and management. India has 25 Marine Protected Areas in the peninsular region and 106 in islands, collectively covering approximately 10,000 square km of the country's geographical areas.

Ensuring Sustainability of Fisheries

India has the highest population of fishing communities globally. These communities are spread over 3,600 fishing villages. More than 14.50 million people depend on fisheries for their livelihood. In order to ensure sustainable development of the sector, a number of measures have been taken by the government, with an emphasis on livelihood creation as well as resource conservation. Some of the measures include establishment of a Potential Fishing Zone Advisory programme, modernization and upgradation of fishing centres as well as banning of mechanized fishing in certain areas.

In alignment with the vision of a "Blue Revolution" a central plan, the Integrated Development and Management of Fisheries was formulated followed by a detailed Integrated National Fisheries Action Plan, 2016. The plan envisages connecting 15 million beneficiaries for livelihood opportunities through various interventions. Further, the government has emphasized maintenance of the ecological integrity of the marine environment, in order to ensure that there are no adverse effects on endangered marine species.

Protection of Coastal Ecosystems

Various national and sub-national legislations are in place for the management and protection of the coastal and marine environment. India has also ratified numerous international conventions related to the use of oceans and their resources, including the United Nations Convention on the Law of the Sea. An online mechanism for predicting the movement of oil spills, the online Oil Spill Advisory System, was launched in 2015. In addition, the revised National Oil Spill Disaster Contingency Plan, 2015 reflects the important national regulations as well as the current international norms.

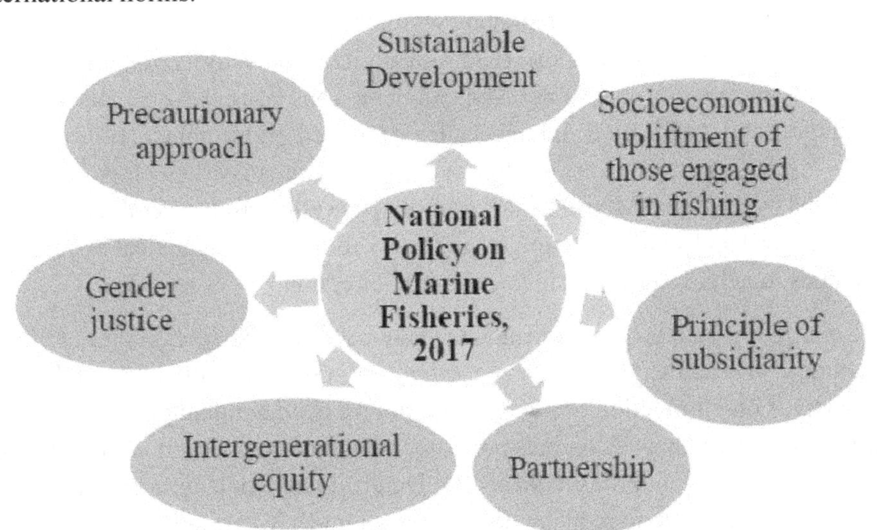

Source: Voluntary National Review Report on the Implementation of Sustainable Development Goals by Niti Ayog, Govt of India for United Nations High Level Political Forum 2017

Further, levels of marine pollution are being monitored by the government at various locations along the country's coastline through the Coastal Ocean Monitoring and Prediction System. India is also setting up a Marine Observation System along the coast to gain a better understanding of coastal processes and monitor water quality.

Holistic Development of Islands and Coastal Areas

India's flagship programme, "Sagarmala" was launched for promoting port connectivity, development and industrialization, in a phased manner during 2015 to 2025. Holistic and sustainable development of coastal communities, especially the population engaged in fishing, is one of the key pillars of the programme. Coastal tourism is also being promoted under the programme for enabling access to better livelihood opportunities.

Indian initiatives towards achieving SDGs

As the fastest growing major economy of the world, India is uniquely placed to deliver on its commitments to inclusive and sustainable development. The country has played a key role in shaping the SDGs and ensuring the balance among its three pillars - economic, social and environmental. It has launched many programs to make progress towards these goals. Notwithstanding its scarce financial resources due to relatively low per-capita income, large population and vast geographical expanse, India is committed to achieving within a short period such ambitious goals as universal rural electrification, road and digital connectivity for all, massive expansions of clean and renewable energy, sanitation and housing for all and universal elementary school education. Taking cue from the memorable phrase "*Sabka Saath, Sabka Vikas*", translated as "Collective Effort, Inclusive Development" and enunciated by the Prime Minister, there is a collective effort from several stakeholders such as central and state governments, industry, civil society, technical experts and academics to achieve progress in the required direction.

Reinforcing India's commitment to the national development agenda and SDGs, the country's Parliament has organized several forums to develop policy and action perspectives on elimination of poverty, promoting gender equality and addressing climate change. Even as it combats poverty, India remains committed to protecting the environment. Under its Nationally Determined Contributions, India has ambitiously committed to reducing the emissions intensity per unit of GDP by 33% -35% by 2030 relative to its 2005 levels. Furthermore, it plans to create an additional carbon sink of 2.5-3 billion tonnes through additional tree cover.

India recognizes that the promotion of global wellbeing requires institutionalizing the concept of 'one world' through partnerships based on solidarity, equity and sharing. In alignment with this philosophy, India's Prime Minister launched the International Solar Alliance at the UN Climate Change Conference in Paris in 2015. The Alliance promotes cooperation among 121 countries with the objective of reducing the price of solar energy. This is to be accomplished through standardization of solar technologies and boosting research and development. Additionally, India has provided platforms like the India-Africa Forum Summit and BRICS Summit for facilitating meaningful dialogue among nations. Beyond bilateral engagements that have been strengthened, focusing on norm setting at multilateral institutions and strengthening the United Nations has also been an important priority. Finally, India continues to play a significant role in development cooperation in its immediate and extended neighbourhood. NITI Aayog, the Government of India's premier think tank, has been entrusted with the task of coordinating the SDGs. It has undertaken mapping of schemes relating to the SDGs and their targets, and has identified the lead and supporting ministries for each target (Annexure 4).

UN Support for Localising the SDGs

The UN Country Team in India supports NITI Aayog in its efforts to address the interconnectedness of the goals, to ensure that no one is left behind and to advocate for adequate financing to achieve the SDGs. In close collaboration with NITI Aayog and partners, the UN has supported thematic consultations on the SDGs to bring together various state governments, central ministries, civil society organisations and academia to deliberate on specific SDGs.

The UN in India currently supports state governments in localising the SDGs to address key development challenges at the state level. State Governments play a key role in visioning, planning, budgeting, and developing implementation and monitoring systems for the SDGs and are a crucial driving force for SDG progress. They are key to India's progress on the SDG Agenda as they are best placed to 'put people first' and to ensuring that 'no one is left behind'. Many of the Government's flagship programmes such as *Swachh Bharat, Make in India, Skill India, and Digital India* are at the core of the SDGs. State and local governments play a pivotal role in many of these programmes. The role of local governments is equally important; 15 of the 17 SDGs directly relate to activities undertaken by local governments in the country. States in turn have prepared roadmaps for implementation of SDGs and role matrix which clearly identifies roles and responsibilities of various stakeholders. Some have identified indicators for each of the 17 SDGs while outlining the baseline, targets, milestones as well as key strategies for realizing the SDGs.

SDG India Index

Given the importance accorded by the Government of India to achieving SDGs, NITI Aayog decided to estimate the progress through a single measurable index that would serve as an advocacy tool and trigger action at the State level. NITI Aayog has constructed the SDG India Index spanning across 13 out of 17 SDGs (leaving out Goals 12, 13, 14 and 17). The Index tracks the progress of all the States and UTs on a set of 62 Priority Indicators, measuring their progress on the outcomes of the interventions and schemes of the Government of India. The SDG India Index is intended to provide a holistic view on the social, economic and environmental status of the country and its States and UTs. The SDG India Index is an aggregate measure which can be understood and used by everyone—policymakers, businesses, civil society and the general public. It has been designed to provide an aggregate assessment of the performance of all Indian States and UTs, and to help leaders and change makers evaluate their performance on social, economic and environmental parameters. It aims to measure India and its States' progress towards the SDGs for 2030 (Annexure 5).

Suggestions for better implementation of SDG 14 in India

1. Regulatory framework to be strengthened; overlap of jurisdiction and inter-state issues to be sorted
2. Information on ecologically sensitive zones such as MPAs should be made available to the public.
3. Ecosystem-based fishery management to be practised
4. Subsidy regime for fishermen needs to be rationalized
5. Development of mariculture/ cage culture
6. Action Plan on effective enforcement and implementation of the MFR Act and its rules by all the respective coastal states.
7. Regular patrolling in terrestrial waters in order to curb the illegal fishing activities in sea
8. Purchase of new patrol vessels with latest technology for patrolling the sea for reporting illegal fishing activity.
9. Fishery related activities specifically inland sector should be suitably included in SDG targets.
10. A clear "National Vision: on SDG-14 may be evolved from which roles and responsibilities of different agencies with time limits and roadmaps may be indicated to track the progress on implementation of SDG.

Conclusion

Sustainable Development Goal No 14 which is "Life Below Water" is the most important SDG for the sustainable development of fisheries and aquaculture. The Goal 14 suggests sustainable development through conservation and sustainable use of oceans, seas and marine resources. The proportion of fish stocks within biologically sustainable levels have declined from 90% (1974) to 67% (2015) globally. At the global level, FAO has developed measures like CCRF; Port state measures agreement to prevent, deter and eliminate IUU; Global Record of fishing vessels, refrigerated transport vessels and supply vessels; Voluntary guidelines on catch documentation schemes; Voluntary guidelines for securing sustainable small-scale fisheries in the context of food security and poverty eradication; The Nansen research vessels; and Guidelines on ecolabelling in fisheries and aquaculture etc for achieving Sustainable Development Goal No 14.

India is one among the key countries in the development of fisheries sector globally. Contribution of India to the global fisheries sector cannot be ignored. However, there is ample potential to grow on a large scale, as India has all the suitable resources along with technology & professionals which can pave the way for growth of the sector in alignment with FAO's sustainable development goals. Govt of India is in the midst of successfully implementing schemes for achieving blue revolution in the country in line with the SDG 14 and other related SDGs through mapping of the same with all associated ministries, departments at national as well as at the local level.

Reinforcing India's commitment to the national development agenda and Sustainable Development Goals, initiatives have been taken to develop policy and action perspectives on elimination of poverty, promoting gender equality and addressing climate change. The NITI Aayog has done detailed mapping and assigning of the 17 Goals and 169 targets to concerned nodal Central ministries. State Governments have in turn carried out similar mapping of the SDGs and targets and assigned responsibilities to the relevant departments in their respective states. An Integrated National Fisheries Action Plan, 2016 envisages connecting 15 million beneficiaries for livelihood opportunities through various interventions and further, the government has emphasized maintenance of the ecological integrity of the marine environment, in order to ensure that there are no adverse effects on endangered marine species. Recently, GoI developed a national indicator framework for SDG 14 detailing targets, national indicator, data source and periodicity, a step towards achieving blue revolution through sustainable fisheries development.

Annexure 4 Mapping of goals and targets under SDG 14

Sustainable Development Goals - National Indicator Framework

Version 2.1 (as on 29.06.2020)

Target	National Indicator	Data Source	Periodicity
Goal 14. Conserve and sustainably use the oceans, seas and marine resources for sustainable development			
14.1 By 2025, prevent and significantly reduce marine pollution of all kinds, in particular from land-based activities, including marinedebris and nutrient pollution	14.1.1 Coastal Water Quality Index	Ministry of Earth Sciences	Annual
	14.1.2 Percentage use of nitrogenous fertilizer to total fertilizer (N,P & K)	DAC & FW, Ministry of Agriculture and Farmer's Welfare	Annual
14.2 By 2020, sustainably manage and protect marine and coastal ecosystems to avoid significant adverse impacts, including by strengthening their resilience, and take action for their restoration in order to achieve healthy and productive oceans	14.2.1 Percentage change in area under mangroves, (similar to 14.5.2)	Ministry of Environment Forest and Climate Change (MoEF&CC)	2 Years
	14.2.3 Percentage change in Marine Protected Areas (MPA)	Ministry of Environment Forest and Climate Change (MoEF&CC)	2 Years
14.3 Minimize and address the impacts of ocean acidification, including through enhanced scientific cooperation at all levels	14.3.1 Average marine acidity (pH) measured at agreed site of representative sampling stations	Ministry of Earth Sciences	Annual
14.4 By 2020, effectively regulate harvesting and end overfishing, illegal, unreported and unregulated fishing and destructive fishing practices and implement science-based management plans, in order to restore fish stocks in the shortest time feasible, at least to levels that can produce maximum sustainable yield as determined by their biological characteristics	14.4.1 Maximum Sustainable Yield (MSY) in fishing (in Million Tonnes/Year)	Department of Fisheries, Ministry of Fisheries, Animal Husbandry, & Dairying	Annual

[Table Contd.

Contd. Table]

Target	National Indicator	Data Source	Periodicity
14.5 By 2020, conserve at least 10 per cent of coastal and marine areas, consistent with national and international law and based on the best available scientific information	14.5.1 Coverage of protected areas in relation to marine areas	Ministry of Environment Forest and Climate Change (MoEF&CC)	Annual
	14.5.2 Percentage change in area under mangroves, (similar to 14.2.1)	Ministry of Environment Forest and Climate Change (MoEF&CC)	2 Years
14.6 By 2020, prohibit certain forms of fisheries subsidies which contribute to over-capacity and overfishing, eliminate subsidies that contribute to illegal, unreported and unregulated fishing and refrain from introducing new such subsidies, recognizing that appropriate and effective special and differential treatment for developing and least developed countries should be an integral part of the World Trade Organization fisheries subsidiesnegotiation	National Indicator is under development		
14.7 By 2030, increase the economic benefits to small island developing States and least developed countries from the sustainable use of marine resources, including through sustainable management of fisheries, aquaculture and tourism	National Indicator is under development		
14.a Increase scientific knowledge, develop research capacity and transfer marine technology, taking into account the Intergovernmental Oceanographic Commission Criteria and Guidelines on the Transfer of Marine Technology, in order to improve ocean health and to enhance the contribution of marine bio-diversity to the development of developing countries, in particular small island developing States and least developed countries	14.a.1 Allocation of budget resources (Budget Estimates) for Ocean Services, Modelling, Applications, Resources and Technology (OSMART) scheme (in Rs Crore)	Ministry of Earth Sciences	Annual

[Table Contd.

Contd. Table]

Target	National Indicator	Data Source	Periodicity
14.b Provide access for small-scale artisanal fishers to marine resources and markets	14.b.1 Assistance to the traditional / artisanal fishers for procurement of Fibre Reinforced Plastic (FRP) boats and other associated fishing-implements, (in Number & in Rs. Lakh)	Department of Fisheries, Ministry of Fisheries, Animal Husbandry, & Dairying	Annual
14.c Enhance the conservation and sustainable use of oceans and their resources by implementing international law as reflected in the United Nations Convention on the Law of the Sea, which provides the legal framework for the conservation and sustainable use of oceans and their resources	14.c.1 Compliance of international laws	Ministry of Earth Sciences	Annual

Source: Ministry of Statistics & Program Implementation, Govt of India, 29.06.2020

Annexure 5 Indicator of SDG India Index 2.0 for SDG Goal 14

S.No.	Target	Indicator	Year	Source
1	14.1	Percentage change in use of nitrogen fertilizers in the coastal states	2015-16	MoSPI
2	14.1	Percentage Ph level of rivers	2015	MoSPI
3	14.1	Percentage of Dissolved oxygen in rivers	2015	MoSPI
4	14.1	Percentage of BOD in river	2015	MoSPI
5	14.1	Coastal Water Quality Index	2015-16	MoSPI
6	14.2	Percentage change in area under mangroves	2017	MoEF&CC

Source: Localising SDGs Early Lessons from India 2019 by Niti Ayog, Govt of India

CHAPTER 8

OVERVIEW OF RECENTLY LAUNCHED PRADHAN MANTRI MATSYA SAMPADA YOJANA (PMMSY) IN INDIA

Background of blue revolution

Fisheries is an important economic activity and a flourishing sector in India with varied resources and potential, engaging over 16 million people at the primary level. The sector has witnessed tremendous growth with an increased fish production from 7.5 lakh tonnes in 1950-51 to 125.9 lakh tonnes during 2017-18. The export earnings of the sector registered Rs. 45,106 crores during 2017-18. It has contributed about 0.92 per cent to the National Gross Value Added (GVA) and 6.16 per cent to the Agricultural GVA (2017-18). India is currently one of the important players in international seafood trade making the position stronger and stronger year on year.

Blue Revolution or the ***Neel Kranti Mission*** was started in 2015 for the duration of 5 years (2015-16 to 2019-20) by Government of India. The vision was to achieve economic prosperity of the country and the fishers/ fish farmers as well as contribute towards food and nutritional security. The focus was on tapping the full production potential and enhance productivity substantially from aquaculture and fisheries resources, both inland and marine besides increasing the share of Indian fisheries in the export area. This was a Centrally Sponsored Scheme (CSS) with an initial outlay of Rs. 3000 crores from the central government. An allocation of Rs. 2767.49 crores was made for implementation of scheme, out of which the Department of Fisheries was able to utilize Rs 2158.13 crores. The implementation of the blue revolution scheme was through NFDB.

Role of National Fisheries Development Board (NFDB) towards achieving blue Revolution in India

The National Fisheries Development Board (NFDB) was established in 2006 as an autonomous organization under the administrative control of the Department of Fisheries, Ministry of Agriculture and Farmers Welfare, to enhance fish production and productivity in the country and to coordinate fishery development in an integrated and holistic manner. Now, the Board works under the Ministry of Fisheries, Animal Husbandry and Dairying. It was set up to realize the untapped potential of fisheries sector in inland and marine fish capture and culture, processing and marketing of fish and accelerate the overall growth of fisheries sector with the application of modern technology backed by research & development. The primary role of NFDB was to channelize Govt. of India funds though activities such as identifying the needs of implementing agencies, providing technical guidance, monitoring physical and financial progress of projects, impact assessment, etc., that have remained as integral components. It has taken up numerous and multifarious developmental activities which have brought visible positive changes in production and productivity as well as post-harvest operations of the fisheries sector. It aims to achieve economic prosperity of fishers and fish farmers that is expected to be achieved by developing fisheries in a sustainable manner keeping in view biosecurity and environmental concerns.

Launch of Pradhan Mantri Matsya Sampada Yojana (PMMSY)

The potential for growth in fisheries sector is immense and the country is on the threshold of massive development in fisheries and aquaculture. Given the abundance of resources with potential and the national importance attributed, aquaculture in India is poised for great expansion in the near future. At the national level, the Govt. of India plans to develop a road map for enhancing fish production through BLUE REVOLUTION with the help of funding model of Pradhan Mantri Matsya Sampada Yojana (PMMSY) that was launched in June 2020. The scheme is expected to revolutionize the fisheries sector, invigorate it with the latest technology, infrastructure and ensure financial assistance.

Vision of PMMSY

"Ecologically healthy, economically viable and socially inclusive fisheries sector that contributes towards economic prosperity and well-being of fishers, and fish farmers and other stakeholders, food and nutritional security of the country in a sustainable and responsible manner"

Objectives of the Scheme

PMMSY has been conceptualized with the following objectives

- Harnessing of fisheries potential in a sustainable, responsible, inclusive and equitable manner
- Enhancing of fish production and productivity through expansion, intensification, diversification and productive utilization of land and water
- Modernizing and strengthening of value chain - post-harvest management and quality improvement
- Doubling fishers and fish farmers' incomes and generation of employment
- Enhancing contribution to Agriculture GVA and exports
- Social, physical and economic security for fishers and fish farmers
- Robust fisheries management and regulatory framework

Beneficiaries of PMMSY

The scheme aims to increase fishermens' income. Beneficiaries include fishers, fish farmers, fish worker, fish sellers, SC / ST / Female / differently-abled persons, fisheries Cooperative Societies / Associations, Fish farmers Producer Organization, Fisheries Development Corporation, Self Help Group (SHG) / Joint Liability Group (JLG) and individual entrepreneur and central government.

Impact of PMSSY on fisheries and aquaculture

- This scheme aims to increase fish production from 13.75 million metric ton (2018-19) to 22 million metric tons by 2024
- The scheme will maintain an average annual growth of about 9% in fish production
- This scheme is expected to increase the contribution of GVA of fisheries to agricultural GVA from 7.28% in 2018-19 to around 9% by 2024-25
- It is expected to double export earnings from Rs 46,589 crore (2018-19) to around Rs 1,00,000 crore by 2024-25
- The scheme aims to improve productivity in aquaculture from the current national average of 3 tonnes to around 5 tonnes per hectare
- Post-harvest crop loss is expected to reduce from 20-25% at present to about 10%
- This scheme aims to increase the consumption of domestic fish from 5-6 kg to about 12 kg per person

- The scheme is expected to generate direct gainful employment opportunities to around 55 lakhs in the fisheries sector along the supply and value chain.

Strategies to achieve blue revolution through implementation of PMMSY Scheme (2020-25)

- Emphasis on harnessing of resources in sustainable and responsible manner
- Addressing critical gaps, infusion of technology and water management
- Enhancing stakeholders' economic returns along the value chain combined with robust fisheries management and regulatory framework
- Cluster/area-based approach to the extent possible

Interventions

Broad interventions suggested in PMMSY fall under three categories (1) Enhancement of production and productivity (2) Infrastructure and post-harvest management (3) Fisheries management and regulatory framework.

(i) Enhancement of production and productivity

- Expansion, intensification, diversification, technological infusion and productive utilization of land and water in a sustainable and responsible manner.
- Marine and inland fisheries sectors including aquaculture and mariculture, input support and facilities like national brood banks, hatcheries, rearing facilities, quality seed units including specific pathogen-free or resistant seed facilities will be provided for enhancing production and productivity. Infrastructure and systems for seed and feed certification, input quality testing, aquatic animal health management including quarantine, and disease diagnostics laboratories and referral laboratories, capacity building and establishment of extension support services will be supported.
- Support under PMMSY will be provided to traditional fishermen for acquiring deep-sea fishing vessels, promotion of technologically advanced fishing vessels and fishing gear for fishermen/fishermen groups through State/UT governments.
- Income generating economic activities like Mariculture including open-sea cage cultivation, Seaweed cultivation and processing, pearl and bivalve cultivation will be supported.
- Thrust will be given for fisheries development in Islands to harness their marine resources.

- Support will be provided for ornamental fish cultivation.
- Inland fisheries would be supported through provisions for fish seed stocking of reservoirs, wetlands like beels, ox-bow lakes etc. Composite fish culture will also be encouraged.
- Under PMMSY support will be provided for stocking of reservoirs with quality fingerlings of Indian Major Carps and other suitable species, creation of in-situ hatcheries and fingerlings rearing units for production of quality fingerlings for stocking, integrated development of reservoirs etc.
- Support will be provided for quality brackish water shrimp farming to ensure sustained income to the marginalized small farmers and for fuelling growth of exports.
- For promoting growth of fisheries sector in the Himalayan states/UTs, Cold-water fisheries (trout, IMCs etc.) will be supported
- Recreational fishing will be supported
- Support will be provided for focused and region-specific interventions for promotion and development of inland fisheries and aquaculture in the north-eastern region.
- Establishment of and handholding of fisheries and aquaculture Start-ups would be a priority intervention under PMMSY
- Integrated Aquaparks would be developed under PMMSY
- Establishment of Fisheries Incubation Centers (FICs) would be supported under PMMSY
- Special focus will be accorded for Training and Capacity building of fishermen, fish farmers and fish workers
- Support for genetic improvement programmes for finfish and shellfish would be provided.

(ii) Infrastructure and Post-Harvest Management

- Under PMMSY it is proposed to develop modern fishing harbours and landing centres and modernize/upgrade the existing ones to suit the present and future needs.
- Post-harvest infrastructure including cold chain for reduction of post-harvest losses will be developed and strengthened.
- Modern wholesale fish markets including supermarkets, retail fish markets and outlets, mobile fish and live-fish markets will be developed.
- Fish marketing mechanisms will be strengthened to protect fishers and fish farmers from the vagaries of middlemen, traders and safeguard the interests of consumers.

- An appropriate IT enabled traceability and labeling system will be supported.
- Integrated modern coastal fishing villages will be developed under PMMSY.
- Production and productivity linked critical infrastructure will be developed.

(iii) Fisheries Management and Regulatory Framework

- Provide need-based support to states/UTs for formulation and implementation of fisheries management plans.
- It is proposed to develop and manage a robust Monitoring, Control and Surveillance (MCS) regime.
- Fisheries documentation and database of fishers both marine and inland will be taken up under PMMSY.
- Support will be provided for sea/ocean ranching, use of satellite technology
- Financial assistance will be provided under PMMSY for livelihood and nutritional support for traditional and socio-economically backward.
- Use of by-catch and juvenile fish excluders and turtle exclusion devices will be promoted.
- Provide requisite regulatory infrastructure including boats, devices, equipment, etc. under PMMSY.
- Disease monitoring and surveillance programme i.e. National Surveillance Programme on Aquatic Animal Diseases (NSPAAD) will be further strengthened.
- Assistance for insurance cover to fishing vessels of marine fishers will be provided.
- Fitment of Bio toilets in mechanized fishing vessels for maintaining hygiene

Implementation of the scheme

The PMMSY expects to bridge critical gaps in fish production and productivity, quality, technology, post-harvest infrastructure and management, modernisation and strengthening of value chain, traceability, establishing a robust fisheries management framework and fishers' welfare. It will be implemented in all the States and Union Territories for a period of five years from FY 2020-21 to FY 2024-25.

The scheme will be operated through two separate Components namely (a) Central Sector Scheme (CS) and (b) Centrally Sponsored Scheme (CSS) with a total estimated investment of Rs. 20,050 crores. It comprises of Central share of Rs. 9407 crores, State share of Rs 4880 crores and Beneficiaries contribution of Rs. 5763 crores. The implementation of CSS will be through NFDB and CS directly through the State governments. The Centrally Sponsored Scheme component is further segregated into Non-beneficiary oriented and beneficiary orientated sub-components/activities under the following three broad heads: (i) Enhancement of Production and Productivity (ii) Infrastructure and Post-harvest Management (iii) Fisheries Management and Regulatory Framework.

Details of Central Sector sub-components/activities with 100% centre funding are given under Annexure 6. Details of Beneficiary oriented sub-components/activities under centrally sponsored scheme is provided in Annexure 7. Details of Non-beneficiary oriented sub-components/activities under centrally sponsored scheme is given in Annexure 8.

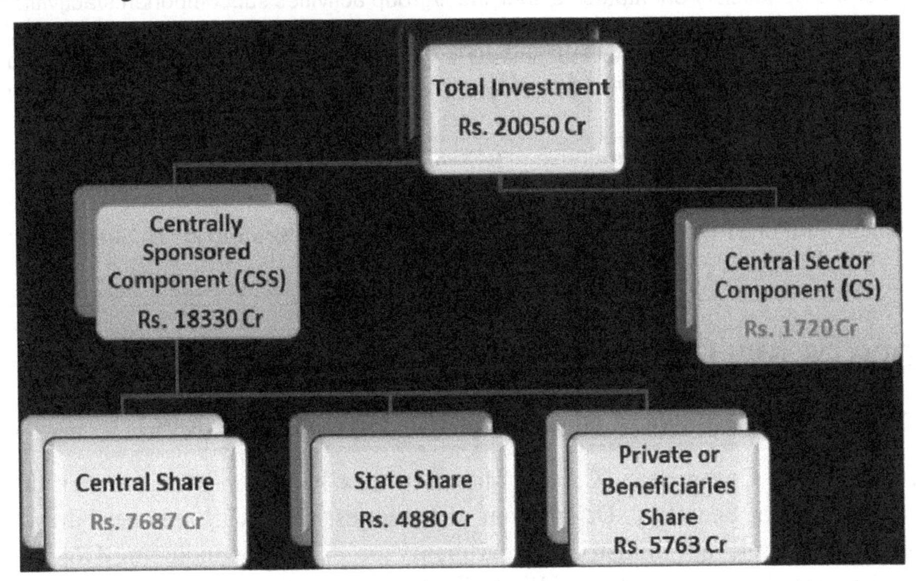

Implementation Schedule - 5 years (2020-21 to 2024-25)

Source: PMMSY Operational guidelines, GoI, Ministry of Fisheries, Animal Husbandry and Dairying, Department of Fisheries, 20 June, 2020

Funding pattern of Central Sector and Centrally Sponsored schemes under PMMSY

Scheme	
(I) Central Sector Scheme (CS)	
(a) The entire project/unit cost will be borne by the Central government	100% central funding
(b) Wherever direct beneficiary oriented i.e. individual/group activities are undertaken by the entities of Central government including National Fisheries Development Board (NFDB)	General category - 40% SC/ST/Women category - 60%. (of the unit/project cost)

[Table Contd.

Contd. Table]

(II) Centrally Sponsored Scheme (CSS)	
A For the Non-beneficiary orientated sub-components/activities under CSS component to be implemented by the States/UTs, the entire project/unit cost will be shared between Centre and State as detailed below	
(a) North Eastern & Himalayan States	90% Central + 10% State share
(b) Other States	60% Central + 40% State share
(c) Union Territories (with legislature and without legislature)	100% Central share
B. For the Beneficiary orientated i.e. individual/group activities subcomponents/activities under CSS component to be implemented by the States/UTs, the Government financial assistance of both Centre and State/UTs governments together will be limited to General category - 40% of the project/unit cost SC/ST/Women - 60% of the project/unit cost	
(a) North Eastern & Himalayan States	90% Central share and 10% State share
(b) Other States	60% Central share and 40% State share
(c) Union Territories (with legislature and without legislature)	100% Central share (No UT Share).

MODE OF IMPLEMENTATION

Institutional framework at the Central Government level

Central Apex Committee (CAC): The CAC has been constituted under the Chairmanship of Secretary, Department of Fisheries, GoI, with members drawn from different relevant ministries/ departments /organizations including DoF for steering overall implementation of PMMSY including monitoring and review.

Project Appraisal Committee (PAC): The PAC comprising Experts and headed by the Chief Executive, NFDB has been constituted for appraisal of Projects/ Proposals under Centrally sponsored scheme submitted by states/UTs after prior approval of State level Approval and Monitoring Committee (SLAMC). After appraisal by PAC, viable projects/proposals under CSS would be recommended to DoF for approval and release of admissible central assistance.

Project Monitoring and Evaluation Unit (PMEU): The DoF under Ministry of Fisheries, Animal Husbandry and Dairying would monitor and evaluate the implementation of PMMSY periodically through the PMEU headed by Joint secretary, DoF with expert members.

Project Monitoring Unit (PMU): THE PMU has been set up in NFDB, constituted comprising experts and headed by Chief Executive, NFDB for monitoring Projects/Proposals on a regular basis.

Institutional framework at the State/UT and District level

District Level Committee (DLC): The DLC headed by the district Collector/ Deputy Commissioner at the district level will be responsible for preparation and approval of annual district fisheries plans, implementation, supervision and monitoring of PMMSY schemes.

State level Approval and Monitoring Committee (SLAMC): The SLAMC constituted in each state headed by the senior most Secretary in charge of the DoF in the states/UTs is responsible for smooth implementation, supervision and monitoring of PMMSY schemes.

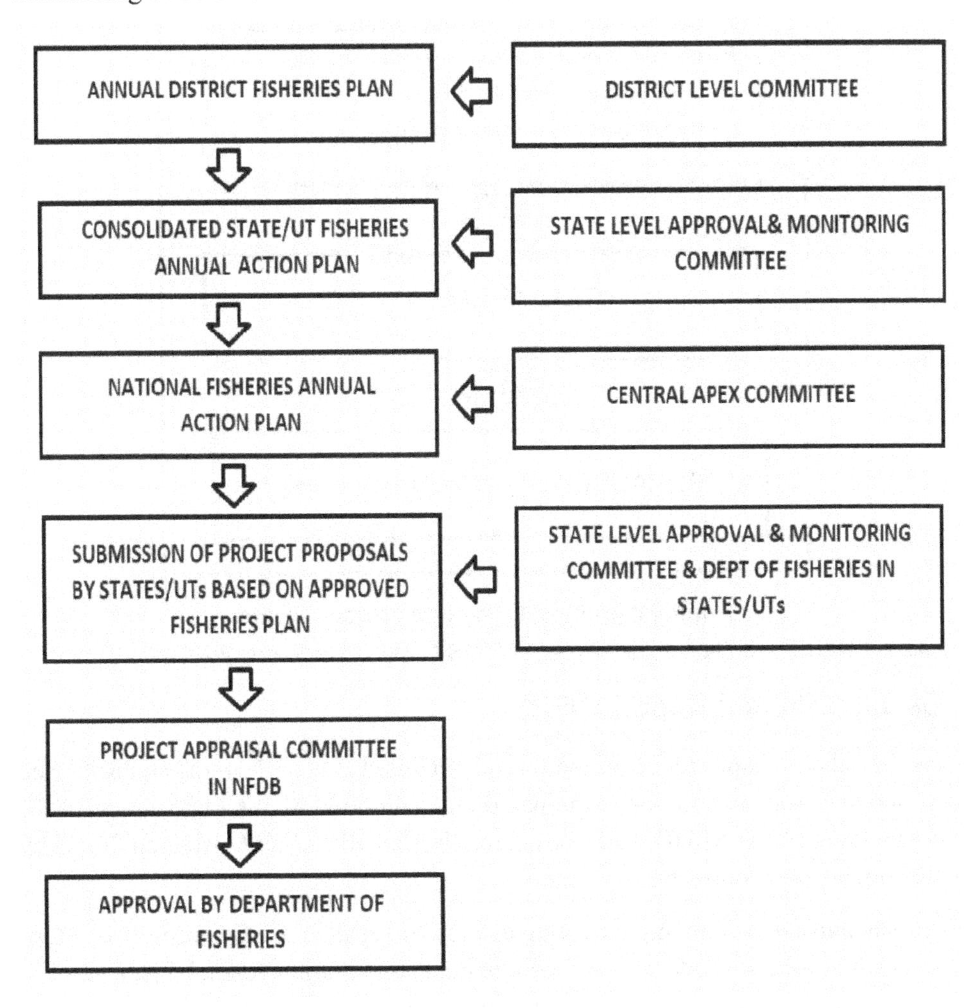

Approval process of Centrally Sponsored Scheme

Source: NFDB operational guidelines of PMMSY, Ready Reckoner, 2020

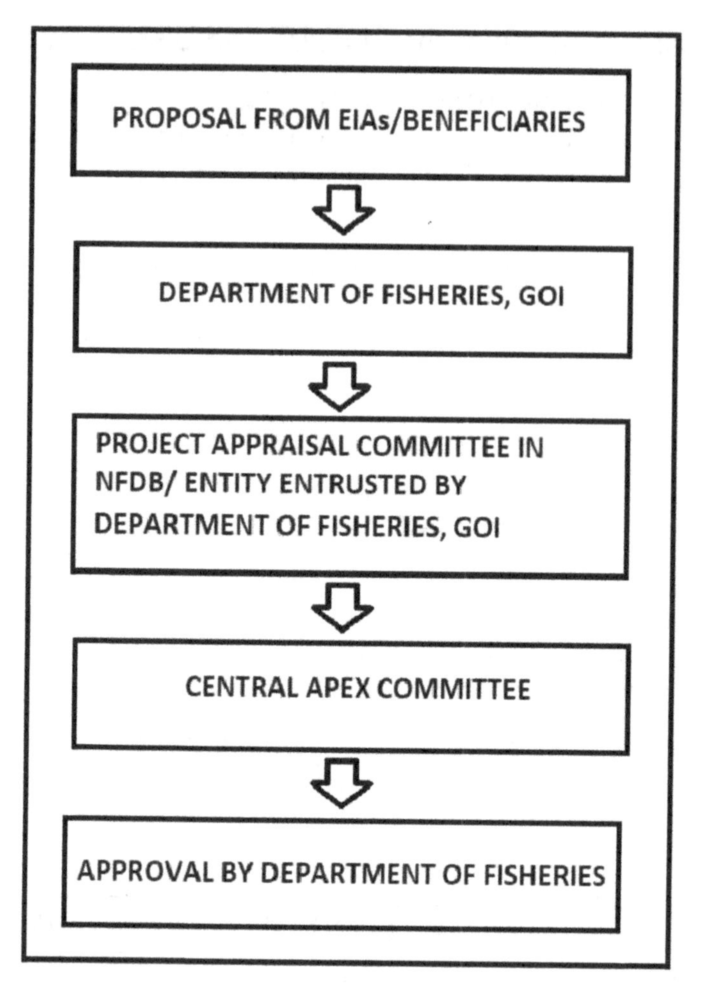

Approval process for central sector scheme

Source: NFDB operational guidelines of PMMSY, Ready Reckoner, 2020

Detailed Project Report (DPR)

As the scheme is primarily project-based, Detailed Project Reports (DPRs) / Self Contained Proposals should be prepared and submitted by the End Implementing Agencies (EIA) to NFDB/ DoF, GoI. Broadly, the DPR / Self Contained Proposals shall include the following:

1. Background of the implementing agency (other than the Department of State/ UT) and their credential and competencies, including financial statement of previous three years in case of autonomous agencies, entrepreneurs
2. Feasibility studies wherever required to assess the demand and supply gaps of intended benefits, particularly in the project locality

3. Project objectives
4. Anticipated benefits in quantifiable terms
5. Cost benefit analysis, wherever required (especially in case of bankable projects)
6. Bio-security and Environment concerns
7. Documentary evidence of availability of land and statutory clearances/ permissions/ licenses, wherever required
8. Sources of funding for implementation of the project (Central assistance, State contribution, own contribution/ bank loan etc. as the case may be)
9. Clear time-lines (in form of a Bar Chart) for completion of the project
10. Undertaking to the effect that there shall be no duplication of central funding or implementation of a similar project by the same agency in the same location
11. Detailed Cost Estimate of the project.
12. Presentation of the project before PAC/NFDB or such entity as decided by the DoF,GoI.

Linkages with Other Schemes

The scheme provides for suitable linkages and convergence with the following schemes/ programmes: Sagarmala Programme of the Ministry of Shipping; Pradhan Mantri Kisan Sampada Yojana of Ministry of Food Processing Industries; Mahatma Gandhi National Rural Employment Guarantee Scheme (MGNREGS); Rastriya Krishi Vikas Yojana and Kisan Credit Card (KCC) of Ministry of Agriculture and Farmers Welfare; National Rural Livelihoods Mission (NRLM); and other schemes of Ministry of Agriculture and Farmers Welfare; Schemes of Department of Commerce; Ministry of Earth Sciences, Ministry of Home; Department of Space; and Ministry of Jal Shakti.

The State Governments, UTs and other agencies shall therefore explore possibility of suitable convergence with the above-mentioned schemes or any other scheme wherever possible, which are being locally implemented. In such proposals, the activities to be covered, quantum of financial sharing proposed under each scheme shall also be clearly indicated in the DPR/Self Contained Proposal.

Inclusive Development

PMMSY encompasses inclusive development by providing higher financial assistance to Women, Scheduled Castes (SCs) and Scheduled Tribes (STs). SLAMC is expected to ensure adequate coverage to small and marginal farmers, SCs, STs and women while selecting beneficiaries under PMMSY.

Pre-Investment Activities

The expenditure towards completion of essential pre-investment activities required for project formulation shall be considered for central assistance on a sharing basis as per the funding pattern of the scheme for activities undertaken only 6 months prior to submission of proposal to SG.

The central funding on completion of essential pre-investment activities shall be restricted to 1% of the total estimated project cost (with a ceiling of Rs. 150 lakhs for multi-crore infrastructure projects), which shall be shared as per the funding patterns of the scheme.

The broad activities involved in project formulation and to be covered for assistance under the scheme are (i) surveys and investigations of all types, (ii) pre-feasibility studies, (iii) preparation of Pre-Feasibility Reports (PFRs), (iv) project planning and designing, (v) preparation of Feasibility Reports (FRs), (vi) Detailed Project Reports (DPRs)/Self Contained Proposals, (vii) Techno Economic Feasibility Reports (TEFRs) and (viii) structural design and detailed cost estimates etc.

The beneficiaries/applicants shall be required to incur the expenditure for completion of the essential pre-investment activities during the course of project formulation. Such expenditure shall be included in the individual DPR with supporting documents/certificates/receipts etc., for consideration under the scheme. The central share of such pre-investment expenditure shall be reimbursed only after the project is approved by the Competent Authority.

In case the proposal is not approved by the authority or fails to meet PMMSY objectives due to its technical viability/ feasibility/ prepared by beyond the scope/ ambit of the Scheme, non- producing of required clearances, non-availability of land, environmental and sustainability concerns or any other reason whatsoever, the expenditure incurred for completion of the pre- investment activities shall not be reimbursed under the scheme. It is the responsibility of the beneficiaries/EIAs to formulate viable/feasible and result oriented proposals etc. The GOI shall not have any commitment to meet the expenditure incurred by the project proponent in formulating unviable and unacceptably proposals.

Fisheries and Aquaculture Infrastructure Development Fund (FIDF)

India is bestowed with varied and huge potential resources in the form of rivers and canals (1.95 lakh km); floodplain lakes (7.98 lakh hectare); ponds and tanks (24.33 lakh hectare); reservoirs (31.50 lakh hectare) and brackish water (14.10 lakh hectare). In addition, the marine fisheries activities spread along the country's long coastline of 8118 km with an EEZ of 2.02 million square km. with export earnings of Rs.45,107 crores from fisheries sector in 2017-18 (US$ 7.08 billion), the fisheries sector has been playing significant role in the national economy.

The Mission Blue Revolution envisioned by the Government primarily focuses to enhance fish productivity, fish production at a growth rate of 6% to 8% and creation of need based infrastructure facilities for fisheries. Through the concerted efforts put in by the Central and State Governments in implementation of various programmes and policies, the country's fish production has increased from 0.75 million tonnes in 1950-51 to 12.61 million tonnes during 2017-18, contributing to the economic development together with food and nutritional security.

Keeping in view that (i) there is limited availability of funds through the normal budgetary process and even these are mostly grant-based without the ability to leverage them for credit based finance, (ii) there is conspicuous lack of credit funding in fisheries sector and (iii) to fill the large gaps in fisheries infrastructure, the Department of Fisheries, Ministry of Agriculture and Farmers Welfare (presently Ministry of Fisheries, Animal Husbandry and Dairying) set up a dedicated Fisheries and Aquaculture Infrastructure Development Fund(FIDF). FIDF envisages creation of fisheries infrastructure facilities both in marine and inland fisheries sectors and augment the fish production to achieve the target of 22 million tonnes by 2025 set under the PMMSY.

A corpus of Rs. 10,000 crores was assigned in 2018 for setting up of a Fisheries and Aquaculture Infrastructure Development Fund (FIDF) for fisheries sector and an Animal Husbandry Infrastructure Development Fund (AHIDF) for financing infrastructure requirement of animal husbandry sector. The proposed Fisheries and Aquaculture Infrastructure Development Fund entails an estimated fund size of Rs 7522.48 Crore comprising of Rs 5266.40 crore to be raised by the Nodal Loaning Entities (NLEs), Rs 1316.60 crore beneficiaries' contribution and Rs 939.48 crore budgetary support from Government of India. It is planned to implement FIDF in all states and union territories.

Objectives of FIDF

1. Creation and modernization of capture & culture fisheries infrastructure
2. Creation of Marine aquaculture infrastructure
3. Creation and modernization of Inland fisheries infrastructure
4. Reduce post-harvest losses and improve domestic marketing facilities through infrastructure support.
5. To bridge the resource gap and facilitate completion of ongoing infrastructure projects.

Development of fishing harbours in Tamil Nadu under FIDF

Fisheries and Aquaculture Infrastructure Development Fund was created with a total of Rs. 7522.48 crore to address the infrastructure requirement for fisheries sector. Government of India urged the coastal states to pay attention to deep sea fishing, post harvesting, cage culture and export promotion. FIDF provides concessional finance to the eligible entities, cooperatives, individuals and entrepreneurs for development of identified fisheries infrastructure. The National Bank for Agriculture and Rural Development (NABARD), National Cooperatives Development Corporation (NCDC) and all scheduled banks are Nodal Loaning entities (NLEs) to provide concessional finance through the FIDF. The Department of Fisheries, Ministry of Fisheries, Animal Husbandry and Dairying under the FIDF provides interest subvention up to 3% per annum for providing the concessional finance by the NLEs at the interest rate not lower than 5% per annum.

The first tripartite Memorandum of Agreement was signed between the department of Fisheries Government of India, NABARD and the Government of Tamil Nadu for the implementation of Fisheries and Aquaculture Infrastructure Development Fund (FIDF) in December 2019. It availed the initial concessional finance of Rs 420 crore from NABARD for development of three fishing harbours in the State namely, (i) Tharangampadi in Nagapattinam District, (ii) Thiruvottriyur Kuppam in Tiruvallur District and (iii) Mudhunagar in Cuddalore District. These will create safe landing and berthing facilities for a large number of fishing vessels plying in the area, augment fish production in the regions, facilitate hygienic post–harvest handling of fish, stimulate growth of fisheries related economic activities and employment opportunities. NABARD as one of the Nodal Loaning Entities (NLEs) provided concessional finance for development of fisheries infrastructure facilities through State Governments/State Entities under the FIDF, after execution of the Tripartite MoA.

Conclusion

The Fisheries sector is gaining importance not only because of its significance in food production but also on account of its value in export earnings and the economic development of the country. It is also instrumental in providing sources of livelihood to a large section of economically backward and resource-poor population of the country. The PMMSY special focus is on fisheries development in islands, northeast, Himalayan states/UTs, and aspirational districts. It expects to bridge critical gaps in fish production and productivity, quality, technology, post-harvest infrastructure and management, modernisation and strengthening of value chain, traceability, establishing a robust fisheries management framework and fishers' welfare. It will be implemented in all the States and Union Territories for a period of five years from FY 2020-21 to FY 2024-25. Government of India is focussing on responsible and sustainable fisheries and aquaculture development to achieve blue revolution through implementation of PMMSY scheme during 2020-21 to 2024-25. Result-oriented joint efforts from all stakeholders (Central Government, State and Local Governments, Entrepreneurs, Processors, Marketers, Exporters, Sellers, Traditional and Modern Businesses) are expected to achieve the targets for fisheries sector under Blue Revolution in India.

Annexure 6

CENTRAL SECTOR SCHEME SUB-COMPONENTS/ACTIVITIES WITH 100% CENTRAL FUNDING UNDER PRADHAN MANTRI MATSYA SAMPADA YOJANA

S.No.	Sub-component	Implementation details	Cost
1	Genetic improvement programs and Nucleus Breeding Centers (NBCs)	DPR mode. The End Implementing Agencies (EIA) must submit Detailed Project Report (DPR)	Unit cost and Central Government financial assistance will be decided by the CAC
2	Innovations and Innovative projects/ activities, Technology demonstration including startups, incubators and pilot projects	Detailed Project Report (DPR) mode.In the case of direct beneficiary oriented i.e. individual/ group activities, project will be done through entities of Central government including NFDB. In such cases the central assistance will be up to 40% of the unit/ project cost for General category and 60% for SC/ ST/Women category.	Unit cost of each project will be evaluated on case to case basis and approved by CAC ● Innovative projects- up to Rs. 1 crore ● Incubation Centres- up to Rs. 3 crores ● Technology Demonstration project- up to Rs. 2 crores ● Startups up to Rs. 50 lakhs ● Pilot projects- up to Rs. 2 crores ● Any other projects as recommended by CAC
3	Training, Awareness, Exposure and Capacity building	Detailed Project Report (DPR) mode	Until further guidelines are formulated the following activities will be supported:

[Table Contd.

Contd. Table]

S.No.	Sub-component	Implementation details	Cost
			• All states have to conduct a One-day state level Awareness cum Training Programme on PMMSY for the stake holders with at least 500 participants and for which central assistance of Rs. 5 lakh will be provided. • One day district level/ regional level (combining 2 to 3 districts, wherever possible) for creating awareness on PMMSY has to be conducted. Awareness cum Training Programme on PMMSY will be conducted for the stakeholders with not less than 500 participants by the state for which central assistance of Rs. 100,000 (for each program) will be provided.
4	Aquatic Quarantine Facilities	The purpose and scope of setting up of Quarantine Stations in fisheries sector is to prevent the ingress of dangerous exotic diseases into the country through imported germ- plasm, live aquatic animal and aquatic products	• Setting up of Aquatic Quarantine Facilities would be supported on 100% Central Assistance. • peration and management (O&M) costs of the AQF should be borne by the concerned sponsoring entity • States/UTs should provide requisite land free of cost for establishing the AQF

[Table Contd.

Contd. Table]

S.No.	Sub-component	Implementation details	Cost
5	Modernization of fishing harbours of central government and its entities	DPR mode	Cost will be as per the actual requirement/need. Site specific DPR may be submitted to DoF
6	Support to NFDB, Fisheries Institutions and Regulatory Authorities of Department of Fisheries, Government of India and need based assistance to State Fisheries Development Boards	Need-based supports in terms of infrastructure etc.	Support will be based on Detailed Project Report (DPR) and the quantum of support would be based on actual need and as decided by the DoF as per recommendations of CAC
7	Support for survey and training vessels for Fisheries Institutes including dredger TSD Sindhuraj owned by the DoF and GoI.	Detailed Project Report (DPR) mode for procurement of survey and training vessels for Fisheries Institutions of FSI involved in fisheries survey and CIFNET involved in imparting training and undertaking course	Unit cost of each project will be evaluated on case to case basis and recommended by CAC
8	Disease Monitoring and Surveillance Network	The disease monitoring and surveillance program i.e. National Surveillance Program on Aquatic Animal Diseases (NSPAAD) will be further strengthened	Second phase of National Surveillance Program on Aquatic Animal Diseases (NSPAAD) would be implemented by DoF on exiting terms and conditions of NSPAAD phase-I given the larger public interest/importance attached to this program
9	Fish data collection, fishers' survey and strengthening of fisheries database	Strengthening of fisheries database which includes survey and regular census of inland and marine fishermen, resource/fish stock assessment (including seaweeds), documentation, etc.	Will be implemented as per actuals by DoF. Wherever EIAs other than DoF undertake these activities, the same would be considered on case to case basis based on a DPR/Self-contained Proposal submitted by the EIAs and subject to recommendations of CAC.

[Table Contd.

Contd. Table]

S.No.	Sub-component	Implementation details	Cost
10	Support to security agencies to ensure safety and security of marine fishermen at sea	To strengthen Security agencies involved in safeguarding safety and security of fishermen at sea by providing requisite regulatory infrastructure including boats, devices, equipment, etc. under PMMSY.	DPR/Self-contained Proposal has to be submitted with justification for the regulatory infrastructure including speed boat for patrolling, security devices and communication equipment etc.
11	Fish Farmers Producer Organizations/ Companies (FFPOs/Cs)	It is proposed to set up 500 Fish Farmers Producer Organizations/Companies (FFPOs/Cs) to economically empower the fishers and fish farmers	The cost norms, guidelines and modalities, etc. for setting up and hand-holding of FFPOs/Cs would be prepared by the DoF and finalized by the CAC
12	Certification, accreditation, traceability and labelling.	1. For setting quality standards for shell fish / fin fish hatcheries/Feed mills in India and ensuring that their production process conforms to norms of quality seed/feed. 2. Economically empower the hatchery owners/ feed mill to ensure the availability and supply of quality fish/shrimp seed and feed to all farmers at a reasonable price 3. To keep traceability of Brood-stocks and documentation of seed production in case of hatcheries and traceability of raw material, process documentation for feed mill.	Will be implemented on DPR/Self Contained Proposal basis on recommendations of CAC and approval of DoF

[Table Contd.

Contd. Table]

S.No.	Sub-component	Implementation details	Cost
13	Administrative Expenses for implementation of PMMSY (to meet expenses of both for Central Sector and Centrally Sponsored Schemes components)		Overall Administrative Expenses for each project/schemes/sub components will not exceed 2.5% of the Central assistance

Annexure 7

Beneficiary oriented sub-components and activities under Centrally Sponsored components of PMMSY

S.No.	Activity	Unit Cost (Rs. in lakhs)	Government al Assistance (Rs. Lakhs)	
			General (40%)	SC/ST/ Women (60%)UTs
	A. ENHANCEMENT OF FISH PRODUCTION AND PRODUCTIVITY			
	Development of inland fisheries and aquaculture			
1	Establishment of New Freshwater Finfish Hatcheries	25	10	15
2	Establishment of New Freshwater Scampi Hatcheries	50	20	30
3	Construction of New Rearing ponds (nursery/seed rearing ponds)	7.00	2.80	4.20
4	Construction of New Grow-out ponds	7.00	2.80	4.20
5	Inputs for fresh water Aquaculture including Composite fish culture, Scampi, Pangasius, Tilapia etc.	4.00	1.60	2.40
6	Establishment of need based New Brackish Hatcheries (shell fish and fin fish)	50	20	30
7	Construction of New ponds for Brackish Water Aquaculture	8.00	3.20	4.80
8	Construction of New ponds for Saline / Alkaline areas	8.00	3.20	4.80
9	Inputs for Brackish Water Aquaculture	6.00	2.40	3.60
10	Inputs for Saline /Alkaline Water Aquaculture	6.00	2.40	3.60

[Table Contd.

Contd. Table]

S.No.	Activity	Unit Cost (Rs. in lakhs)	Government al Assistance (Rs. Lakhs)	
			General (40%)	SC/ST/ Women (60%)UTs
11	Construction of Biofloc ponds for Brackish water/Saline/ Alkaline areas including inputs of Rs 8 lakhs/Ha	18.00	7.2	10.8
12	Construction of Biofloc ponds for Freshwater areas including inputs of Rs 4 lakhs/Ha	14.00	5.6	8.4
13	Stocking of Fingerlings in Reservoirs @1000FL/ha (3.0/1lakh FL)	Rs.3/- Fingerlings	Rs.1.2/- Fingerlings	Rs.1.8/- Fingerlings
14	Stocking of Fingerlings in Wet lands @1000FL/ha (3.0/1lakh /FL)	Rs.3/- Fingerlings	Rs.1.2/- Fingerlings	Rs.1.8/- Fingerlings
Development of marine fisheries including mariculture and seaweed cultivation				
15	Establishment of Small Marine Finfish Hatcheries	50	20	30
16	Construction of large Marine Finfish Hatcheries	250	100	150
17	Marine Finfish Nurseries	15	6	9
18	Establishment of Open Sea cages (100-120 cubic meter volume)	5	2	3
19	Establishment of Seaweed culture rafts including inputs (per raft).	0.015	0.006	0.009
20	Establishment of Seaweed culture with Monoline/ tube net Method including inputs (one unit is approximately equal to 15 ropes of 25m length)	0.08	0.032	0.048
21	Bivalve cultivation (mussels, clams, pearl etc.)	0.20	0.08	0.12
Development of fisheries in North-eastern and Himalayan States/UTs				
22	Establishment of Trout Fish Hatcheries.	50	20	30
23	Construction of Raceways of minimum of 50 cubic meter	3.0	1.20	1.80
24	Inputs for Trout Rearing Units	2.50	1.00	1.50
25	Construction of New Ponds	8.40	3.36	5.04
26	Establishment of Medium RAS for Cold water Fisheries. (with 4 tank of minimum 50 m^3 /tank capacity and fish production capacity of 4 ton/crop)	20	8	12

[Table Contd.

Contd. Table]

S.No.	Activity	Unit Cost (Rs. in lakhs)	Government al Assistance (Rs. Lakhs)	
			General (40%)	SC/ST/ Women
27	Establishment of large RAS for cold water fisheries (with 10 tanks of minimum 50 m^3/ tank capacity and fish production capacity of 10 ton/crop)	50	20	30
28	Input support for Integrated fish farming (paddy cum fish cultivation, livestock cum fish, etc).	1.00	0.40	0.60
29	Establishment of Cages in cold water regions	5.00	2.00	3.00
Development of ornamental and recreational fisheries				
30	Backyard Ornamental fish Rearing unit (both Marine and Fresh water)	3.00	1.20	1.80
31	Medium Scale Ornamental fish Rearing Unit (Marine and Freshwater Fish)	8.00	3.20	4.80
32	Integrated Ornamental fish unit (breeding and rearing for fresh water fish)	25	10	15
33	Integrated Ornamental fish unit (breeding and rearing for marine fish)	30	12	18
34	Establishment of Fresh water Ornamental Fish Brood Bank	100	40	60
35	Promotion of Recreational Fisheries	50	20	30
Technology infusion and adaptation				
36	Establishment of large RAS (with 8 tanks of minimum 90 m3/tank capacity 40 ton/crop)/ Biofloc (50 tanks of 4m dia and 1.5 high) culture system.	50	20	30
37	Establishment of Medium RAS (with 6 tank of minimum 30m3/tank capacity 10ton/crop)/ Biofloc culture system(25 tanks of 4m dia and 1.m high)	25	10	15
38	Establishment of small RAS (with 1 tank of 100m^3 capacity/Biofl oc (7 tanks of 4m dia and 1.5 high) culture system	7.50	3.00	4.50
39	Establishment of Backyard mini RAS units	0.50	0.20	0.30
40	Installation of Cages in Reservoirs	3.00	1.20	1.80
41	Pen culture in open water bodies	3.00	1.20	1.80

[Table Contd.

Contd. Table]

S.No.	Activity	Unit Cost (Rs. in lakhs)	Government al Assistance (Rs. Lakhs)	
			General (40%)	SC/ST/ Women
	B. INFRASTRUCTURE AND POST-HARVEST MANAGEMENT			
Post harvest and cold chain infrastructure				
Construction of Cold Storages/Ice Plants				
42	Plant/storage of minimum 10-ton capacity.	40	16	24
43	Plant/storage of minimum 20-ton capacity	80	32	48
44	Plant/storage of minimum 30-ton capacity	120	48	72
45	Plant of minimum 50-ton capacity	150	60	90
46	Modernization of Cold storage /Ice Plant	50	20	30
47	Refrigerated vehicles	25	10	15
48	Insulated vehicles	20	8	12
49	Motor cycle with Ice Box	0.75	0.30	0.45
50	Cycle with Ice Boxes	0.10	0.04	0.06
51	Three wheeler with Ice Box including rickshaws for fish vending	3.00	1.20	1.80
52	Live fish vending Centres	20.00	8.00	12.00
Fish feed mills				
53	Fish Feed MillsMini Mills of production Capacity of 2 ton /Day	30.00	12.00	18.00
54	Fish Feed MillsMedium Mills of production Capacity of 8 ton /Day	100.00	40.00	60.00
55	Fish Feed MillsLarge mills of production Capacity of 20 ton /Day	200.00	80.00	120.00
56	Fish Feed Plants of production Capacity of at least 100 ton /Day	650.00	260.00	390.00
Markets and marketing infrastructure				
57	Construction of fish retail markets including ornamental fish/aquarium markets.	100.00	40.00	60.00
58	Construction of fish kiosks including kiosks of aquarium/ornamental fish	10.00	4.00	6.00
59	Fish Value Add Enterprises Units	50.00	20.00	30.00
60	E-platform for e-trading and e-marketing of fish and fisheries products	–	–	–

[Table Contd.

Contd. Table]

S.No.	Activity	Unit Cost (Rs. in lakhs)	Government al Assistance (Rs. Lakhs)	
			General (40%)	SC/ST/ Women
Development of deep-sea fishing				
61	Support for acquisition of Deep-sea fishing vessels for traditional fishermen	120.00	48.00	72.00
62	Up gradation of existing fishing vessels for export Competency	15.00	6.00	9.00
63	Establishment of Bio-toilets in mechanized fishing vessels	0.50	0.20	0.30
Aquatic health management				
64	Establishment of Disease diagnostic and quality testing labs	25.00	10.00	15.00
65	Disease diagnostic and quality testing Mobile labs/clinics	35.00	14.00	21.00
C. FISHERIES MANAGEMENT AND REGULATORY FRAMEWORK				
Monitoring, Control and Surveillance (MCS)				
66	Communication and /or Tracking Devices for traditional and motorized vessels like VHF/DAT/NA VIC/Transpo nders etc.	0.35	0.014	0.21
Strengthening of safety and security of fishermen				
68	Support for providing safety kits for fishermen of Traditional and motorized fishing vessels (other than Communication and/or Tracking Device mentioned above)	1.00	0.40	0.60
69	Providing boats (replacement) and nets for traditional fishermen	5.00	2.00	3.00
70	Support to Fishermen for PFZ devices and network including the cost of installation and maintenance etc.	0.11	0.044	0.066
Fisheries extension and support services				
71	Extension and support Services.	25.00	10.00	15.00
Livelihood and nutritional support for fishers for conservation of fisheries resources				
72	Livelihood and nutritional support for socio-economically backward active traditional fishers' families for conservation of fisheries resources during fishing ban/lean period.	Total assistance - Rs 3000 per annum per enrolled beneficiary Share pattern a) North Eastern & the Himalayan States 80% Central + 20% State share b) Other States50% Central + 50% State share c) Union Territories (with legislature and without legislature) 100% Central		

[Table Contd.

Contd. Table]

S.No.	Activity	Unit Cost (Rs. in lakhs)	Government al Assistance (Rs. Lakhs)	
			General (40%)	SC/ST/ Women
Each enrolled beneficiary under this component is required to contribute Rs. 1500 annually. The beneficiary fishers will save Rs. 1500 over a period of 9 months during fishing season annually towards their contribution with a bank designated by the State/ UT Department of Fisheries. States/UTs will devise suitable modalities to ensure transparency and smooth implementation of this activity. Depositing of beneficiary contribution on a lump sum basis in a period of one or two months may be avoided				
Insurance of Fishing Vessels and Fishermen				
73	Insurance to fishers	The fishers shall be eligible for insurance and insurance coverage under the PMMSY as below: (a) Rs.5.00 lakh against death or permanent total disability (b) Rs. 2.50 lakh against permanent partial disability The insurance cover shall be for a period of 12 months and premium shall be paid annually.		
74	Insurance premium subvention for fishing vessels	The government assistance will be up to 40% of the annual premium amount for General category and 60% for SC/ST/Women and rest of the premium will be borne by the beneficiary.		

Annexure 8

Non-beneficiary-oriented activities under Centrally Sponsored components of PMMSY

S.No.	Activity	Unit Cost (Rs. in lakhs)	Central Assistance (Rs. Lakhs)		UTs
			General States	Northeastern & Himalayan States	
	A. ENHANCEMENT OF FISH PRODUCTION AND PRODUCTIVITY				
Development of inland fisheries and aquaculture					
1	Establishment of Brood Banks (including seed banks for seaweeds)	500	300	450	500
2	Integrated Development of Reservoirs (Large) (Area: more than 5000 ha	600	360	540	600
3	Integrated Development of Reservoirs (Medium) (Area: 1000 to 5000 hectares)	400	240	360	400
4	Integrated Development of Reservoirs (Area: less than 1000 hectares)	300	180	270	300
5	Integrated Aqua Parks (IAPs)	10000	6000	9000	10000
6	Development of fisheries in the Himalayan and North-Eastern States/ UTs(Support to states for import of germplasm.)	Need based – On project mode. The State Government will submit a Detailed Project Report (DPR) with justification, fish species, techno financial details.			
	B. INFRASTRUCTURE AND POST-HARVEST MANAGEMENT				
7	Development of fishing harbours and fish landing centres	20000	12000	——	20000
8	Modernization/ Up-gradation of existing Fishing Harbours	5000	3000	——	5000
9	Modern Integrated Fish Landing Centres	2500	1500	2250	2500
10	Maintenance of Dredging of EXISTING fishing harbours	500	300	—	500
Markets and marketing infrastructure					
11	Construction of state of art wholesale fish market	5000	3000	4500	5000

[Table Contd.

Contd. Table]

S.No.	Activity	Unit Cost (Rs. in lakhs)	Central Assistance (Rs. Lakhs)		UTs
			General States	Northeastern & Himalayan States	
12	Organic Aquaculture Promotion and Certification	**DPR/SCP based** Submit Detailed Project Report or Self Contained Proposals with cost estimates and techno-financial details. The projects would be approved on DPR/SCP. The unit cost would be decided by DoF on case to case basis as per essentiality.			
13	Promotion of Domestic fish consumption, branding, Fish mark, GI in fish, Himalayan Trout-Tuna branding, Ornamental fishes promotion and branding etc.				
Development of deep-sea fishing					
14	Promotion of technologically advanced vessels to marine fishermen/fisher men groups through State/UT Governments	5000	3000	—	5000
Integrated modern coastal fishing villages					
15	Integrated modern coastal fishing villages	750	450	—	750
Aquatic Health Management					
16	Aquatic Referral Labs for Quality Testing and Disease Diagnostics.	1000	600	900	1000
C. FISHERIES MANAGEMENT AND REGULATORY FRAMEWORK					
Monitoring, Control and Surveillance (MCS)					
17	Common Infrastructure for MCS including Hub stations, towers, IT based software, peripherals, networks and operations etc	The State Government/UT administration shall submit Detailed Project Report (DPR) indicating the infrastructure requirements, viability assessment, detailed cost estimate, operational modalities, timeline for completion of the project. The State/UT shall also confirm availability of land and clearance and state budgetary resource			

Table Contd.

Contd. Table]

S.No.	Activity	Unit Cost (Rs. in lakhs)	Central Assistance (Rs. Lakhs)		UTs
			General States	Northeastern & Himalayan States	
Fisheries extension and support services					
18	Multipurpose Support Services – Sagar Mitra (performance-based incentives along with requisite IT/Communication support like Tablet/mobile telephony etc. would be provided to Sagar Mitras) (It is envisaged to engage a total of 3477 Sagar Mitras and deploy them at the rate of one Sagar Mitra per marine coastal fishing village in the maritime States/UTs. Sagar Mitra purely on contractual basis. Sagar Mitra purely on contractual basis interface between the Government and fishers)	12.4	7.44	—	12.4

CHAPTER 9

CLIMATE CHANGE AND IMPACT ON FISHERIES AND AQUACULTURE

Introduction

Since 1988 the Intergovernmental Panel on Climate Change (IPCC) has been providing regular, evidence-based updates on climate change and its political and economic impacts. These updates comprehensively synthesize the internationally accepted consensus on the science of climate change, its causes and consequences.

The 2015 Paris Climate Agreement recognizes the need for effective and progressive responses to the urgent threat of climate change, through mitigation and adaptation measures, while taking into account the particular vulnerabilities of food production systems. The climate change will lead to significant changes in the availability and trade of fish products, with potentially important geopolitical and economic consequences, especially for those countries most dependent on the sector. The interaction between ecosystem changes and management responses is crucial to minimize the threats and maximize the opportunities emerging from climate change. Production changes are partly a result of expected shifts in the distribution of species, which are likely to cause conflicts between users, both within and between countries (Barange *et al.,* 2018).

The conference of parties (COP21) of the United Nations Framework Convention on Climate Change highlighted the vital need to reverse the current trend of overexploitation and pollution to restore aquatic ecosystem services and the productive capacity of the oceans. The resultant Paris Agreement aims to keep the increase in global average temperature below 2^{0}C (SOFIA, 2016).

COP21 for the first time featured the role of oceans, inland waters and aquatic ecosystems for temperature regulation and carbon sequestration. The Sendai Framework for Disaster Risk Reduction, an international document was adopted by UN member states between 14th and 18th of March 2015 at the World Conference on Disaster Risk Reduction held in Sendai, Japan and endorsed by the UN General Assembly in June 2015 (FAO,2016.)

The atmosphere and oceans have warmed and sea level has risen. The uptake of additional energy in the climate system is caused by the increase in the atmospheric concentration of carbon dioxide (CO_2) and other greenhouse gases (GHGs). CO_2 concentrations have increased by 40 percent since pre-industrial times, primarily from fossil fuel emissions. The ocean has absorbed 93 percent of this additional heat and sequestered 30 percent of the emitted anthropogenic CO_2. Over the period 1901 to 2010, global mean sea level also rose by 0.19 m. Aquatic systems that sustain fisheries and aquaculture are undergoing significant changes as a result of global warming. Temperature of water bodies is increasing across the globe, which results in more pronounced stratification of the water column. An increase in SST of 0.2 °C to 0.3 °C over the previous 45 years was reported along the coast of India. Dissolved oxygen levels decrease with increased temperature, and oxygen minimum zones in the oceans have expanded over the last decades, both in coastal and offshore areas. The absorption of increasing amounts of anthropogenic CO_2 by the oceans results in acidification of waters, with potentially detrimental impacts on shell-forming aquatic life (Barange,2018).

Impacts of climate change on marine fisheries

Shifts in the distribution of species of fish of importance to fisheries are one of the most widely recognized and acknowledged impacts of climate change on the oceans. Sea level rise is a phenomenon driven by global warming that is being experienced in many regions at different rates. At the ecosystem level, common impacts emphasized in the different regions are shifts in distribution by fish species and other taxonomic groups, increasing incidences of coral bleaching with serious implications for the affected ecosystems as a whole, and increasing frequency in outbreaks of harmful algal blooms. Increasing frequency and intensity of such events is expected to lead to substantial reductions in the extent of live coral cover, and could lead to a loss of coral reef species and changes in the dominant species assemblages. These changes will lead to significantly altered ecosystem services.

Climate change is leading to changes in the temperature, pH and salinity of water and the incidence and intensity of extreme weather events, all of which can have impacts on food safety. For example, the growth rates of pathogenic bacteria that occur in the marine environment have been found to increase at higher water temperatures, while changes in seasonality and other environmental conditions can influence the incidence of parasites and some foodborne viruses. Changes in the environment can also modify the population dynamics of the hosts of food-borne parasites.

Hazards in food safety and aquatic animal health

Coping with climate-driven changes will require giving greater attention to monitoring of key environmental parameters, including water and air temperature, pH and salinity, to enable advance prediction of imminent problems related to food safety such as the incidence of toxins, pathogens and contaminants in bivalve molluscs and the fish species that are more susceptible to such threats. Implementation of effective early warning systems will need collaboration between the relevant sectors and stakeholders, including those responsible for aquatic animal health, the marine environment and food safety and public health, at both national and international levels. Climate change also brings increased risks for animal health. Aquaculture development is leading to more intense production so as to ensure economic sustainability but this has the effect of increasing the probabilities of disease outbreaks as well as the challenges in controlling them.

Challenges to fishing communities particularly small-scale fishers

Billions of people around the world depend on fisheries and aquaculture for food, essential nutrients and livelihoods. The sector is already under stress from pollution, habitat degradation, overfishing and harmful practices; climate variability, climate change and ocean acidification represent additional threats to the sector and dependent communities. The impacts of climate change are expected to be heaviest for small-scale fishers in several regions. Small-scale fishers are also considered to be among the most vulnerable groups especially in the developing countries and they tend to have lower adaptive capacity to cope with and therefore to be more vulnerable to climate change.

Relocation of resources and replacement with less commercially valuable species requires diversification of fishing operations and markets; changes in the timing of fish spawning and recruitment will need adjustments to management interventions; increases in the frequency and severity of storms may affect infrastructure, both at sea and on shore; in areas where production is already limited by temperature, traditional productive areas may be reduced and dependent communities will need to diversify their livelihoods; and the impact of ocean acidification may be locally significant, for example in activities dependent on coral reefs are major challenges to fishing communities posed by climate change.

Impact of climate-driven disasters

Climate-related disasters now account for more than 80 percent of all disaster events, with large social and economic impacts, including both short- and long-term displacement of people and populations (UNISDR, 2015). Fisheries and aquaculture are sectors frequently exposed to the impacts of climate change and therefore, face serious threats from extreme events. With the increased and increasing number of extreme events and the likelihood of resulting disasters, there is an urgent need to invest in coherent and convergent disaster risk reduction and adaptation measures and preparedness for climate disaster response and recovery in the fisheries and aquaculture sectors. This should lead to a shift from reactive management after disasters have occurred to proactive management and risk reduction of climate risks and hazards.

Adaptation to climate change

Efforts towards adaptation to climate change include implementation of the FAO Code of Conduct for Responsible Fisheries and related instruments, ecosystem approaches to fisheries, spatial planning including effective systems of marine protected areas, ensuring participatory systems of governance and strengthening control and enforcement in the fisheries sector.

Small-scale and artisanal fisheries and fishers are identified as being particularly vulnerable to the impacts of climate change and a number of the adaptation options are aimed primarily at them. They include implementation of the FAO *Voluntary guidelines for securing sustainable small-scale fisheries* (FAO, 2015), and the *Voluntary guidelines on the responsible governance of tenure of land, fisheries and forests* (FAO, 2012) to promote secure tenure rights and equitable access to fisheries as a means of eradicating hunger and poverty and supporting sustainable development. Other specific options include wider use of community based approaches to fisheries governance, flexibility to enable switching of gears and target species in response to changes, creation of alternative livelihoods, capacity-building to enhance resilience in different ways, and improving the economic stability of small-scale fishers and those involved in associated activities through improved access to credit, microfinance, insurance services and investment.

Measures to be considered include improvement of early warning systems, safety at sea and protection of fisheries-related infrastructure such as safer harbours, landing sites and markets. Disaster risks and losses in fisheries and aquaculture can be prevented and minimized by implementing the Sendai Framework.

Greenhouse emissions in marine capture fisheries

Carbon footprint has become a widely used term and concept in the public debate on responsibility and abatement action against the threat of global climate change. It had a tremendous increase in public appearance over the last few years and is now a buzzword widely used across the media, the government and in the business world.

The carbon footprint is a measure of the total amount of carbon dioxide emissions that is directly and indirectly caused by an activity or is accumulated over the life stages of a product (Wiedmann and Minx, 2008). Carbon dioxide is a major component of the Greenhouse gases warming our planet. Fishing is considered as the most energy-intensive food production method and is dependent on fossil fuels. However, emissions due to fishing received less attention compared to the direct impact that fishing had on stocks and associated marine ecosystems (Tyedmers *et. al.*, 2005). While the use of fossil fuels has increased the availability of fish to fisheries, the dependence of the fishing sector on fossil fuels raises concerns related to climate change, ocean acidification and economic vulnerability.

The world's marine capture fish landing had been stagnant from 1990 to 2011 but the emissions from fishing fleet had increased by 28% (Parker *et al.*, 2018). In Indian marine fisheries, the enhanced fishing effort and efficiency in the last five decades has resulted in substantial increase in diesel consumption, equivalent to CO_2 emission of 0.30 million tonnes (mt) in the year 1961 to 3.60 mt in 2010. For every tonne of fish caught, the CO_2 emission has increased from 0.50 to 1.02 t during the period. There is scope to reduce CO_2 by setting emission norms and improving fuel efficiency of marine fishing boats (Vivekanandan *et al.*, 2013). India contributed 2.7% to the global marine fisheries CO_2 emission (134 mt). Harvest phase of marine fisheries life cycle contributed the highest emission (Ghosh *et al.*, 2014).

Since, fishing vessels are the largest contributors to carbon footprint, replacing the fuel-intensive fishing method with alternate methods can reduce the emissions from fishery sector (Tan and Culaba, 2009; Ghosh *et al.*, 2014). Carbon footprint reduction is feasible with profitable fishery (Tan and Culaba, 2009).

CO_2 emissions from India's marine fleets are relatively low, and they present the country with both challenges as well as an opportunity. It is necessary to use energy efficient fishing gears to reduce C and CO_2 emissions. Some alternative ways should be found out to replace the fuel use with green fishing methods. Solar and electric energies might be used instead of fuel and excess greenhouse gases can be reduced (Ghosh *et al.*, 2014).

Marine fishing in India can be made further eco-friendly by using alternate fishing techniques which use less fuel to reduce CO_2 emissions. Fuel cost accounts for 50-55% of the operating cost of mechanised boats. Mechanised crafts use fuel for reaching the fishing ground as well as for operation of fishing gear. Passive and semi-active fishing methods could reduce the amount of energy used per kg of landed fish (Thrane, 2004).

Mitigation measures

Although a small global contributor, capture fisheries have a responsibility to limit GHG emissions as much as possible. Eliminating excessive fishing capacity and over-fishing; improving fisheries management; reducing post-harvest losses and increasing waste recycling will decrease the sectors' CO_2 emissions and improve the resilience of aquatic ecosystems. Other technical solutions to reduce fuel use might include shifting towards static fishing technologies and to more efficient vessels and gears. In some cases, win-win conditions could be identified, where reduced fuel-use strategies would link with reducing fishing effort, improving returns to vessels, safeguarding stocks and improving resilience to climate change (FAO, 2008).

A significant reduction of GHG emissions in capture fisheries and aquaculture can be achieved by reduction in energy consumption; use of fuel-efficient fishing methods; enhancing fuel efficiency in fisheries and aquaculture operations; integrated farming; and better feeds and feed management.

Reducing energy consumption

In capture fisheries, the vessel and gear are two main sources of energy consumption. Stationary gears consume less energy compared to mobile gears as energy consumption is limited to cruising up to the fishing ground in the former while it is for both propulsion and fishing in the case of latter. The choice of vessel design, size of engine and type of propeller determines fuel efficiency. Proper vessel length to width ratio, smooth hull painting and fairing, bulbous bows, high efficiency internal combustion engines, and larger diameter propellers with nozzles are some of the important features for a highly fuel-efficient vessel. For fishing gears, especially towed fishing gears, the use of efficient otter boards, off-bottom fishing, high-strength materials, large mesh sizes, and smaller diameter twines are some of the measures that reduce fuel consumption. Opportunities for reducing dependence on fossil fuel exist in usage of renewable energy systems such as wind and solar-powered generation of electricity.

Use of low impact fuel efficient fishing methods and gears

Fishing gears have varying degrees of impact on marine ecosystems. Towed gears like dredges and otter board trawls affect the bottom where a contact with the fishing gear components and the bottom occurs. Seines interact with the bottom surrounded by the net and herding ropes. Stationary fishing gears have minor effect on the bottom habitats. Off-bottom fishing gears like pelagic trawls and purse seines have little or no bottom impact. Gill nets, long lines, pots and traps comprise fishing gears with low energy use.

Enhancing fuel efficiency of fisheries and aquaculture operations

Fisheries management has a great impact on all aspects of fish production, especially in capture fisheries and can therefore affect efficiency of fuel use. Management measures that reduce overall fishing effort and improve stock abundance make significant contributions to fuel efficiency in capture fisheries. Fuel efficiency and GHG emissions from fisheries should be considered as an integral part of fisheries management to sustainably reduce fuel use and GHG emission in fisheries. Area closures are a widely applied measure that can contribute to ensuring high and sustainable stock biomass, and therefore fuel efficiency.

In the case of capture fisheries, through use of efficient engines and larger propellers in fishing vessels, as well as through improving vessel shapes and other hull modifications and by reducing the mean speed of vessels, 10-30 % reduction can be achieved. In the case of fishing gears, switching from highly fuel intensive methods to low impact fuel efficient methods that require less fuel would reduce GHG emissions.

Integrated farming

In aquaculture, GHG emissions are greatest in intensive production of finfish and crustaceans, which is heavily reliant on feeds and aeration. Integrated food production systems such as that of fish with rice, and shrimp-mangrove cultivation can substantially reduce overall GHG emissions from aquaculture production systems.

Better feeds and feed management

There are also opportunities to reduce GHG emissions in aquaculture, which include improved technologies to increase efficiency in the use of inputs, greater reliance on energy from renewable sources, improving feed conversion rates, and switching from feed-based on fish to feed made from crop-based ingredients. The integration of pond aquaculture with agriculture is also a potential option for reducing fuel consumption and emissions.

Aquaculture in the context of Blue Dimensions

Aquaculture has been responsible for the supply of fish for human consumption. In order to meet the demand for food from a growing global population, aquaculture production must be increased to fill the gap between supply and demand with rapidly expanding global fish demand and relatively stable capture fisheries. global aquaculture production is projected to reach 109 million tons in 2030, with a growth of 37% over 2016 (FAO, 2018). Long-term growth of the aquaculture industry needs both environmentally sound practices and sustainable resource management (Naylor *et al.*, 2000).

The rapid development of aquaculture has been considered the blue revolution (Costa-Pierce, 2002; Simpson, 2011), which refers to the significant emergence of aquaculture as a vital and very productive agricultural activity (McGinn, 1998; Movik *et al.*, 2005). The blue revolution of aquaculture is becoming increasingly an attractive approach to enhance food production (White *et al.*, 2004). In fact, aquaculture is increasing food production and contributing to human nutrition and food security (Béné *et al.*, 2016). Small-scale freshwater rural aquaculture contributes to food security, poverty alleviation, and socioeconomic benefits in many countries (ADB, 2005; Bondad-Reantaso and Subasinghe, 2013). In recent years, eco-friendly bio-floc for shrimp farming technology with microorganisms has been developed for maintaining water quality and reducing feed application (Emerenciano *et al.*, 2013).

Rainfall that does not absorb the soil is blue water (Sood *et al.*, 2014). Blue water implies groundwater and surface runoff in lakes, ponds, and rivers that can be withdrawn for irrigation and human utilization (Chapagain and Hoekstra, 2011; Falkenmark and Rockström, 2006; Hoff *et al.*, 2010; Menzel and Matovelle, 2010; Rockström *et al.*, 2009). Globally, 39% of rainfall contributes to blue water of which 36% goes to seas and oceans (Molden, 2007). Thus, water in aquifers, seas, and oceans is also known as blue water. Aquaculture primarily depends on blue water due to the availability of surface and groundwater. Access to blue water is vital for aquaculture as a considerable amount of water is required to grow fish. Globally, aquaculture used 201 km^3 of freshwater in 2010 (Mungkung *et al.*, 2014). Aquatic ecosystems fully depend on blue water, which support and maintain aquaculture, aquatic biodiversity, and fisheries (Molden, 2007; Rockström *et al.*, 2007). Blue water in ponds, lakes, rivers, estuaries, and seas are used for

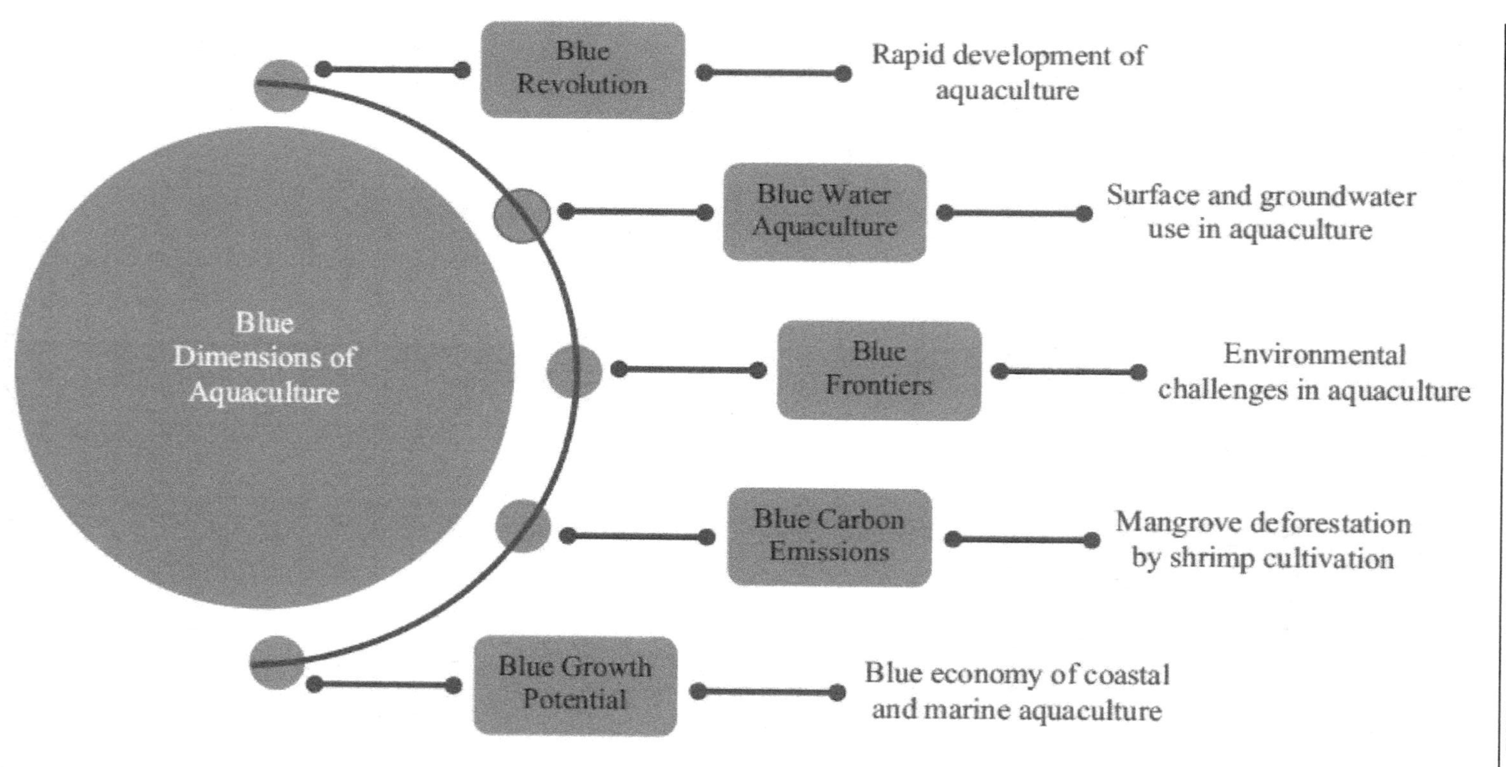

Fig. 1: Framework of Blue Dimension of Aquaculture (Source: The blue dimensions of global aquaculture: A global synthesis, Science of the Total Environment 652, 2019)

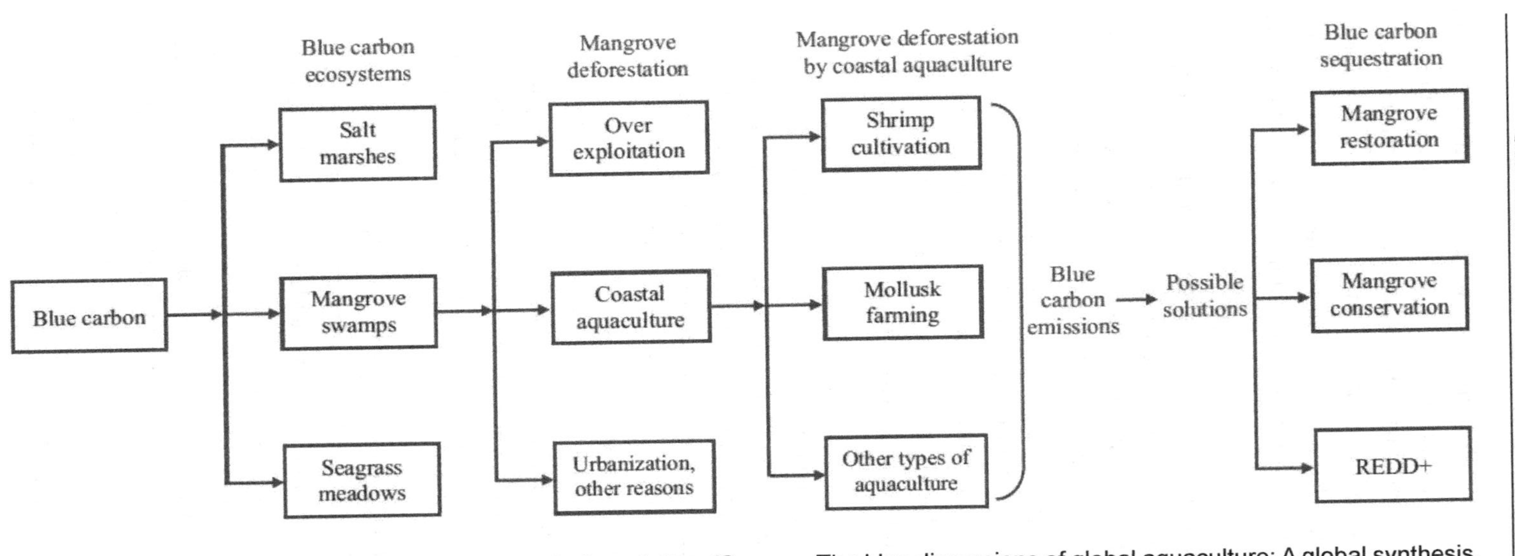

Fig. 2: Blue Carbon emissions from Mangrove Deforestation (Source: The blue dimensions of global aquaculture: A global synthesis, Science of the Total Environment 652, 2019)

inland, coastal, and marine aquaculture. Utilizing blue water in ponds for inland freshwater aquaculture is significant. Irrigation return water is also applied for aquaculture, which is known as "Integrated Irrigation Aquaculture (IIA)" (Halwart and van Dam, 2006). There is a substantial contribution of IIA where fish farms have access to irrigation return water from canals and tributaries. Cage culture, floodplain aquaculture, and pen culture are also important for using blue water resources.

Aquaculture must take an ecosystem approach to reduce environmental impacts of the blue revolution (Costa-Pierce, 2002; Neori *et al.*, 2007). To avoid environmental hazards in aquaculture, sustainable intensification with efficient use of land and water, technological developments, Better Management Practices (BMP), better site selection within the carrying capacity of ecosystems, and an Ecosystem Approach to Aquaculture (EAA) must be followed (Edwards, 2015). Fish producers need to understand and comply with environmental regulations through BMP and EAA. Attention should be paid to management interventions that could provide improved environmental performance in aquaculture (Hall *et al.*, 2011).

Carbon emissions are greatly increasing due to devastating effects on mangroves by coastal aquaculture including shrimp farming (Kauffman *et al.*, 2017; Pendleton *et al.*, 2012). The conversion of mangroves to shrimp farms releases considerable amounts of blue carbon and reduces storage facilities. On average, blue carbon emissions from the conversion of mangroves to shrimp farms is 554 tons/ha (Kauffman *et al.*, 2017). Comparatively, carbon emissions from 1 ha of mangrove forest conversion to shrimp farm are equal to the emissions of 5 ha of tropical evergreen forest conversion and 11.5 ha of tropical dry forest conversion (Kauffman *et al.*, 2014). On average, blue carbon emissions from deforested mangroves for each kilogram of Shrimp production is 437 kg or 1603 kg of CO_2e (Kauffman *et al.*, 2017).

Carbon emissions with other GHG, including CH_4 and N_2O, are the main reason for climate change (IPCC, 2014). Reducing blue carbon emissions from mangrove deforestation by shrimp cultivation is vital to tackle anthropogenic climate change. Preventing mangrove loss and the conservation of mangrove forests can help to reduce blue carbon emissions for climate change mitigation (Duarte *et al.*, 2013; McLeod *et al.*, 2011; Murdiyarso *et al.*, 2015; Pendleton *et al.*, 2012; Siikamäki *et al.*, 2012). Mangrove restoration and conservation could increase blue carbon sequestration as mangroves sequester blue carbon at the rate of 1.15–1.39 tons/ha annually (Bouillon *et al.*, 2008; Nellemann *et al.*, 2009; Siikamäki *et al.*, 2012). According to Alongi (2014), the global mean carbon sequestration rate for mangroves is 1.74 tons/ha. Integrated mangrove-shrimp cultivation can help to reduce blue carbon emissions through mangrove restoration, which in turn sequesters blue carbon (Ahmed *et al.*, 2018).

Globally, onshore and offshore mariculture potential is large. Although most mariculture is practiced inshore, 0.3 million km of global coastline (44% of maritime nations) are not utilized for mariculture (Kapetsky *et al.*, 2013). Mariculture is an approach to expand seafood production which could be a possible solution to feed the growing global population (Duarte *et al.*, 2009). Seafood production is expected to be predominantly sourced through mariculture by 2050 (Diana *et al.*, 2013). A combination of climatic factors, including ocean acidification, sea level rise, and sea surface temperature could have severe impacts on coastal and marine aquaculture. Ocean acidification has adverse impacts on calcifying species, which affect shell formation that undermine seafood production (Clements and Chopin, 2017; De Silva and Soto, 2009). Fish are highly sensitive to ecological conditions and changes in coastal and marine ecosystems due to sea level rise and sea surface temperature could have adverse effects on finfish and shellfish production. Considering the vulnerability of climate change, adaptation strategies must be developed to cope with the challenges. Adaptation to climate change is an essential component of a blue economy approach (World Bank and UNDESA, 2017).

In order to meet the demand for food from a growing global population, aquaculture production must increase sustainably while at the same time its environmental impacts must reduce considerably. Producing fish in an environmentally sustainable way is essential. EAA with BMP must be considered for reducing environmental impacts of the blue revolution of aquaculture.

Green House Gas (GHG) Emissions and Aquaculture

Climate change is having profound effects on aquaculture ecosystems around the world. While contributing to increased GHG emissions, aquaculture is also affected by the environmental impacts of climate change. Different geographies and production systems are affected in different ways and certain production systems, such as cage and pond operations, are especially vulnerable to the effects, as many of their environmental parameters are beyond farmers' controls.

Impact of climate change on Aquaculture

Aquaculture is making an increasing contribution to global production of fish, crustaceans and molluscs and thereby to the livelihoods, food security and nutrition of millions of people. By helping to meet the growing demand for these products, aquaculture also alleviates the price increases that would otherwise result from any escalating gap between supply and demand. Aquaculture no longer enjoys the high annual growth rates of the 1980s and 1990s (11.3 percent and 10.0 percent, excluding aquatic plants), but remains the fastest growing global food production system. Average annual growth declined to 5.8 percent during the period 2000 to 2016, although double-digit growth still occurred in a small number of individual countries, particularly in Africa from 2006 to 2010. Overall, between 1950 and 2015 global aquaculture production grew at a mean annual rate of 7.7 percent and by 2016 had reached 80.0 million tonnes of food fish and 30.1 million tonnes of aquatic plants (FAO, 2018), equivalent to 53 percent of global production of fish for food by capture fisheries and aquaculture combined.

Climate change can have direct and indirect impacts on aquaculture, and in the short and long-term. Some examples of short-term impacts include losses of production and infrastructure arising from extreme events such as floods, increased risk of diseases, parasites and harmful algal blooms, and reduced production because of negative impacts on farming conditions. Long-term impacts include reduced availability of wild seed as well as reduced precipitation leading to increasing competition for freshwater. Climate-driven changes in temperature, precipitation, incidence and extent of hypoxia and sea level rise, amongst others, will have long-term impacts on the aquaculture sector at scales ranging from the organism to the farming system to national and global.

It is clear that these changes will potentially have both favourable and unfavourable impacts on aquaculture, but unfavourable changes are likely to outweigh favourable ones, particularly in developing countries where adaptive capacity is typically weakest. The threats of climate change to aquaculture have been recognized by some countries and, as of June 2017, of the 142 countries that had submitted their NDCs (Nationally Determined Contributions), 19 referred to aquaculture or fish farming. Nine of those included a focus on adapting aquaculture to climate change, while ten included proposals to use the development of aquaculture as an adaptation and/or mitigation measure in their efforts to address climate change.

Assessments at national scale provide useful guidance for governments and decision-makers at global and national levels but there is also usually high diversity within countries. Vulnerability assessments and therefore adaptation planning also needs to be to be conducted at finer, localized scales where the specific practices, stakeholders and communities, and local environmental conditions can be taken into account. The global assessments considered sensitivity, exposure and adaptive capacity as the components of vulnerability. For freshwater aquaculture, that study found Asia to be the most vulnerable area, influenced strongly by the high production from the continent, with Viet Nam being the most vulnerable country in Asia, followed by Bangladesh, the Lao People's Democratic Republic and China. Belize, Honduras, Costa Rica and Ecuador were assessed as being the most vulnerable countries in the Americas, while Uganda, Nigeria and Egypt were found to be particularly vulnerable in Africa.

In the case of brackish water production, Viet Nam, Egypt and Thailand emerged as having the highest vulnerabilities & the countries with lowest adaptive capacity to cope with the impacts of climate change included Senegal, Côte d'Ivoire, the United Republic of Tanzania, Madagascar, India, Bangladesh, Cambodia and Papua New Guinea. For marine aquaculture, Norway and Chile were identified as being the most vulnerable, reflecting the high production of those countries in comparison to others. China, Viet Nam, and the Philippines were found to be the most vulnerable countries in Asia, while Madagascar was the most vulnerable country in Africa. Mozambique, Madagascar, Senegal and Papua New Guinea were identified as countries with particularly low adaptive capacity.

Options for adaptation and building resilience in aquaculture emphasize that they should be applied in accordance with an ecosystem approach to aquaculture (EAA). They include:

- improved management of farms and choice of farmed species
- improved spatial planning of farms that takes climate-related risks into account
- improved environmental monitoring involving users
- improved local, national and international coordination of prevention and mitigation

According to the fifth assessment report of the IPCC (Jimenez Cisneros, Oki, Arnell *et al.*, 2014), climate change is projected to result in a significant reduction in renewable surface water and groundwater resources in most of the dry subtropical regions, which can be expected to lead to greater competition between different types of agriculture and between agriculture and other sectors. As with inland fisheries, this expected trend, and other inter-sectoral interactions, means that focusing only on adaptation within aquaculture is unlikely to be sufficient and effective reduction of vulnerability in the sector requires the integration of aquaculture into holistic, multi-sectoral watershed and coastal zone management and adaptive planning. Few of the potential measures for aquaculture in context of climate change are

- The expected increase in frequency and intensity of extreme weather events could lead to an increase in the number of escapees from aquaculture farms.

This impact could be minimized by measures such as regulating the movement of aquatic germplasm originating from outside the area, certification of cage equipment that minimizes the risk of escapes, modifying pond systems, and developing the capacity of farmers and implementing management measures to address the problem.

- Consumption of water by aquaculture will add to competition for the resource in places where availability and quality of freshwater is reduced by climate change. This problem can be alleviated by reducing water consumption by aquaculture through technological and managerial improvements. Measures such as these need to be complemented by similar actions across sectors and users, and should be a part of integrated, multi-sectoral policies, legal and regulatory frameworks and actions.
- Aquaculture could be negatively affected if the impacts of climate change on the availability of fishmeal and fish oil are negative, as these are important components of feeds for some species. This problem could be reduced in the longer-term by increasing the use of fish processing wastes in feeds as well as by use of new feedstuffs, the production of which is already increasing.
- Aquaculture can also contribute to climate change adaptation in other sectors. For example, culture-based fisheries could be used to alleviate the effects of reduced recruitment in capture fisheries as a result of change. Aquaculture is also frequently seen as a promising alternative livelihood for fishers and other stakeholders when capture fisheries can no longer support them because of climate change, over-exploitation and other factors. A common message on aquaculture is that there are important gaps in current knowledge and understanding of scientific, institutional and socio-economic aspects of the sector and the likely impacts of change.

Contribution of aquaculture to climate change

While estimates are scarce, emerging research suggests the rapid development of industrial scale aquaculture, as a result of plateauing fisheries stocks, is contributing to increasing global GHG emissions (Yuan *et al.*, 2019). GHG emissions are the leading cause of climate change, which poses risks to the aquaculture sector and is expected to bring severe ecological and biological changes in terms of productivity, species abundance, ecosystem stability and pathogen levels worldwide (FAO, 2016: B), all of which will have an impact on the operating costs of aquaculture companies. Overall, industrial aquaculture is contributing to climate change,while also being exposed to significant risks because of it.

Few estimates of the global aquaculture sector's GHG emissions exist. FAO documents indicate that the "paucity of data" on GHG emissions in seafood supply chains is inhibiting the development of measures to reduce energy use in the sector (FAO, 2017). Further, the sector's emissions are difficult to calculate as they are usually accounted for across several national GHG inventories, including agriculture, waste, fuel combustion and refrigeration. Carbon emissions released by land conversions for aquaculture are rarely considered when calculating aquaculture emissions. For example, the impact of converting wetland and mangrove areas into fishponds is underestimated. From 2000 to 2015, 122 million tonnes of carbon was released due to mangrove deforestation, which is equivalent to Brazil's annual emissions (Erickson-Davies, 2018). Wetlands also play an important role in the global carbon cycle; they cover just 6% of the world's land surface but contain 12% of its global carbon pool (Erwin, 2009). In China, paddy fields are increasingly being converted to aquaculture ponds (currently these conversions make up over 50% of Chinese inland fish ponds) which is contributing to global warming from methane emissions (Yuan *et al.*, 2019). Finally, GHG emissions from aquaculture operations vary widely, depending on the species cultured and the type of production system used. However, a 2019 study found that freshwater aquaculture in the top 21 producing nations is responsible for 1.82% of global methane and 0.34% of global nitrous oxide emissions (Yuan *et al.*, 2019).

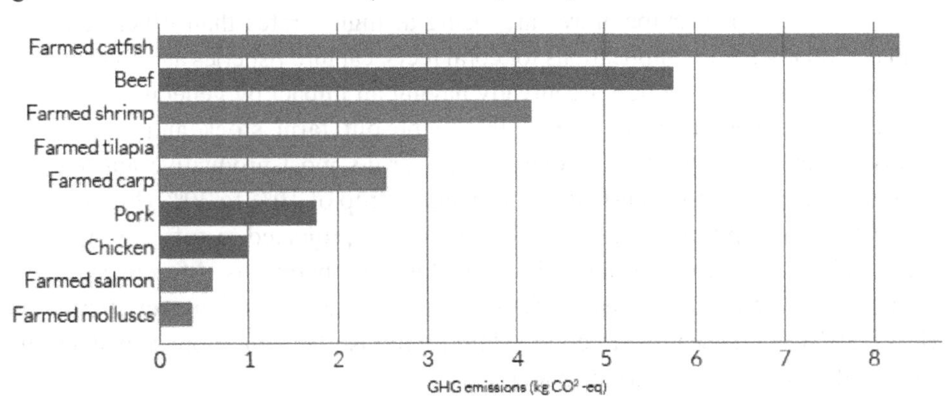

Fig. 3: Average GHG Emissions per 40 G / Protein

Source: Adapted from Hilborn *et al.* (2018)

For most species, the most emission-intensive stages of fish farming are in feed production. In Atlantic salmon and rainbow trout production, feed accounts for around 87% of total GHG emissions (Rasenberg *et al.*, 2013). Shrimp is an exception, as on-farm energy use is the most emission-intensive stage. Intensive shrimp farms require the use of aerators and pumping systems to maintain water quality, which drives up the operational energy expenditure.

Compared to land animal proteins, average farmed salmon production has relatively low GHG emissions. However, shrimp aquaculture GHG emissions are comparable to those arising from beef production: the data in Figure suggest shrimp production emissions are around 20% lower than beef production, however another study suggests they are as high as in beef production (Poore and Nemecek, 2018).

HOW CLIMATE RISK IMPACTS AQUACULTURE

Europe and North America

In northern parts of Europe and North America, where temperatures are rising fastest, severe impacts are expected to make it more challenging to farm salmon (Froehlich *et al.*, 2018). However, climate impacts on the salmon sector are still largely uncertain and require research. For example, warmer water and salinity changes will likely result in increased salmon diseases and parasites, while pathogens associated with a colder climate may be reduced (Troell *et al.*, 2017). Overall, the lack of clarity exposes investors in the European and North American salmon aquaculture sector to risks.

Asia

In Asia, sea surface temperatures are rising at higher rates than offshore areas, with potentially severe implications for coral reefs, capture fisheries and aquaculture. Warmer waters in the region are already having an impact by generating stronger and more frequent tropical storms that wipe out farm stock and erase farm infrastructures. Southeast Asia, one of the world's most productive aquaculture regions, is expected to experience a production drop of 10% to 30% (Froehlich *et al.*, 2018). Shrimp farms located in the tropics are projected to suffer from rising sea levels, which could flood inland freshwater operations. Moreover, ocean acidification is expected to lead to declines in mollusc production over 2020-2060 (FAO, 2016: A). Diseases affecting salmon will also become more prevalent due to climate change and warmer waters.

Other regions

Some areas, however, especially in Africa and South America, are predicted to see aquaculture productivity increase by up to 30%. Some species, such as tilapia production in the tropical Pacific, may benefit from climate change (FAO, 2016: A). Aquaculture companies such as Tassal have reported that they are currently working on adaptation measures in response to rising sea temperatures and extreme weather events, including geographic diversification of its marine farm portfolio.

Conclusion

Climate change affects communities and livelihoods in fisheries and aquaculture, and efforts to adapt to and mitigate climate change must therefore be human-centred. Climate adaptation strategies must emphasize the need for poverty eradication and food security, in accordance with the Paris Agreement, the United Nations 2030 Agenda for Sustainable Development and other international instruments, such as the *Voluntary guidelines for securing sustainable small-scale fisheries in the context of food security and poverty eradication.* Climate change poses several challenges to fishing communities. Some of them are diversification of fishing operations and markets due to relocation of resources, need of dependent communities to diversify their livelihoods in areas where production is limited on account of temperature, adjustments to management interventions due to changes in spawning and recruitment pattern and efforts to rebuild infrastructure, both at sea and on shore due to the frequent and severe storms.

COP21 for the first time featured the role of oceans, inland waters and aquatic ecosystems for temperature regulation and carbon sequestration. Disaster risks and losses in fisheries and aquaculture can be prevented and minimized by implementing the Sendai Framework. The Sendai Framework for Disaster Risk Reduction, an international document was adopted by UN member states in March 2015 at the World Conference on Disaster Risk Reduction held in Sendai, Japan and endorsed by the UN General Assembly in June 2015. Supporting sustainable fisheries and healthy ecosystems will help mitigate and reduce impacts from climate change and variability and unlock the Blue Growth potential of aquatic systems in line with the principles of the Code of Conduct for Responsible Fisheries and related instruments.

CHAPTER 10

SUSTAINABLE BLUE ECONOMY

Introduction

Blue economy covers all economic activities related to oceans, seas and coasts. The term "blue economy" has been used in different ways, comprising range of economic sectors and related policies that together determine whether the use of oceanic resources is sustainable. An important challenge of the blue economy is to understand and better manage the many aspects of oceanic sustainability, ranging from sustainable fisheries to ecosystem health to pollution. A second significant issue is the realization that the sustainable management of ocean resources requires collaboration across nation-states and across the public-private sectors.

The "blue economy" concept seeks to promote economic growth, social inclusion, and the preservation or improvement of livelihoods while at the same time ensuring environmental sustainability of the oceans and coastal areas. At its core it refers to the decoupling of socio-economic development through oceans-related sectors and activities from environmental and ecosystems degradation. It draws from scientific findings that ocean resources are limited and that the health of the oceans has drastically declined due to anthropogenic activities. These changes are already being profoundly felt, affecting human well-being and societies, and the impacts are likely to be amplified in the future, especially in view of projected population growth.

The blue economy has diverse components, including established traditional ocean industries such as fisheries, tourism, and maritime transport, but also new and emerging activities, such as offshore renewable energy, aquaculture, seabed extractive activities, and marine biotechnology and bioprospecting. A number of services provided by ocean ecosystems, and for which markets do not exist, also contribute significantly to economic and other human activity such as carbon sequestration, coastal protection, waste disposal and the existence of biodiversity.

A study by core group of experts on blue economy in December 2019 highlighted the need to accord priority to five areas: Fisheries and aquaculture; Sea port and Shipping including seaport development; Tourism including Island Development for tourism; renewable ocean energy; and Mining including offshore hydrocarbons and seabed minerals (FICCI-KAS,2019).

The blue economy aims to move beyond business as usual and to consider economic development and ocean health as compatible propositions. It is generally understood to be a long-term strategy aimed at supporting sustainable and equitable economic growth through oceans-related sectors and activities. The blue economy is relevant to all countries and can be applied on various scales, from local to global. In order to become actionable, the blue economy concept must be supported by a trusted and diversified knowledge base, and complemented with management and development resources that help inspire and support innovation.

A blue economy approach must fully anticipate and incorporate the impacts of climate change on marine and coastal ecosystems—impacts both already observed and anticipated. Understanding of these impacts is constantly improving and can be organized around several main "vectors": acidification, sea-level rise, higher water temperatures, and changes in ocean currents. These different vectors, however, are unequally known and hard to model, in terms of both scope—where they will occur, where they will be felt the most—and severity. For instance, while not as well understood as the other impacts, and more difficult to measure, the impacts of acidification are likely to be the most severe and most widespread, essentially throughout any carbon-dependent ecological processes. Likewise, the effects of sea-level change will be felt differently in different parts of the world, depending on the ecosystems around which it occurs. Most importantly, however, and unlike in terrestrial ecosystems, further uncertainty results from the complex interactions within and between these ecosystems. In spite of this uncertainty, the current state of knowledge is sufficient to understand that these impacts will be felt on critical marine and coastal ecosystems throughout the world and that they fundamentally affect any approach to the management of marine resources, including by adding a new and increasing sense of urgency.

Healthy oceans and seas can greatly contribute to inclusiveness and poverty reduction, and are essential for a more sustainable future. In spite of all its promises, the potential to develop a blue economy is limited by a series of challenges. First and foremost is the need to overcome current economic trends that are rapidly degrading ocean resources through unsustainable extraction of marine resources, physical alterations and destruction of marine and coastal habitats and landscapes, climate change, and marine pollution.

The second set of challenges is the need to invest in the human capital required to harness the employment and development benefits of investing in innovative blue economy sectors. The third set of challenges relates to strengthening the concept and overcoming inadequate valuation of marine resources and ecosystem services provided by the oceans; isolated sectoral management of activities in the oceans, which makes it difficult to address cumulative impacts; inadequate human, institutional, and technical capacity; underdeveloped and often inadequate planning tools; and lack of full implementation of the 1982 United Nations Convention on the Law of the Sea (UNCLOS) and relevant conventions and instruments. While stimulating growth in individual oceanic sectors is comparatively straightforward, it is not always clear what a sustainable blue economy should look like and the conditions under which it is most likely to develop.

Components of blue economy

The blue economy includes established ocean industries such as fisheries and tourism as well as emerging and new activities—such as offshore renewable energy, aquaculture, seabed mining, and marine biotechnology. Larger industries such as coastal development, shipping, and port infrastructure and services also rely on the oceans, seas, and coasts. Underlying the need for diversified economies are demographic trends such as population growth and rapid coastal urbanization, which fuel the search for food and job security and for alternative sources of minerals and energy, as well as seaborne trade. At the same time, new technologies can offer significant opportunities to tap into new and previously unexploited resources (UNEP 2015; UNDESA 2014a; Economist Intelligence Unit 2015). The environmental and social impacts of each of these industries, as well as their potential economic benefits, are unique. And their contribution to the blue economy will need to be weighed against principles and policies that have been established globally, regionally, and/or nationally. For an activity to contribute to the blue economy, it would need to include at least two of the four elements of resource efficiency: reducing food loss and waste along the value chain, energy efficiency (reducing the carbon footprint), decent employment, and innovative financing or technologies. In addition, the project would need to provide environmental, social, and economic benefits (FAO 2014c; WWF 2015; UNDESA 2014a).

The components of blue economy are broadly given under five categories

- Harvesting and trade of marine living resources
- Extraction and use of marine non-living resources (non-renewable)
- Use of renewable non-exhaustible natural forces (wind, wave & tidal energy)

- Commerce and trade in and around oceans
- Indirect contribution to economic activities and environments

Details of activities, related industries/sectors and drivers of growth are given in Annexure 9.

The Transition to a Blue Economy

Sustainable development implies that economic development is both inclusive and environmentally sound, and to be undertaken in a manner that does not deplete the natural resources that societies depend on in the long term. The need to balance the economic, social, and environmental dimensions of sustainable development in relation to oceans is a key component of the blue economy. It is also a difficult balance to reach in practice, given that ocean resources are limited and the health of the oceans has drastically declined due to human activities—ranging from damage caused by carbon dioxide emissions to nutrient, chemical, and plastics pollution, unsustainable fishing, habitat degradation and destruction, and the spread of invasive species. The scientists and experts who prepared the First Global Integrated Marine Assessment (also known as the World Ocean Assessment) warned that the world's oceans face major pressures simultaneously with such great impacts that the limits of their carrying capacity are being reached—or in some cases have been reached—and that delays in implementing solutions to the problems that have already been identified as threatening to degrade the world's oceans will lead, unnecessarily, to greater environmental, social, and economic costs.

The importance of oceans for sustainable development is widely recognized by the international community and was embodied in, among others, Agenda 21, the Johannesburg Plan of Implementation, various decisions taken by the Commission on Sustainable Development, the Rio+20 outcome document The Future We Want, and the 2030 Agenda for Sustainable Development. The 1982 United Nations Convention on the Law of the Sea (UNCLOS), together with its implementing agreements—the 1994 Agreement relating to the implementation of Part XI of UNCLOS and the 1995 United Nations Fish Stocks Agreement—sets out the legal framework within which all activities in the oceans and seas must be carried out and is of strategic importance as the basis for national, regional, and global action and cooperation in the marine sector. This includes the conservation and sustainable use of all areas of the oceans and their resources. The concept of a blue economy came out of the 2012 Rio+20 Conference and emphasizes conservation and sustainable management, based on the premise that healthy ocean ecosystems are more productive and form a vital basis for sustainable ocean-based economies (UNDESA 2014a).

Under "business as usual," the costs of marine ecosystem degradation from human uses should be high, but they are not quantified or accounted for. At the same time, the economic contribution of the ocean to humankind has been significantly undervalued in particular where the value of non-market goods and services, such as carbon sequestration, coastal protection and recreation, and cultural and spiritual values, are concerned. In contrast, a new form of understanding the oceans, and which incorporates environmental and social dimensions, requires a paradigm shift acknowledging and valuing all ocean benefits (UNEP 2015).

The blue economy moves beyond business as usual to consider economic development and ocean health as compatible propositions. It comes from a realization that humanity cannot continue, let alone accelerate, human-induced changes to ocean ecosystems. In a blue economy, the environmental risks of and ecological damage from economic activity are mitigated or significantly reduced. Thus, economic activity is in balance with the long-term capacity of ocean ecosystems to support this activity and remain resilient and healthy. It is generally understood to be a long-term strategy aimed at supporting sustainable economic growth through oceans-related sectors and activities, while at the same time improving human well-being and social equity and preserving the environment (UNEP 2013; UNCTAD 2016).

A blue economy is low-carbon, efficient, and clean (UNDESA 2014a). It is also an economy that is based on sharing, circularity, collaboration, solidarity, resilience, opportunity, and interdependence. Its growth is driven by investments that reduce carbon emissions and pollution, enhance energy efficiency, harness the power of natural capital— such as the oceans—and halt the loss of biodiversity and the benefits that ecosystems provide (UNEP 2013).

Blue growth, or environmentally sustainable economic growth based on the oceans, is a strategy of sustaining economic growth and job creation necessary to reduce poverty in the face of worsening resource constraints and climate crisis. The World Bank defines green growth as "growth that is efficient in its use of natural resources, clean in that it minimizes pollution and environmental impacts, and resilient in that it accounts for natural hazards and the role of environmental management and natural capital in preventing physical disasters" (World Bank 2012a).

Each country will thus need to draft its vision for a sustainable oceans' economy, including how to balance growth and sustainability to enable optimal use of ocean resources with maximum benefit (or at least minimal harm) to the environment. The vision could be supported by development of plans and policies, referred to as blue economy plans, for the maritime zones of each country, which would support the attainment of the agreed-upon vision. The vision must further be anchored in the provisions of UNCLOS, which provides the necessary legal certainty with respect to maritime rights and obligations of states, including with regard to maritime space and resources.

An important dimension of the blue economy involves how established ocean industries are transitioning to more environmentally responsible practices. An early example of this comes from the fisheries sector. The Blue Growth Initiative of the Food and Agriculture Organization of the United Nations (FAO) will assist countries in developing and implementing blue economy and growth agendas.

A blue economy is supported by a trusted and diversified knowledge base and complemented with resources, which helps to inspire and support innovation (UNEP 2015). In too many instances, clear policy frameworks are not developed at the national level, yet are essential, as is an engaged process of stakeholder consultation and co-creation of a common vision for the blue economy nationally.

Effective implementation of UNCLOS, its two implementing agreements, and other relevant conventions and instruments

UNCLOS is widely considered to serve as "the constitution for the oceans," thereby providing for a legal order in the oceans, the very foundation of sustainable economies. As such, it is accepted as the framework that provides for the rights and obligations of States in the oceans, including defining the various maritime zones of jurisdiction and corresponding rights and obligations of states within them. It accommodates different uses of the oceans for economic and social development, balanced with the need to protect and preserve the marine environment. UNCLOS and related agreements serve to guide the management of the oceans and seas and the activities that take place on and within them, and they contribute to international peace and security, the equitable and efficient use of ocean resources, the protection and preservation of the marine environment, and the realization of a just and equitable economic order.

An assessment of the value of marine resources and their corresponding ecosystem services

Not only are marine living resources poorly measured and understood, they are also rarely valued properly. Measuring the blue economy gives a country a first-order understanding of the economic importance of the oceans and seas. In Mauritania, a study showed that the value of fisheries and renewable marine resources was much greater than that of the minerals that were the basis for most government decisions on marine resource management. Understanding that, in comparison with mineral resources, marine living resources are of much higher total value and renewable, the government adopted an alternative approach to development based on realizing the long-term potential for blue growth (Mele, 2014).

Increased reliance on evidence-based decision making

Countries increasingly recognize that they need more knowledge on biophysical characteristics, carrying capacity, and synergies or trade-offs between oceans-related sectors to ensure an efficient and sustainable management of different activities. Better scientific and economic data are required to understand these activities and their environmental costs. Marine and coastal spatial planning and integrated maritime surveillance can give authorities, businesses, and communities a better picture of what is happening in this unique space. Digital mapping of maritime and coastal space and natural assets in turn can form the basis for cross-sectoral analysis and planning in order to prevent conflicts and avoid externalities. Similarly, the growing science of data-limited stock assessments can provide critical information needed for improved fisheries management. In South Africa and Indonesia, mobile technology is being tested to gather previously unavailable data, for example on fish landings and fish stock health (World Bank 2016a).

A framework for ecosystem-based management

Historically, economic activity in the oceans has been managed on a sectoral basis, with only limited coordination between ministries, regulatory bodies, and industry when overseeing, among other things, overlap of property rights (particularly licenses for the exploration of extractive materials), shipping routes, and fishing grounds. Governing a sustainable blue economy will be far more complex. Unlike sectoral approaches that can lead to disconnected decisions, inefficient resource use, and missed opportunities, ecosystem-based management in which both the economy and ecosystems thrive will use tools such as integrated coastal zone management, marine spatial planning (MSP), and marine protected areas (MPAs) as part of the transition to a blue economy. Ecosystem-based management can facilitate a coordinated approach to the application of different policies affecting the coastal zone and maritime activities, from traditional ocean sectors to new businesses focused on ocean health. For ecosystem-based management to be successful, several supporting conditions must be met, including clear laws and regulations, strong institutions and inter-ministerial cooperation, inclusive decision-making processes involving all stakeholders, evidence-based support, and credible arbitration mechanisms. This institutional support and capacity vary greatly among countries.

Improved governance to grow a blue economy

This is essential for sustainable use of oceans, seas, and marine resources, for biodiversity conservation, for improved human well-being, and for ecosystem resilience. The use of science, data and technology is critical to underpin governance reforms and shape management decisions. Traditional knowledge and practices can also provide culturally appropriate approaches for supporting improved governance.

New data that can sway decision makers

Well managed, the goods and services produced from marine ecosystems could make a much greater contribution to reducing poverty, building resilient communities, fostering strong economies, and feeding over 9 billion people by 2050. The World Bank's 2016 Sunken Billions Revisited report shows that fisheries properly managed, with a significant reduction in overcapacity and overfishing, could provide additional economic benefits to the global economy in excess of US$80 billion each year (World Bank 2016b). That is almost 30 times the annual net benefits currently accruing to the fisheries sector in spite of prevalent overfishing.

Broad and resilient partnerships for coordination and collaboration of blue economy projects and initiatives

According to an analysis of case studies by the United Nations Environment Programme (UNEP), the blue economy makes its strongest gains when leveraging existing institutional relationships to address strategic gaps that affect multiple sectors and players and that catalyse visible benefits for them in the long term (UNEP 2015). A shift to a blue economy requires dedicated short- and long-term efforts, which can seize existing opportunities to bring together stakeholders. In addition, the blue economy requires the building of inclusive processes, including a concerted effort to identify and involve marginalized groups. Improving market infrastructure and access for small-scale and artisanal fishers can create more-sustainable outcomes that benefit the poor, for example through building on buyer demand for sustainable seafood.

Innovative financing to direct investments into economic activities that can enhance ocean health

Many public and private economic activities that could serve to restore ocean health will carry higher upfront costs and returns that will not immediately accrue to investors (Economist Intelligence Unit 2015). This suggests the need for new and innovative financing mechanisms, more capital than is currently being deployed, and a greater degree of collaboration between the public and private sectors. The private sector can play a key role in the blue economy, especially in SIDS. Business is the engine for trade, economic growth, and jobs, which are critical to poverty reduction.

Indicators to measure and track progress

Indicators used to track progress toward social and ecological sustainability are largely ignored in standard economic metrics such as gross domestic product (GDP) and will be needed to measure key transformation changes in different sectors of the blue economy. Thus, in countries like Mauritius, an important step in developing a blue economy has been the exploration of alternative economic indicators, based on the recognition that well-being is supported by a variety of economic, social, cultural, and natural assets and processes. Such initiatives are fundamental to developing more diversified, country-specific goals and progress indicators (UNEP 2015). These, in turn, are crucial to formulating policies that can halt ecosystem losses and thereby provide clearer pathways to sustained blue economy prosperity in the long term. Sector-specific monitoring is also necessary to fully understand the economic, environmental, and social impacts of each sector on local and national levels. For example, the International Network of Sustainable Tourism Observatories of the United Nations World Tourism Organization (UNWTO) monitors these impacts on the destination level, fostering the evidence-based management of tourism.

In the context of its initiative on Wealth Accounting and the Valuation of Ecosystem Services, the World Bank has led a joint effort to improve the availability of natural solutions to managing coastal areas. This initiative resulted in the publication of "Managing Coasts with Natural Solutions—Guidelines for Measuring and Valuing the Coastal Protection Services of Mangroves and Coral Reefs." This newly developed methodology could in turn be expanded to other, non-tropical marine and coastal ecosystems, leading to much improved decision making based on accurate assessment and valuation of ecosystems and the natural services they provide.

Challenges to the Blue Economy

The potential to develop the blue economy is limited by a series of challenges. For much of human history, aquatic ecosystems have been viewed and treated as limitless resources and largely cost-free repositories of waste. These resources, however, are far from limitless, and the world is increasingly seeing the impacts of this approach. The narrow coastal interface is oversubscribed by myriad sectors and is increasingly affected by climate change. Rising demand, ineffective governance institutions, inadequate economic incentives, technological advances, lack of or inadequate capacities, lack of full implementation of UNCLOS and other legal instruments, and insufficient application of management tools have often led to poorly regulated activities. This in turn has resulted in excessive use and, in some cases, irreversible change of valuable marine resources and coastal areas. In this increasingly competitive space, the interests of those most dependent and vulnerable (for example, small-scale artisanal fishers) are often marginalized, mostly for the benefit of other, more visible sectors (such as coastal tourism), where the actual economic benefits—while more clearly apparent at first—may actually be ephemeral or directly exported to foreign investors. The major human impacts include, among others, the following:

- Unsustainable extraction from marine resources, such as unsustainable fishing as a result of technological improvements coupled with poorly managed access to fish stocks and rising demand. FAO estimates that approximately 57 percent of fish stocks are fully exploited and another 30 percent are over-exploited, depleted, or recovering (FAO 2016). Fish stocks are further exploited by illegal, unreported, and unregulated fishing, which is responsible for roughly 11–26 million tons of fish catch annually, or US$10–22 billion in unlawful or undocumented revenue.
- Physical alterations and destruction of marine and coastal habitats and landscapes due largely to coastal development, deforestation, and mining. Coastal erosion also destroys infrastructure and livelihoods. Unplanned and unregulated development in the narrow coastal interface and nearshore areas has led to significant externalities between sectors, suboptimal siting of infrastructure, overlapping uses of land and marine areas, marginalization of poor communities, and loss or degradation of critical habitats.
- Marine pollution, for example in the form of excess nutrients from untreated sewerage, agricultural runoff, and marine debris such as plastics.
- Impacts of climate change, for example in the form of both slow-onset events like sea-level rise and more intense and frequent weather events. The long-term climate change impacts on ocean systems are not yet fully understood,

but it is clear that changes in sea temperature, acidity, and major oceanic currents, among others, already threaten marine life, habitats, and the communities that depend on them.

- Unfair trade: Exclusive Economic Zones, areas in which a state has sovereign rights over exploration and use of marine resources, are crucial to the economies of small island developing states and often dwarf their corresponding land mass and government's administrative capacity. In the case of fishing agreements allowing access to an EEZ, there is usually a low appropriation of fisheries export revenues by national operators and insufficient transfer to national stakeholders of specific fishing knowledge by foreign fishing companies, so the potential for national exploitation of those resources is reduced in the long run.

Despite a range of actors and large investments, current attempts to overcome these challenges have mostly been piecemeal, with no comprehensive strategy. Even when one sectoral policy achieves some success, these results are often undermined by externalities from activities in another sector. For example, coastal zone management efforts, or support to coastal fishers, tend to be undermined by unbridled sand mining, ill-sited ports or aquaculture farms, or unregulated tourism development. In coastal zones, declines in mangrove forest habitat resulting from habitat conversion, wood harvest, sea-level rise, destruction of dune systems from sand mining, and changes in sediment and pollutant loading from river basins combined with land reclamation for agriculture or infrastructure have serious negative impacts on fisheries by reducing or degrading spawning and feeding habitats. Loss of mangrove forests, for instance, threatens profits from seafood harvests in excess of US$4 billion per year; in Belize, mangrove-rich areas produce on average 71 percent more fish biomass than areas with few mangroves.

In view of the challenges facing SIDS and coastal LDCs, partnerships can be looked at as a way to enhance capacity- building. Such partnerships already exist in more established sectors, such as fisheries, maritime transport, and tourism, but are found less in newer and emerging sectors. There is thus an opportunity to develop additional partnerships around newer economic activities, such as marine biotechnology and renewable ocean energy.

A list of partnerships relating to the blue economy and its diverse sectors is available on the UN DESA website (https:www.un.org/development/desa/en/). The list of partnerships is dynamic and can be updated by users.

Carbon Sequestration (Blue Carbon)

"Blue carbon" is the carbon captured in oceans and coastal ecosystems (Herr, Pidgeon, and Laffoley 2012). The carbon captured by living organisms in oceans is stored in the form of biomass and sediments from mangroves, salt marshes, sea grasses, and—potentially—algae. Several key coastal habitats, such as sea grasses and mangroves, fix carbon at a much higher rate than comparable terrestrial systems (FAO 2014b; Pendleton *et al.* 2012). These "blue carbon sinks" can sequester up to five times the amounts of carbon absorbed by tropical forests, and they present an important opportunity for ecosystem-based climate mitigation, which also preserves the essential ecosystem services of these habitats (FAO 2014b).

Globally, the areas of wetland ecosystems declined 64–71 percent in the twentieth century, and wetland losses and degradation continue worldwide (Ramsar, 2015). Adverse changes to wetlands and coral reefs are estimated to result in more than US$20 trillion in losses of ecosystem services annually (Ramsar, 2015). The full social cost of carbon released into the atmosphere as a result of mangrove clearance has been estimated at between US$3.6 billion and US$18.8 billion per year at a price of US$41 per ton of carbon dioxide (that is, the true "social" cost) (Pendleton *et al.* 2012). Conserving marine and coastal ecosystems that sequester carbon could therefore lead to significant emissions reduction.

Arguably, and based on the untapped value of the carbon sequestered thus, if the value of the services provided by these coastal ecosystems in storing carbon could be quantified, payments could theoretically also be extracted, and paid to communities involved in managing and conserving these habitats through a "carbon market approach." In theory, at least, blue carbon could be traded and handled much like green carbon currently is (such as forest carbon under the UN collaborative initiative on Reducing Emissions from Deforestation and Forest Degradation) and entered into emission and climate mitigation protocols along with other carbon-binding ecosystems (FAO 2014b).

In practice, however, it should be noted that blue carbon has not yet been fully included in emissions accounting and that standards for blue carbon markets are still in their infancy. Some blue carbon pilot projects are currently under way around the world, including in SIDS and coastal LDCs, and research on carbon storage capacity of coastal ecosystems is being undertaken, but much uncertainty still prevails. It is already understood that carbon absorption capacity will differ based on geography and the physical interactions at play, so it would be almost impossible to come up with one proxy value, based on area of coverage. Nevertheless, carbon absorption is clearly one of the critical ecosystem services that must be quantified and valued (along with other services such as shelter, resilience to erosion, source of food), with the potential for a whole blue carbon sector to emerge eventually, particularly as carbon prices on the voluntary or compliance markets increase.

Blue Bonds

Blue bonds are modeled after green bonds. They are issued to raise capital and investment for existing and new projects with environmental benefits. The Seychelles plan to issue blue bonds, the first trial of this instrument among SIDS. Bond sales, facilitated by multilateral institutions including the World Bank and the African Development Bank, will fund the implementation of a fisheries management plan to develop the Seychelles' semi-industrial and artisanal fisheries (Rustomjee 2016). A number of projects are considered by the World Bank in which blue bonds provide the means to fund blue economy and fisheries development.

Steps for Future Action

The different pathways toward the blue economy depend on national and local priorities and goals. Nevertheless, there are common steps that will be required by all countries aiming to adopt this approach to managing their oceans.

- Countries must accurately value the contribution of natural oceanic capital to welfare, in order to make the right policy decisions, including with regards to trade-offs amongst different sectors of the blue economy.
- Investment in, and use of the best available science, data, and technology is critical to underpinning governance reforms and shaping management decisions to enact long-term change.
- Each country should weigh the relative importance of each sector of the blue economy and decide, based on its own priorities and circumstances, which ones to prioritize. This prioritization can be carried out through appropriate investments and should be based on accurate valuation of its national capital, natural, human and productive.
- Anticipating and adapting to the impacts of climate change is an essential component of a blue economy approach. National investments to that end must be complemented by regional and global cooperation around shared priorities and objectives.
- Ensuring ocean health will require new investment, and targeted financial instruments—including blue bonds, insurance and debt-for-adaptation swaps—

can help leverage this investment in order to ensure that it maximizes a triple bottom line in terms of financial, social, and environmental returns.

- The effective implementation of the United Nations Convention on the Law of the Sea is a necessary aspect of promoting the blue economy concept worldwide. That convention sets out the legal framework within which all activities in the oceans and seas must be carried out, including the conservation and sustainable use of the oceans and their resources. The effective implementation of the Convention, its Implementing Agreements and other relevant instruments is essential to build robust legal and institutional frameworks, including for investment and business innovation. These frameworks will help achieve SDG and NDC commitments, especially economic diversification, job creation, food security, poverty reduction, and economic development.
- Realizing the full potential of the blue economy also requires the effective inclusion and active participation of all societal groups, especially women, young people, local communities, indigenous peoples, and marginalized or underrepresented groups. In this context, traditional knowledge and practices can also provide culturally appropriate approaches for supporting improved governance.
- Developing coastal and marine spatial plans (CMSP) is an important step to guide decision making for the blue economy, and for resolving conflicts over ocean space. CMSP brings a spatial dimension to the regulation of marine activities by helping to establish geographical patterns of sea uses within a given area.
- The private sector can and must play a key role in the blue economy. Business is the engine for trade, economic growth and jobs, which are critical to poverty reduction.
- In view of the challenges facing SIDS and coastal LDCs, partnerships can be looked at as a way to enhance capacity building. Such partnerships already exist in more established sectors, such as fisheries, maritime transport, and tourism, but they are less evident in newer and emerging sectors. There is thus an opportunity to develop additional partnerships to support national, regional, and international efforts in emerging industries, such as deep-sea mining, marine biotechnology, and renewable ocean energy. The goal of such partnerships is to agree on common goals, build government and workforce capacity in the SIDS and coastal LDCs, and to leverage actions beyond the scope of individual national governments and companies.

Conclusion

The "blue economy" concept seeks to promote economic growth, social inclusion, and the preservation or improvement of livelihoods while at the same time ensuring environmental sustainability of the oceans and coastal areas. An important challenge of blue economy is to understand different facets of sustainable oceans from fisheries to marine pollution. The blue economy aims to move beyond business as usual and to consider economic development and ocean health as compatible propositions. A blue economy approach must fully anticipate and incorporate the impacts of climate change on marine and coastal ecosystems—impacts both already observed and anticipated. Healthy oceans and seas can greatly contribute to inclusiveness and poverty reduction, and are essential for a more sustainable future. While stimulating growth in individual oceanic sectors is comparatively straightforward, it is not always clear what a sustainable blue economy should look like and the conditions under which it is most likely to develop.

Annexure 9 Components of Blue Economy

Sr No	Type of Activity	Activity Subcategories	Related Industries/Sectors	Drivers of Growth
1	Harvesting and trade of marine living resources	Seafood harvesting	Fisheries (primary fishproduction)	Demand for food and nutrition, especially protein
			Secondary fisheries and related activities (e.g. processing;net and gear making; ice production & supply; boat construction & maintenance;manufacturing of fish-processing equipment; packaging, marketing & distribution)	Demand for food and nutrition, especially protein
			Trade of seafood products	Demand for food and nutrition,especially protein
			Trade of non-edible seafood products	Demand for cosmetic, petand pharmaceutical products
			Aquaculture	Demand for food and nutrition, especially protein
		Use of marine living resources for pharmaceutical products and chemical applications	Marine biotechnology & bioprospecting	

[Table Contd.

Contd. Table]

Sr No	Type of Activity	Activity Subcategories	Related Industries/Sectors	Drivers of Growth
2	Extraction and use of marine non-living resources (non-renewable)	Extraction of minerals	Seabed mining	Demand for minerals
		Extraction of energy sources	Oil and gas	Demand for (alternative) energy sources
		Freshwater generation	Desalination	Demand for freshwater
3	Use of renewable non-exhaustiblenatural forces (wind,wave & tidal energy)	Generation of (off-shore) renewable energy	Renewables	Demand for (alternative) energy sources
4	Commerce and tradein and around oceans	Transport and trade	Shipping and shipbuilding Maritime transport	Growth in seaborne trade, transport demand, international regulations, maritime transport industries (shipbuilding, scrapping, registration, seafaring, portoperations etc.)
			Ports and related services	
		Coastal development	National planning ministries and departments, private sector	Coastal urbanization, national regulations
		Tourism and recreation	National tourism authorities, private sector, other relevant sectors	Global growth of tourism
5	Indirect contribution toeconomic activities andenvironments	Carbon sequestration	Blue carbon	Climate mitigation
		Coastal protection	Habitat protection, restoration	Resilient growth
		Waste disposal for land-based industry	Assimilation of nutrients, solid waste	Wastewater Management
		Existence of biodiversity	Protection of species, habitats	Conservation

Source: World Bank and United Nations Department of Economic and Social Affairs. 2017. The Potential of the Blue Economy

CHAPTER 11

PLANT AND CELL BASED SEAFOOD: AN EMERGING ERA

Introduction

Development of innovative solutions is the need of the hour to address the growing global demand for seafood without causing any damage to environment and ecosystems. Development of plant and cell-based or cultivated seafood is a new concept taking shape and if it takes off will help to provide safe, nutritious, tasty, delicious, easily available and affordable seafood product to consumers.

Though the seafood is more popular and famous food across the globe; supply of seafood from capture fisheries has limitations as most of the commercial fisheries are overexploited or on the verge of overexploitation. Because of this, the annual global production from capture fisheries is almost stagnant. There are many factors which are creating pressure on ecosystems mainly marine ecosystems. Reducing pressure on global fisheries is critical to allow ocean ecosystem to recover to its original state. On the other side; aquaculture industry is growing very fast contributing more than half of global total fisheries production. However, aquaculture practices are not always responsible which may have adverse impacts on the environment. Commercial production of plant-based seafood (using plant-derived ingredients to replicate the flavor and texture of seafood) and Commercial production of cell-based seafood (produced by cultivating cells from marine animals) are two innovative approaches in meeting the global demand for seafood.

Plant-based and cell-based meat exhibit fundamentally higher efficiencies than cycling caloric value through animals, and they offer the unique opportunity to level the trophic playing field within seafood production. In other words, the raw materials and resources to create plant-based or cell-based versions of a top predator like tuna are essentially the same as those required for plant-based or cell-based versions of species at the bottom of the food chain.

Cell-based meat has become the industry-preferred term for products made from animal cell culture as a neutral term that is conducive to conversations with the conventional meat industry and with regulatory agencies. Other terms include clean meat — as a nod to the environmental benefits (akin to clean energy) as well as the reduced risk of bacterial contamination — and cultured meat and *in vitro* meat have historically been used as well. The term "cultured" is particularly problematic when referring to cell-based meat as applied to seafood, as cultured seafood is widely understood to mean seafood produced through aquaculture. Thus, the terms cell-based meat and cell-based seafood are more appropriate to use.

Accelerating the development and commercialization of scalable plant-based and cell-based seafood products that compete on taste, price, accessibility, and nutritional quality with their ocean-derived counterparts should comprise a core component of global strategies to maintain the vitality and, ultimately, the survival of our oceans. In the last decade, the market has seen massive shifts in consumer demand and product innovation for plant-based alternatives to products of terrestrial animal agriculture.

These trends are likely to reflect a similar forthcoming transformation within the seafood industry. In fact, there is reason to believe that the transition of seafood toward plant-based and cell-based meat solutions will occur with more urgency than for products like meat, poultry, and dairy, for which production has largely kept pace with increasing demand. The rapidly growing unmet demand for seafood coupled with the looming collapse of many global fisheries is likely to accelerate this shift. Furthermore, factors like the high incidence of seafood allergies and the high price points of several seafood products — especially products that are consumed raw and thus pose special consumer risks — generate a sizeable number of highly motivated early-adopters and market entry points for plant-based and cell-based seafood products.

The transition to plant-based and cell-based seafood can be further accelerated by concerted efforts to apply insights from the development, commercialization, and generation of demand for plant-based and cell-based versions of terrestrial animal agriculture products. While many of these insights can be translated directly to plant-based and cell-based seafood, the seafood sector does pose some unique technical challenges for both plant-based and cell-based approaches.

Consumer research providing a more nuanced understanding of seafood purchasing behaviour across diverse consumer segments and cultures is also needed, to enable refinement of marketing and product development strategies. While plant-based and cell-based seafood products will ultimately be produced and supplied through the private sector, the underlying technologies and their path toward commercialization will require a robust innovation ecosystem. Given that virtually no dedicated funding outside of a few companies' R&D budgets has been expended in this area and that the estimated total global R&D expenditure to date across all forms of plant-based and cell-based seafood is on the order of $10-20 million, this industry exhibits tremendous potential to benefit from concerted public and private resource allocation. To accelerate the process from early product development through to widespread market adoption, activities must be coordinated across start-up companies, multiple sectors of established industries, private and public funders and investors, governments, trade associations, and academic and other research institutions. All of these entities – and any individual who envisions a future with sustainable oceans of abundance – should consider this a call to contribute to the development and growth of the plant-based and cell-based seafood industry. (The Good Food Institute, 2019)

Coller FAIRR Protein Producer Index

The FAIRR Initiative has developed an index to analyse the largest global meat, dairy and aquaculture producers by combining nine environmental, social and governance (ESG) risk factors with the Sustainable Development Goals (SDGs). The benchmark will be primarily a resource for institutional investors and other actors interested in the livestock sector. The important point to note here is that the initiative has given larger importance to alternative sustainable proteins by associating maximum number of SDG's to it.

Plant-Based and Cell-Based Seafood As a New Solution

As incomes rise and population increases, the United Nations projects an increase in demand for seafood of more than 45 million tons between the mid-2010s and early-2020s, even after accounting for higher prices. Wild fisheries are already harvested at maximum capacity, and they are increasingly yielding species that are of low value for human consumption, which are instead processed into fishmeal and fish oil. Coupled with projections for a slowed rate of growth of the aquaculture industry in coming years, this creates a severe demand-supply gap. There is an urgent and sizable need for altogether new approaches to meet increasing global demand for seafood. (The Good Food Institute, 2019)

The production of plant-based and cell-based seafood is not limited by considerations like wild population productivity or geographical restrictions. Instead, these production platforms rely on consistent manufacturing and raw material inputs with robust supply chains and unconstrained supply. Some new and established companies are developing seafood product lines made from highly efficient protein sources such as fungi (see Terramino Foods and Quorn) that can potentially utilize byproduct streams and residual biomass from other agricultural or biological industries as feedstocks. Manufacturing facilities for plant-based and cell-based seafood need not be constructed near sensitive, expensive, and overburdened coastal areas and can instead be situated for most efficient logistical access for raw materials and final product distribution.

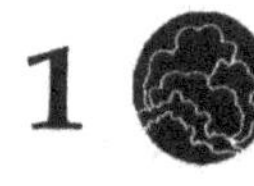

1 **GREENHOUSE GAS EMISSIONS**
Disproportionate amount of GHGs generated by livestock makes companies engaged in factory farming vulnerable to transition and physical risks

2 **DEFORESTATION AND BIODIVERSITY LOSS**
Global movements tracking forest loss target factory farming companies and can lead to shareholder divestment and / or weaken customer loyalty

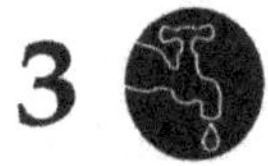

3 **WATER USE AND WATER SCARCITY**
Beef, pork, dairy, and poultry companies consume large quantities of water both directly and indirectly via their purchase of animal feed

4 **WASTE AND WATER POLLUTION**
Companies are facing greater scrutiny about the impact of waste on surrounding communities and the environment, meaning potential fines and regulation

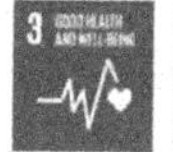

5 **ANTIBIOTICS**
Drug-resistant infections are a serious public health threat which will likely impact productivity on a national scale

6 **WORKING CONDITIONS**
Operational risks, which can involve worker injuries and reputational risk, as well as food product contaminated by sick workers

7 **ANIMAL WELFARE**
Poor animal welfare presents operational and reputational risks for companies

8 **FOOD SAFETY**
A series of high profile food safety incidents in meat and dairy have focused consumer concerns on threat of food contamination and foodborne illnesses

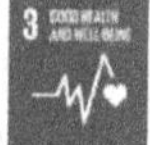

9 **SUSTAINABLE PROTEINS**
Reduced reliance on animal protein sources is key to sustainable development

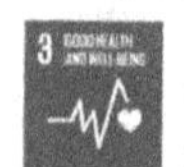

*SDG 2: Zero Hunger; SDG 3: Good Health and Well-being; SDG 6: Clean Water and Sanitation; SDG 8: Decent Work and Economic Growth; SDG 12: Responsible Consumption & Production; SDG 13: Climate Action; SDG 14: Life Below Water; SDG 15: Life on Land

Source: Coller FAIRR Protein Producer Index 2019

Plant-based and cell-based seafood producers are able to generate products in direct response to consumer demand rather than being dictated by availability, in sharp contrast to both wild-caught seafood and farmed seafood. Even though cultured species are purposefully farmed, the availability of a given farmed species is partly a matter of consumer demand but also partly derives from the relative ease of culturing that species. For example, we have already seen how species that are relatively amenable to aquaculture have come to dominate farmed fish consumption, despite consumers' very high demand for higher-value species like large predatory fish that are not well suited for aquaculture.

By contrast, the resource requirements and raw material inputs for producing cell-based tuna meat are virtually identical to those required to produce an equal mass of cell-based tilapia meat because fish muscle cells grown in cultivators will exhibit essentially the same metabolic requirements regardless of the species of origin. In other words, cell-based seafood production eliminates the multiple compounding layers of energy loss that occur to produce wild tuna due to its higher position on the food chain (trophic level). Likewise, the raw materials and production processes for making plant-based tuna (or any other high-value species or product) are virtually identical to those required to make plant-based tilapia (or any other low-value species or product). The differences in resource requirements and production processes from one species or product to another reside in subtle changes in the formulation and manufacturing process to develop unique flavors and textures mimicking each species.

Plant-based and cell-based seafood producers are able to generate products in direct response to consumer demand rather than being dictated by availability, in sharp contrast to both wild-caught seafood and farmed seafood. As a result, high quality and highly desirable products will become more accessible to consumers without the need to monetize low-value species or byproducts. Furthermore, plant-based and cell-based seafood solve the so-called carcass-balancing problem, thereby reducing waste across the entire food system. This term is most often used in terrestrial animal farming, referring to the need to monetize all parts of an animal carcass while consumers do not demand various cuts in the precise ratio in which they are found on the carcass. Seafood industry processes low-value byproducts of fillet preparation into minced products like fish sticks or compressed cakes made from deboned fish proteins. These products can be made through plant-based and cell-based approaches, but they will no longer flood the market as low-value byproducts from the production of more desirable cuts. As a result, high-quality and highly desirable products such as whole filets will become more accessible to consumers without the need to monetize low-value waste products. (The Good Food Institute, 2019)

Plant Based Seafood Products

While the plant-based seafood industry can learn from and build upon the success of terrestrial meat and poultry alternatives, plant-based seafood has lagged in growth and diversity of products and brands. Many of the fundamental production techniques used to structure plant proteins into fibrous food products resembling animal muscle tissue exhibit cross-applicability to many types of meat, including seafood. Thus, there is ample opportunity for existing plant-based meat companies to pivot a portion of their formulation and product development effort to adapt their recipes and protocols to seafood product lines. However, seafood also presents novel challenges regarding structure (in the case of particularly segmented or flaky forms of fish meat), appealing flavor profiles, and unique ingredient sourcing (such as cost-effective sources of animal-free omega-3 fatty acids). Some of these latter challenges may best be addressed through open-access, publicly funded research.

Plant-based seafood products comprise a very small fraction of the global seafood market, leaving almost unlimited growth potential in all segments. To date, only a handful of brands carry any plant based seafood product lines, and these lines cover fewer than a dozen of the hundreds of species of marine animals that are regularly consumed around the globe.

Notable emerging plant-based brands and companies include Good Catch, Terramino Foods, and Ocean Hugger, all of which have formed since 2016. Good Catch has developed flaked fish products such as tuna, crab cakes, and fish burgers. Terramino Foods debuted their prototype salmon burger in April 2018, produced using a fungi-based fermentation platform. Ocean Hugger uses the concept of biomimicry to replicate the texture and flavor of sushi using intact fruits and vegetables where their native flavors are replaced with those invoke fish, such as savory umami. This latter approach is fairly novel among plant-based meats because it does not require intensive processing methods and does not seek to achieve the protein levels of animal-derived meat. Using fibrous plant materials like marinated jackfruit to mimic the shredded texture of meat is another example of this approach, which has recently been used by Tofuna Fysh for a flaked tuna fish product.

Although the last two years have witnessed the launch of several exciting plant-based seafood brands, there is still a great deal of untapped market potential. Most of the plant-based seafood products currently on the market are ground or minced products rather than whole filets. With more sophisticated manufacturing methods, it may be possible to create the layers of fat, collagen, and protein that give fish its desirable cooking properties like flakiness. Additionally, since far more species of fish are consumed compared to species of land animals, there are nearly endless opportunities to develop novel products. (The Good Food Institute, 2019)

Fig.: Established and emerging plant-based seafood brands
(**Source:** GFI, 2019)

Plant-based seafood can leverage growth trends in the seafood category and the plant-based alternatives category to potentially exceed growth rates in the broader plant-based meat category. There is ample opportunity for existing and new brands to expand their distribution within retail channels. While a variety of plant-based beef and chicken products are available in mainstream grocery stores throughout the U.S., only two plant-based seafood products — Gardein's Fishless Filets and Crabless Cakes — are widely available as of early 2019. These products are almost never displayed alongside their conventional counterparts in retailers. Plant-based seafood manufacturers may experience a sales boost by utilizing similar strategies to those employed by other plant-based meat brands such as placement alongside conventional options.

While recent growth of the plant-based seafood industry has proven that commercially successful products are within reach today, further commercialization of high-quality products can be accelerated through open-access research across many areas spanning food science, biotechnology, and social science. The plant-based seafood industry can build upon insights in formulation, structuring, and marketing of other plant-based products, but several research areas are unique to the seafood sector. (Good Food Institute, 2019)

Cell-Based Seafood

While plant-based meat has improved drastically, current evidence indicates that many consumers are likely to continue to demand genuine meat even if presented with highly compelling plant-based options. Furthermore, many plant-based meats are minced or processed products rather than complex structures resembling intact animal muscle tissue. While a large fraction of seafood exists in the form of minced or processed products such as fish sticks, crab cakes, and surimi, many high-value products are intact muscle tissue of fish, shellfish, molluscs, and crustaceans. These products may prove more difficult to recapitulate with plant-based ingredients and thus may necessitate an approach that can produce the sophisticated structures associated with animal muscle tissue.

The cell-based meat production process relies on providing animal cells with molecular and environmental cues that govern developmental organization of the skeletal musculature, adipose, and connective tissues. Regardless of species of origin, the fundamental biological requirements of these cell types are largely similar across species, as developmental pathways are conserved by evolution in organisms ranging from annelids to vertebrates. Thus, the general process workflow for cell-based meat production should look similar for marine animals as for more established mammalian or avian systems.

Modifications to the general process and optimization for each final product will be required because the repertoire of growth factors and the precise nutrient requirements of different cells will vary slightly.

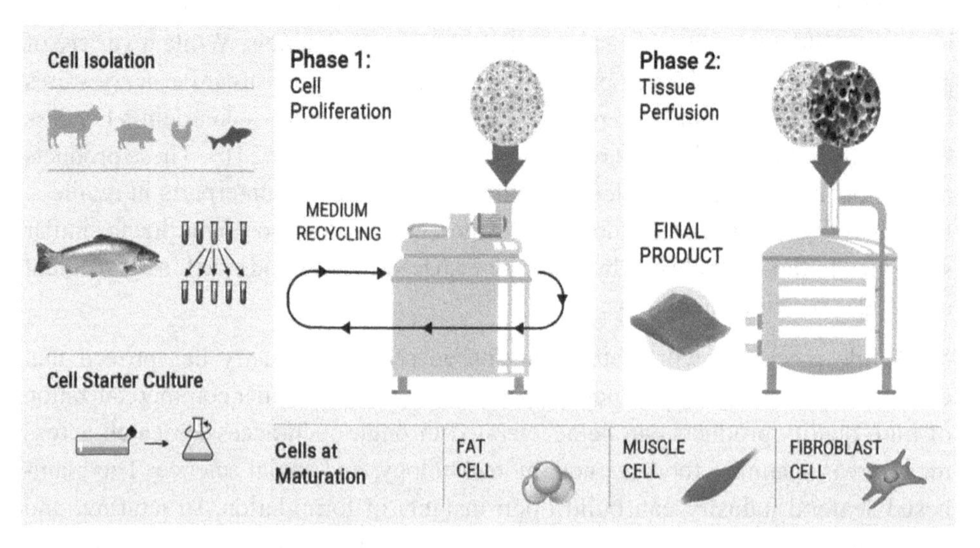

Fig. : Cell-based meat production schematic for seafood
(**Source:** GFI, 2019)

Since 2014, more than two dozen start-up companies have emerged to commercialize cell-based meat. Of these, only four companies have indicated an exclusive or predominant focus on seafood products.

These species comprise only a small portion of the multi-billion-dollar market across all marine animals, leaving immense opportunity for fledgling companies to pioneer work on additional cell-based seafood products from other species. Because the procedures for producing a cell-based seafood product should be relatively transferable across species once optimized, there is potentially a large first-mover advantage for early entrants. Thus, the application of cell-based meat bioprocessing to seafood is a very young endeavor and the allocation of even modest levels of additional resources toward this effort is likely to contribute substantially to the technological maturity and commercial readiness of the field.

Table : Cell-based meat companies with a partial or dedicated focus on cultivating cells from fish and other aquatic animal species*

Company	Location	First Target Product	Capital Raised
Finless Foods	USA (Berkeley, California)	Bluefin Tuna	Seed ($ 3.5 million)
BlueNalu	USA (San Diego, California)	Species not announced	Seed ($ 4.5 million)
Wild Type	USA (San Francisco, California)	Salmon	Seed ($ 3.5 million)
Shiok Meats	Singapore	Crustaceans	Pre-seed (Undisclosed)
Seafuture	Canada (Calgary, Alberta)	Species not announced	Not announced
Avant Meats	China (Hong Kong)	Species not announced	Not announced

*Note that this landscape is rapidly evolving and this represents a snapshot as of January 2019.

(**Source:** GFI, 2019)

The mission to create cell-based seafood products exhibits unique advantages and challenges relative to cell-based meat for other terrestrial farmed species. The most notable challenge is that cells from fish and other marine animals are not routinely cultured in most research labs, so protocols optimized for these cell types are not readily available in most cases. Additionally, resources such as sophisticated genome annotations for seafood-relevant species are limited compared to common laboratory species (e.g. mouse, rat, fly) or even common livestock species (e.g. cow, pig, chicken), and species-specific reagents like validated antibodies are not generally commercially available for marine animals. Thus, this field lacks established protocols and a rich scientific literature from which to draw, thereby requiring significant up-front investment in basic R&D by companies entering this space.

Cell-based seafood exhibits several potential advantages over mammalian or avian cell culture. Cells grown in culture perform best when growing conditions mimic the natural environmental conditions for that particular animal. In contrast to mammalian cell culture (typically conducted at 37°C), fish cell culture can be performed at appreciably lower temperatures (4-24°C for saltwater and 15-37°C for freshwater species). Many fish species also undergo muscle hyperplasia as juveniles, leading to rapid expansion in muscle cell number and biomass. This rapid growth ability may offer higher yield of skeletal muscle tissue in a shorter time when translated to a cell-based meat production environment. Additionally, many fish and crustaceans retain high expression of the enzyme telomerase in multiple tissue types, which may enable long-term proliferative capacity or facilitate the establishment of immortalized cell lines for research use and, ultimately, commercial cell-based seafood production. Many consumers place a premium on fresh seafood that has not been frozen, but product loss due to spoilage is a major concern within the seafood industry. Because cell-based seafood will be produced in aseptic cultivators, product shelf life may be dramatically improved without having to resort to freezing. (Good Food Institute, 2019)

Cell culture of marine species is not prevalent in academic research. There is a substantial need to develop the tools and resources that are already well established for mammalian cell culture — such as cell lines, robust protocols, commercial reagents, transformation vectors and reporters, full genome sequences, and biomolecular ("-omics") datasets for cells derived from marine species. These data and research tools would contribute greater mechanistic insights into the metabolism, growth, and developmental cell biology of these species, which may differ from the canonical pathways that are well characterized in mammalian species. Because of the development of less expensive, higher-throughput techniques for performing all of this work, the required investment in terms of funding, effort, and time to develop comprehensive data sets for each marine species will be orders of magnitude smaller than historical investment to develop these resources for the mammalian research community.

To facilitate all of this work, a public repository of validated cell lines derived from specific marine animal species representing diverse animal genera (such as bony fish, cartilaginous fish, shellfish, crustaceans, and molluscs) is needed. At present, the difficulty of obtaining such cell lines is a significant barrier to entry for both academic researchers and commercial ventures.

Research endeavors provide multiple opportunities for proactive engagement with the existing seafood industry and other entities that protect fisheries and ocean ecosystems. Obtaining access to high-quality primary tissue may require partnering with marine research or conservation organizations, aquariums, aquaculture facilities, or even industrial or recreational fishers. Collaborations involving aquaculture research institutes may prove particularly valuable because the aquaculture industry is experienced in handling aquatic species at all stages of maturity including embryos, and it routinely uses fish cell culture for advanced breeding and to monitor stocks for pathogens.

Key Challenges and Opportunities Unique to the Seafood Space

While many of the insights from the explosive growth of plant-based alternatives for terrestrial animal products can be applied directly to plant-based and cell-based seafood products, the seafood sector does pose some unique challenges and an even greater number of distinct opportunities. Efforts to accelerate the development and adoption of plant-based and cell-based seafood should incorporate these strategic considerations.

Challenges

Seafood is often seen as a healthier alternative to other terrestrial animal Proteins. About 90% of U.S. consumers of all ages associate seafood with positive health benefits. The negative health and environmental effects of red meat have driven many chefs and consumers to switch to inexpensive lean proteins such as chicken or fish. Thus, consumers may be less inclined to seek alternatives to seafood on health grounds, and plant-based and cell-based seafood products will have to emphasize their nutritional equivalence to conventional seafood. Because very few plant-based seafood products (and no cell-based seafood products) exist in the market, it is difficult to forecast projections of consumer demand for these products with a high degree of confidence.

In addition to challenges posed by consumer perceptions of seafood as healthy relative to other types of meat, structural differences between the seafood industry and the terrestrial meat industry may complicate attempts to garner support from key stakeholders to influence the entire sector. The seafood industry is historically more disaggregated than the heavily consolidated terrestrial meat industry. For example, while 13 companies control about 15% of the global seafood catch, just four companies control over 60% of the world's pork production and about 70% of the world's cattle. Additionally, harvest is often geographically dispersed from the point of consumption for both wild caught seafood and farmed seafood, making supply chains more opaque for seafood and requiring complex import and export considerations. This adds a layer of complexity when attempting to exert influence over the major stakeholders in the seafood industry and may present challenges for involving the existing industry as active participants in a wholesale shift toward plant-based and cell-based seafood. This challenge might be partially addressed by working with importers and processors rather than producers, as importation and processing represent points of consolidation within the supply chain.

Opportunities

Despite the overall perception of seafood as healthy, there are notable exceptions and this perception is swiftly changing. A significant portion of the population is excluded from seafood consumption due to health concerns. Fish and shellfish are two of the eight most common food allergens. They are responsible for more than 90% of food allergic reaction episodes in the U.S. Over seven million Americans are allergic to seafood, with shellfish allergies representing the most common food allergy in the U.S. Thus, plant-based seafood products have a built-in potential early-adopter market even larger than that of plant-based dairy. Doctors often advise avoiding all seafood if allergic to either fish or shellfish, since many products like imitation crab often contain other fish or shellfish.

Additionally, some people limit seafood consumption due to concern for high levels of mercury and other toxins, and the FDA advises those who are pregnant or breastfeeding to avoid certain species of fish completely. Seafood has received considerable press for contamination scares and parasitic infections resulting from consuming uncooked fish. As noted above, fish and shellfish can contain high levels of mercury, PCBs, dioxins, and other health contaminants, and are frequently fraudulently labelled either in species or in origin to evade these concerns. Furthermore, meta-analyses have recently called into question some of the supposed health benefits of seafood products, including those associated with omega-3 fatty acid consumption. Thus, the positive association between seafood and health may weaken in coming years.

Beyond these health-related opportunities to drive consumer demand for plant-based and cell-based seafood, these products also exhibit notable advantages to the industry in terms of increased efficiency and reduced loss throughout the supply chain. Nearly half of the edible U.S. seafood supply was lost due to consumer food waste, discarding of bycatch, or distribution spoilage from 2009 to 2013. Seafood products are highly perishable foods, which presents challenges producers and distributors as well as consumers, as many consumers feel they do not feel comfortable assessing whether seafood is fresh and safe to eat. Plant-based items have a longer shelf life and reduce the need for costly refrigerated transportation while providing an attractive opportunity for local production in landlocked areas.

Furthermore, the production process for both plant-based and cell-based seafood is more controllable and predictable, allowing for better real-time response to demand and for much more customized end products that precisely answer this demand. More valuable cuts, product formats, and species of seafood products could be produced without generating low-value byproduct waste. These increases in efficiency create an opportunity for plant-based seafood products to provide a healthier and ultimately less expensive alternative to conventional seafood.

Encouraging Innovation Ecosystem for Sustainable Seafood Products

While plant-based and cell-based seafood products will ultimately be produced and supplied through the private sector, the underlying technologies and their path toward commercialization will require a robust innovation ecosystem. To accelerate maturation from early product development through to widespread market adoption, activities must be coordinated across startup companies, multiple sectors of established industries, private and public funders and investors, governments, trade associations, and academic and other research institutions. All of these entities should consider this a call to action to contribute to the development and growth of these sustainable alternatives to conventional seafood.

Given the global nature of the seafood industry, it is imperative to formulate appropriate regional strategies for bolstering this innovation ecosystem. The seafood industry at present is dominated by Asia in terms of both production and consumption, but much of the innovation in plant-based and cell-based seafood is currently occurring in North America. Regional differences in the aquaculture industry are even more striking: the Asia Pacific region accounts for over 90% of global aquaculture production, while North America is responsible for less than 1%. Because of this significant regional skew, the strategy for advancing plant-based and cell-based seafood in each region should account for unique considerations such as policy, availability of investment capital, nutritional needs, impact on livelihoods, projected demand growth, and other locally relevant factors. It is especially critical to maintain a global perspective regarding consumer attitudes, product/species selection, and the involvement of governments and NGOs in order to maximize the impact of plant-based and cell-based seafood options for meeting global demand and thus alleviating pressure on overtaxed fisheries and aquaculture systems.

Resource allocation toward plant-based and cell-based seafood development is in alignment with many philanthropic foundations and government agencies whose mission or scope is not directly related to ocean sustainability or preservation. Governments and private foundations across many sectors are encouraged to consider how their missions – whether to serve the public good, improve global food security and safety, create a sustainable food supply, or conserve marine ecosystems – may be advanced through efforts to support plant-based and cell-based seafood.

Governments whose constituents are heavily economically dependent upon seafood production or vibrant oceans (for example, for ocean-related tourism) should be especially motivated to support innovation in plant-based and cell-based seafood. For example, countries like Singapore and Israel have demonstrated leadership in supporting alternative protein innovation motivated in part by food security and independence due to their high fraction of imported food. (Good Food Institute, 2019)

Government action can exert influence over the future of plant-based and cell-based seafood in several ways beyond committing additional public research support. Two approaches are: (1) ensuring that plant-based and cell-based seafood products are not handicapped through labeling laws or standards of identity that privilege conventional seafood, and (2) updating the dietary guidelines to specifically include these products as suitable alternatives to conventional seafood.

Trade associations and non-profits with lobbying capacity should prepare to advocate for legislation and regulatory policies that allow plant-based and cell-based seafood to compete on a level playing field with conventional seafood. Start-up incubators and philanthropic foundations with the capacity to make private investments can also solicit specific solutions. A systematic solicitation for letters of interest from various stakeholders such as retailers, distributors, and foodservice outlets would also provide insight into desirable product types and the volume and pricing targets required to secure these contracts. These letters of interest would also de-risk the investment by demonstrating market receptivity and unmet demand despite the fact that many of the companies in this sector are currently pre-revenue. (Good Food Institute, 2019)

Forging strategic partnerships with the existing seafood industry and beyond strategic relationships with the meat, consumer packaged goods, agriculture, industrial biotechnology, and life science industries have proven critical to the success of plant-based and cell-based versions of terrestrial animal products. Plant-based and cell-based seafood manufacturers can employ a similar strategy by partnering with traditional seafood companies, ingredient suppliers, and non-profit organizations early in the emergence of this industry. Companies already involved in seafood harvesting, production, processing, and distribution are well poised to invest in plant-based and cell-based seafood. They can leverage their existing assets, diversify their supply chains, and position their product portfolios to appeal to future consumers.

Ingredient suppliers such as ADM, Cargill, Axiom, Ingredion, Givaudan, and others can make excellent strategic investors or partners for emerging plant-based brands. These companies sell the commodities that comprise plant-based seafood, including proteins, flavorings, fragrances, coloring agents, and lipids, as well as many of the raw materials that will contribute to cell-based seafood nutrient medium. The size, market insight, scientific expertise, and distribution networks of these suppliers make them strong partnership candidates even for early-stage plant-based and cell-based seafood start-ups.

Restaurant foodservice is among the likeliest market entry points for plant-based and cell-based seafood due to higher margins relative to grocery channels and the strong association between seafood and fine dining. Plant-based and cell-based seafood companies could partner with chefs and restaurant concepts to create splashy product launches and drive critical initial sales. This model has been employed with great success by companies like Impossible Foods, which generated tremendous consumer interest and garnered culinary respect by first launching their burger in an exclusive set of high-end restaurants with chefs who are unapologetic about their discerning standards for high-quality meat.

Aquatic and oceanographic research institutes can serve as valuable partners for directing resources toward species that are particularly threatened or whose harvesting tends to be most highly disruptive to ocean ecosystems. Plant-based and cell-based seafood companies could partner with certifying bodies such as the Marine Stewardship Council, the Aquaculture Stewardship Council, the Global Aquaculture Alliance, Friend of the Sea to obtain existing or new certifications for their products, as plant-based and cell based seafood would clearly meet all the metrics for reduced environmental impact. Environmental certifications (such as seals that can be added to product packaging) would serve to familiarize consumers with novel plant-based and cell-based seafood products.

Products like New Wave Foods' plant-based shrimp and Ocean Hugger's plant-based tuna have been featured by several prominent Silicon Valley technology companies' dining services. These contracts tend to garner significant positive press for the plant-based seafood companies as well, thereby elevating the status and visibility of these products. (Good Food Institute, 2019)

Plant-based and cell-based seafood producers may benefit from creating a coalition to respond to regulatory, labeling, and anti-competitive legislative challenges. Such challenges have arisen for other segments of the plant-based and cell-based meat industries. For example, the state of Missouri passed a law in 2018 making it unlawful to misrepresent a product as meat if it is not from slaughtered livestock or poultry. Currently there are no trade associations dedicated solely to plant-based or cell-based seafood companies, but plant-based seafood companies can join the Plant Based Foods Association, which represents the interests of companies that make a variety of plant-based foods.

Systems for regulatory oversight and maintaining global supply chains are significantly different for seafood products than for land-based animal agriculture. Thus, although the production platforms for manufacturing plant-based seafood and cell-based seafood are rather similar to their terrestrial animal agriculture counterparts, dedicated consortia and associations for these industries may be necessary to effectively advocate for their unique considerations. Plant-based and cell-based seafood offer a unique opportunity for ocean-focused companies and organizations to demonstrate their commitment to preserving the marine ecosystems that they seek to share with their clientele.

These endeavors often also incorporate workforce development and training programs to ensure the availability of a highly skilled workforce for new manufacturing methods. The plant-based and cell-based meat sectors have already experienced a shortage of technical talent trained to enter these industries. The establishment of concerted training programs should prepare skilled technical workers for the burgeoning plant-based and cell-based seafood industry before workforce limitations become a bottleneck for companies' growth.

Investors have been among the most active participants in the early-stage emergence of plant-based and cell-based seafood because seasoned investors in disruptive technologies recognize the harbingers of a wholesale disruption of the seafood industry. Immense pressures on a finite resource — wild fish stocks — arise not only from overfishing and increased demand but are substantially compounded by less predictable but equally disruptive influences like climatic shifts or severe weather patterns.

Overfishing of keystone species, pollution, and climate perturbations can collapse marine ecosystems in an irreversible manner, which generates substantial financial risk for investors with commercial interests in the seafood industry. The task of articulating these risks can again take a page from the playbook of terrestrial animal agriculture. A UK-based initiative called FAIRR (Farmed Animal Investment Risk and Return) is a collaborative network of investors that raises awareness of the risks associated with investment in animal farming and highlights protein diversification as a key strategy for responsible asset management.

Developing a similar coalition among investors with financial interests in the seafood industry to position investment opportunities in plant-based and cell-based seafood as critical de-risking components of their portfolio could facilitate a substantial shift in financial resources. Early-stage investors in plant-based and cell-based seafood companies include not only mission-aligned firms such as New Crop Capital, Stray Dog Capital, and Blue Horizon Ventures but also traditional Silicon Valley venture capital firms that have historically had little or no involvement in food technology or consumer brands. Concerted efforts to educate these investors about the current state and existing challenges of this nascent industry will ensure that expectations are aligned and that projections are realistic and achievable, which ultimately positions early-stage companies to successfully achieve milestones to unlock larger follow-on investments.

Strategic investors from the food, ingredients, and seafood industries can also serve a crucial role. These types of investors will be particularly important for accelerating growth through leveraged infrastructure and distribution networks, and they should be brought into the capitalization table early in a plant-based or cell-based seafood company's funding trajectory. Investors have been among the most active participants in the early-stage emergence of plant-based and cell-based seafood because seasoned investors in disruptive technologies recognize the harbingers of a wholesale disruption of the seafood industry.

Vision for The Future of Plant Based/Cell Based Seafood

The rate of adoption of new technologies is entirely dependent upon the amount of resources — including financial capital and human capital as well as political support that can spur additional effort — devoted to their development and deployment. The present moment presents a pivotal opportunity to leapfrog over stopgap measures such as fishery management strategies and intensive aquaculture to develop a truly sustainable, scalable means of satisfying growing global seafood demand.

Globally, humankind's current relationship to seafood is akin to our relationship to terrestrial animal meat several thousand years ago, when domesticated (farmed) animals and wild (hunted) animals contributed roughly equally to meat consumption. Now, with farmed meat accounting for virtually all terrestrial meat consumption — on a scale far greater than wild animal populations could have ever sustained — our civilization is realizing that even the most efficient, intensive animal farming is simply too inefficient to feed a population nearing 10 billion people by 2050. Livestock now constitute far more biomass than all wild terrestrial vertebrate animals combined , and animal agriculture's high resource burden has squeezed terrestrial habitats — and the biodiversity they once contained — to the brink. Sufficient arable land to cultivate the inputs for terrestrial animal agriculture on the scale required to meet projected demand this century simply does not exist. Even if feasible, the amount of animal waste generated through such a system would wreak catastrophic environmental consequences.

The resource allocation toward plant-based and cell-based seafood development is doubly beneficial for the oceans. Just as the plant-based and cell-based seafood sector can leverage insights from alternatives to terrestrial farmed animal products, learnings from plant-based and cell based seafood product development will likewise accrue benefits to the larger animal product alternatives landscape. Thus, investment in plant-based and cell-based seafood benefits oceans both directly and indirectly: it directly alleviates demand for wild-caught and farmed seafood, and it indirectly supports a shift away from industrialized *terrestrial* animal farming, which is intimately intertwined with ocean health through waste runoff and greenhouse gas contributions. Therefore, the oceans benefit immensely from a wholesale shift away from animal product consumption — aquatic and terrestrial — accelerated by widespread availability of cost-competitive plant-based and cell-based meat and seafood.

In the face of growing global demand for seafood, efficient and scalable plant-based and cell-based seafood production offers a new approach for maintaining vibrant and abundant oceans without compromising food security, complementing existing strategies in fisheries management and sustainable aquaculture. Accelerating the development, commercialization, and widespread availability of plant based and cell-based seafood should constitute a core pillar of the strategic plan of all entities whose vision includes responsible stewardship of both land and sea while ensuring human prosper) nonprofit organization dedicated to creating a healthy, humane, and sustainable food supply.

Development of Plant Based/Cell Based Food/Meat/Seafood in India

Why India is Important?

India is second largest country in population in the World. Though we are one of the largest seafood producers in the world, the per capita consumption is yet low compared to global average. India is among one of the countries where the proportion of youngsters in total population is quite good and hence larger young force is therefore available from now till next couple of years. However, the rate of malnutrition is among one of the highest in the world for India hence plant-based and clean meat / seafood present an unprecedented opportunity to address debilitating nutritional deficits sustainably and humanely, without the harms of conventional food sector.

India can be a model for improving nutritional status in a massive population efficiently and humanely through plant-based and clean meat innovation. Encouragement, co-operation and enthusiasm from the government, scientific institutes, and large corporations are making the ways to look seriously in this sector. With thorough involvement from the stakeholders, a huge impact on the future of protein industry in India will progress in right direction. India's young, English-speaking workforce and majority religious culture rooted in vegetarianism lay the groundwork for rapid plant-based and clean meat innovation. Taken with India's strong biopharma industry and research institutes, significant government support for R&D, and growing venture capital sector, good food companies have a promising launch pad.

Simultaneously, there is a clear opportunity for Indian research institutes and biopharma companies to enter the clean meat sector—particularly as part of the supply chain, optimizing critical technology elements: cell lines, cell culture media, scaffolds, and bioreactors. For plant-based products, millets and other indigenous Indian crops provide an excellent opportunity to diversify the global inputs for plant-based meats, seafood, eggs, and dairy. Using these crops would create lucrative markets for farmers and strengthen value chains into which the government is already pouring significant investment.

Good Food Institute India

Good Food Institute is Washington based NGO working in the field of plant and cell-based food. The Indian arm of Good Food Institute signed an MoU in 2019 with Institute of Chemical Technology (ICT), Mumbai to set up a "Centre for excellence in cellular agriculture". Good Food Institute will offer expertise and guidance in setting up the lab through its network of scientists. The concept of cell-based meat is still in an early stage in the country. However, the size and demand of the meat industry is growing rapidly. Two types of research will be conducted at the Lab: open access research to be utilised by the entire sector and contract research opportunities with entrepreneurs.

Conclusion

Plant-based and cell-based seafood is an emerging industry but at a very initial stage today. The traditional ways of culturing seafood produce high emissions of carbon whereas professionals involved in the protein sector assure of reducing those carbon emissions to a large extent with the production of plant-based and cell-based seafood. There are companies in the world which are working in the direction of production, distribution, marketing and sales of these plant-based and cell based seafoods. Ample scope exists in this sector which is expected to reduce the load on the environment by providing the seafood and meat substitutes through the new technologies.

CHAPTER 12

IMPACT OF COVID-19 ON FISHERIES AND AQUACULTURE

Introduction

The impacts of COVID-19 on the fisheries and aquaculture food systems vary and the situation is rapidly evolving. Fish and fish products that are highly dependent on international trade suffered quite early in the development of the pandemic from the restrictions and closures of global markets, whereas fresh fish and shellfish supply chains were severely impacted by the closure of the food service sectors namely hotels, restaurants and catering facilities, including school and workplace canteens. The processing sector also faced closures due to reduced/lost consumer demand. This has had a significant impact, especially on women, who form the majority of the workforce in the post-harvest sector. In some areas, an increase in retail sales has been reported due to the closure of the food service industry. Canned and other preserved seafood products with a longer shelf life have profited from panic buying at the beginning of the crisis. In some markets, suppliers have developed ways to provide direct supplies to consumers (e.g. box schemes) to replace lost fresh fish sales from established retailers. There are still many uncertainties ahead, particularly with regard to the duration and severity of the pandemic, but a prolonged market downturn is likely to introduce long-term transformations to the sector.

The lockdowns implemented by some countries have resulted in logistical difficulties in seafood trade, particularly in relation to transportation and border restrictions. The salmon industry, in particular, suffered from increased air freight costs and cancellation of flights. The tuna industry has reported movement restrictions for professional seafarers, including at-sea fisheries observers, and marine personnel in ports, thereby preventing crew changes and repatriation of seafarers. Shortage of seeds, feeds and related aquaculture items (e.g. vaccines) have also been reported, due to restrictions on transportation and travel of personnel, with particular impacts on the aquaculture industry. As a result of the drop in demand, and resulting price drops, capture fishery production in some countries has been brought to a halt or significantly reduced, which may positively influence wild fish stocks in the short term. In aquaculture, there is growing evidence that unsold produce will result in an increase of live fish stocks, and therefore higher costs for feeding as well as greater risk of fish mortalities.

Although COVID-19 does not infect aquatic species (Bondad-Reantaso *et al.*, 2020), it has affected the fisheries and aquaculture food systems like no other shock before. The protection measures taken by governments to contain the spread of the disease, while necessary, have impacted each step of the seafood supply chain, from fishing and aquaculture production, to processing, transport, and wholesale and retail marketing. Yet, in this period of global pandemic, fish remains an essential source of animal proteins, micronutrients and omega-3 fatty acids, which are vital in low-income food-deficit countries (LIFDCs) and Small Island Developing States (SIDS), where diets are heavily reliant on fish. It is therefore important that these countries continue to have access to fish products.

In addition to being important for the livelihoods of many fish-dependent communities, fish and fish products are among the most highly traded food products in the world, with 38 percent of total fish production entering international trade. The measures necessary to contain the spread of COVID-19 have caused disruption in all segments of both domestic and international supply chains. Protecting each stage of the supply chain is fundamental to avoid global and local food crises, and protect fish-dependent economies (FAO,2020).

FAO summary of Impact of Covid-19 on fisheries and aquaculture (Excerpts reproduced from FAO,2020)

Reduction in Fishing Activities

Fishing activities have decreased in both artisanal and industrial sectors during the pandemic. According to Global Fishing Watch, global industrial fishing activity had fallen globally by about 6.5 percent as at the end of April 2020, compared with previous years, as a result of restrictions and closures related to COVID-19 (Clavelle, 2020). Limited supplies (e.g. ice, fuel, gear and bait), due to suppliers being closed or unable to provide inputs on credit, have also constrained fishing activities. Labour shortages have also had a severe impact on fishing activities, particularly where crews are made up of migrant workers. As a consequence, their families in home countries had to cope with the reduction or halting of remittance flows (World Bank, 2020). In some regions, signs of improvement have been evident in some fisheries (e.g. changes in target species and in marketing strategies in line with changes in demand), with some small-scale fisheries potentially adjusting more quickly to market demand.

Varied Impacts in Aquaculture Production

Effects on aquaculture have varied with regions, species, markets and financial capacity of farms. Following disruptions, many farmers who have been unable to sell their harvest have had to maintain large quantities of live fish. Others have not been able to complete all necessary seasonal tasks such as fish breeding. This has increased costs and risks, in particular when the supply of inputs has also been disrupted, and it is also likely to delay restocking and subsequent harvests. Species grown for export have been severely affected by the disruption of international transport. Although some government or financial institutions have provided financial support, the risk of bankruptcy remains. However, preliminary projections by a few companies indicate that they may have the capacity to recover once the crisis eases.

Aquaculture farms supplying the live-fish market or high-end food services (i.e. restaurants, tourism and hotels) have also been dramatically impacted. Their capacity to recover will depend largely on their ability to re-direct their sales to other markets, especially supermarkets and retail, including their use of digitization tools, which have emerged as a major innovation during the crisis. Small and medium-sized enterprises and farms struggle with cash-flow issues, as the crisis has not only slashed their incomes but also created new expenses associated with the cost of maintaining live stocks in production facilities. The availability of labour and aquaculture inputs needed for production (e.g. medicines, fingerlings and feed) have also been impacted by restrictions on cargo movements, precautionary measures and borders closures.

Processors, Markets & Trade

The main adverse effects are for producers supplying the food services sector, e.g. hotels, restaurants and catering. Some have started direct sales and delivery service to households in an effort to compensate for lost hotel and restaurant demand. Exports have been severely affected owing to transport disruptions. However, food retail sales have remained stable or increased for frozen, canned, marinated and smoked fish with a longer shelf-life.

Processing has been affected by worker health issues and labour shortages due to COVID-19 illness and quarantining of staff. Changes in demand are also affecting storage, resulting in increased food loss and waste. Many wholesale and retail fish markets are often congested and crowded, presenting risks to traders, most of whom are women, as well as to consumers, who take significant risks to maintain their livelihoods and to buy fresh fish to eat. It is the informal supply chains that are facing greater impacts due to the lack of formal contractual relationships (no established cold chain or insurance, among others). Another consequence of the virus outbreak, linked to global trade, is the cancellation of key seafood trade events across the world.

Problems in working conditions, especially for women and other vulnerable workers, along the supply chain

While some small-scale fisheries have been able to adapt (e.g. providing direct sales to consumers), in general small-scale fishers and fish workers have been hit hardest because they lack capital to weather the storm, depend on fishing for their daily income/food, and do not have access to health services, among other reasons. In some parts of the Mediterranean and the Black Sea, more than 90 percent of small-scale fishers have been forced to stop fishing during lockdowns, despite being a primary food production sector, as they have been unable to sell their catches and/or as the prices of fish have dropped below a profitable level (Euronews, 2020). Women, who represent 50 percent of the labour force in fisheries and aquaculture, have been particularly affected by reduced landings and the closure or reduction of processing and marketing activities (CFFA, 2020). In addition, fishers, and fish processors and vendors (many of whom are women), are exposed to a greater risk of infection, as they have close contact with others at all stages of the value chain. Moreover, the widespread informal activities in the sector constitute an added barrier for fishers, women fish workers and fish farmers in accessing protection offered under labour market policies and contributory social protection mechanisms. Working conditions and the safety of fishers in both small and industrial sectors have been affected owing to having to work longer periods, which increases fatigue and stress (FAO, 2020a).

Impact on Management of Fisheries

The impacts of COVID-19 have affected fisheries management processes. While closing fishing operations will offer respite to some overexploited fish populations, similar constraints will also apply to science and management support operations. For example, fish assessment surveys may be reduced or postponed, obligatory fisheries observer programmes may be temporarily suspended, and the postponement of science and management meetings will delay implementation of some necessary measures and the monitoring of management measures. Lockdowns could lead to reduced capacity in Fisheries Monitoring Centres (FMCs), as was the case in West Africa during the 2013-2016 Ebola outbreak – not only were staff not available, but limited national resources were directed to fund emergency activities and this left the FMCs unable to function effectively. Fishers who are able to continue fishing at sea know this and they may adapt their operations to engage in illicit activities and benefit from the MCS shortcomings. A lack of monitoring and enforcement of shared stocks may encourage some states fishing on these stocks to revert to a less responsible level of management, monitoring and control of fishing operations (FAO, 2020a).

A lack of monitoring and enforcement may encourage a less responsible level of management, monitoring and control of fishing operations and there is a risk that levels of illegal, unreported and unregulated (IUU) fishing will increase. Crew safety is an additional concern. However, the crisis has triggered unprecedented responses by governments across the world. Policies and actions taken include measures to protect public health, to safeguard fishers' and fish workers' safety, and to strengthen social protection to support the most vulnerable and avoid a socio-economic crisis. Social protection measures relate to social assistance (e.g. cash transfers), social insurance (e.g. health insurance) and labour market programmes (e.g. unemployment benefits), and steps to ensure the continuity of food supply.

Fear of consumption of fisheries and aquaculture products

Fish and fish products are a key component to a healthy diet and are safe to eat. Misleading perceptions in some countries have led to decreased consumption of these products. Yet, corona virus cannot infect aquatic animals (finfish, reptiles, amphibians and invertebrates such as crustaceans and molluscs), therefore these animals do not play an epidemiological role in spreading COVID-19 to humans.

While there is no evidence of viruses causing respiratory illnesses are transmitted via food or food packaging, fishery and aquaculture products can become contaminated if handled by people who are infected with COVID-19 and who do not follow good hygiene practices. For this reason, as before COVID-19, it is important to emphasize the need to implement robust hygiene practices to protect fishery and aquaculture products from contamination.

Key impacts on global and local seafood-dependent economies and livelihood

It is not yet clear whether the sector will experience a quick or slow recovery after the pandemic is over. While some seafood companies may manage or even benefit from the crisis, a level of industrial consolidation is to be expected, as well as re-sourcing. Digital innovation, accelerated shifts towards Web-based applications, online services and improved product traceability and sustainability are some of the results likely to emerge from the crisis.

At a local level, fishers and fish workers are adapting by changing fishing gears, targeting different species or selling their products to the domestic market. Some fishers, fish farmers and fish workers have started selling directly to the consumer. While these innovations will support communities, especially women operating in the post-harvest sector, domestic markets have limits both in terms of demand and price. In the short term, possible disruptions to economies and livelihoods could come from labour shortages (travel barriers, labour lay-offs, etc.); direct boat-to-consumer sales; aquaculture input shortages (feed, seed, vaccines); as well as fishing (e.g. bait, ice, gear, etc.); competition for sourcing and transport services and a lack of finance and cash flow (delayed payment of past orders).

Implications of COVID-19 on vulnerable and women

The pandemic has created an unprecedented economic, social and health crisis with impacts on the most vulnerable groups including women (harvesters, processors and vendors), migrant fishers, fish workers, ethnic minorities and crew members. Many individuals are not registered, operate in the informal labour market with no labour market policies, including no social protection and no access to relief package/aid. These conditions might exacerbate the secondary effects of COVID-19, including poverty and hunger. The small-scale fisheries sector is trying to make ends meet, to continue fishing and provide locally-caught fresh fish, while experiencing great difficulties due to the closure of markets, limited storage facilities, falling wholesale fish prices and new sanitary requirements and physical distancing measures. Because of these difficulties, many activities have been reduced. The reduction of fishing and fish farming activities will reduce the amount of fish available for processing and trade. Furthermore, mobility restrictions will adversely affect the transfer of fish to markets. This will particularly impact women, who are mostly in charge of these activities. Food loss and waste could also increase if processors do not have access to appropriate storage and cold chain facilities. Frontline employees who are processing seafood suffer from a lack of protective equipment and clothing, which highlights the general lack of access to hygiene and protective equipment for the vulnerable workers of the seafood industry.

In the current situation, migrant fishers and fish workers, including ethnic minorities, are unable to return to their native villages due to lockdowns. They require assistance including food and transportation (where movement restrictions permit) to reach their villages. Working conditions and the safety of fishers at sea will be negatively affected should the number of fishers available to crew vessels be reduced. The availability of crew may be reduced for various reasons including inter alia contracting COVID-19, restrictions on movements or wider lockdowns. In addition, it is difficult for fishermen to maintain physical distancing measures of a metre apart on-board fishing vessels. Should fishing vessels be forced to operate with fewer crew members, this may result in working longer hours, which will compromise safety measures and thereby put the well-being and health of fishers at risk.

Crew on large-scale industrial vessels (pelagic trawlers, purse seiners), that work in rotations of several weeks before being replaced by another crew during their work-break, are unable to travel home due to flight restrictions and quarantine periods. As a consequence, they work longer periods on-board, which increases the likelihood of on-board accidents, fatigue and stress (also relevant to the health of family members back home). Large-scale fishing vessels of distant water fishing fleets also risk outbreaks of COVID-19 cases among crew members while away at sea. COVID-19 may spread rapidly among crew members of a vessel, and medical assistance is not always readily available. Also, when trying to enter a port where the crew are not nationals of the port state, access may be denied.

Self-isolation and restriction of mobility reduce demand for fish and fish products, which has negative economic impacts on women's livelihoods and income (harvesting, processing and trading). In addition to lack of economic opportunities, women fish vendors may be exposed to a greater risk of infection, since markets are a place of close contact and have limited sanitation and hygiene facilities. This is all the more fundamental in view of women's decreased job security, especially those informally employed in the fisheries and aquaculture sectors and migrant workers in seafood processing factories. They are thus unlikely to be eligible for, or have access to, social protection benefits offered by some governments to handle the COVID-19 outbreak. Moreover, lockdowns and mobility restrictions may modify the dynamics and power relationships between men and women within fisherfolk households and communities. It is therefore recommended that special attention and support be given to women and children who are particularly vulnerable to sexual exploitation and abuse in times of crises. As was also observed during the 2013–2016 Ebola outbreak in West Africa, a surge of domestic and family violence cases has already been observed in Australia, China, Indonesia, Italy, Malaysia, Singapore, the United Kingdom of Great Britain and Northern Ireland, and the United States of America as a result of confinement measures (FAO, 2020a).

FAO's Response to COVID-19 Pandemic

FAO's first objective is to ensure food security and nutrition for all. In response to the COVID-19 pandemic, the organization has led an unprecedented response that includes: dedicated COVID-19 pages on the FAO website, with targeted analysis and solutions across food value chains, sectoral and cross-sectoral policy briefs, advice on planting and harvesting plan. FAO also hosts weekly COVID-19 planning meetings with regional and sub-regional offices, and holds regular meetings with Members to update them on the state of and responses to the pandemic. In addition, the FAO Director-General continues to brief national leaders and decision-makers as well as the broader community through interventions within international fora such as the G20, the World Economic Forum, and the United Nations Economic and Social Council (ECOSOC), and by participating in various international exchanges and through bilateral meetings with Members (FAO, 2020).

In the context of fisheries and aquaculture, the response from FAO has been primarily focused on supporting, restarting and strengthening the sector's supply chains and livelihoods, while focusing on the most vulnerable groups and regions. To facilitate this work, the Fisheries and Aquaculture Department established a dedicated COVID-19 Task Force to coordinate departmental initiatives in response to the pandemic and provide coordinated support to measures and interventions addressing the impact of COVID-19 on fisheries and aquaculture. These include:

- Working with Members, industry and civil society representatives, and other stakeholders to monitor the situation and provide policy, management and technical advice, as well as technical support to innovate and adapt practices along the supply chain.
- Coordinating information and responses with international and regional partners, such as regional fishery bodies (FAO, 2020c), intergovernmental economic organizations, research centres and civil Society organizations.
- Continuing to improve its understanding of the virus, to assess any potential risks to global, regional and national food systems – as new information and knowledge become increasingly soundly based (on, for example, international standards, expert opinion, peer-reviewed studies) – and to mobilize resources for coordinated COVID-19 mitigation measures.
- Working with financial institutions and donors to develop comprehensive and coordinated intervention packages to address the most urgent priorities to reactivate supply chains.

Initial Assessment of an Impact of COVID-19 on Global Fisheries & Aquaculture

Regional fisheries management organizations (RFMOs) and regional fisheries advisory bodies (RFABs), collectively referred to as regional fishery bodies (RFBs), have an important role in contributing to fisheries management and scientific research of many important fisheries around the globe. In seeking to better understand the impact of COVID-19 on the functioning of RFBs, questionnaires were sent from FAO to the secretariats of both RFMOs and RFABs. Their responses provide an overview of the current known impacts to the work of RFBs and guidance as to the impact of fisheries products supply and employment, and may provide guidance on possible mitigation actions and measures to be considered. In order to understand the full impact of COVID-19 on fisheries and aquaculture, further assessments both at regional and country level will be necessary (FAO, 2020c).

Several RFMOs and the majority of RFABs concerned with capture fisheries expect that the impact of COVID-19 will have negative consequences on the management of fisheries. The vast majority of RFMOs and RFABs with an MCS role expect that the impact of COVID-19 will have negative consequences on the MCS of fishing activities and the fight against IUU fishing. RFMOs reported impacts of COVID-19 on regular operations while the RFABs mostly reported on issues in relation to markets, migration, restrictions in production and economic crisis in general. One positive impact – the reduction of carbon footprint – was reported. RFABs reported that in capture fisheries, employment in the harvest sector will be most affected whilst in aquaculture it was believed that employment within both the harvesting and post-harvest sectors will be equally significantly affected. With respect to the demand, supply and price of fisheries products, they reported that whilst it is expected that demand for exports in both capture fisheries and aquaculture sectors will be the worst affected by the impact of COVID-19, the demand in the domestic markets is also foreseen to be significantly impacted in both sectors. Therefore, supply to both domestic and import markets in both capture fisheries and aquaculture sectors is also envisaged to be negatively affected by the impact of COVID-19. The impact of COVID-19 on fish prices currently remains uncertain.

RFABs plan to undertake, COVID-19 impact mitigation measures related to teleworking, including development, provision and access to online training materials. RFABs are also considering measures oriented towards policy-making and promotion of fish consumption such as formalizing networks for robust timely analyses on impacts; development of targeted policy and management guidance, such as regional contingency plans; measures to facilitate transport of artisanal fisheries products; promotion of fishery products through printed communications and on social networks; promoting procedures to authorize the export of fishery and aquaculture products; encouraging the search for new fisheries alternatives; and promoting emergency food aid by governments based on fish products.

Global/Regional initiatives amid COVID-19

Food and Agriculture Organization (FAO) FAO has organised virtual dialogues, policy briefs, Q & A section, summary of Impacts of COVID-19 on fisheries and aquaculture as an addendum to SOFIA 2020 report, webinars, on-line conferences among many others.

International Collective in Support of Fishworkers (ICSF) As part of efforts to understand how countries are responding to the COVID-19 crisis, it created an interactive map that presents information gathered in relation to COVID-19 and small-scale fisheries around the world. The information is divided into three different categories: impacts of COVID-19, relief initiatives and community initiatives to donate to.

Friend of the Sea has launched a new audit format known as "Sustainable Augmented Reality Audits" (SARA) in COVID-19 crisis. SARA allows the qualified auditor to carry the onsite inspection from a control panel commanding remote "eyes" and recording a complete video of an audit, following a strict inspection procedure. The complete audio is then saved on blockchain to prevent any possible editing. This initiative has reduced the health concern of not only an auditor but also the other functions without much coming in contact and at the same time reduced the cost burden on companies by more than 70%.

The **Sustainable Fisheries Partnership** shares resources from different organizations and institutions concerning small-scale fisheries in the context of COVID-19.

World Forum of Fisher Peoples (WFFP) released a statement to highlight the importance of small-scale fisheries in achieving food security and nutrition around the world during the COVID-19 pandemic.

WorldFish gathered a comprehensive set of COVID-19-related resources, including news, articles, and publications to inform about the state of fish and aquatic food systems during the COVID-19 pandemic in Africa and Asia.

Blue Ventures gathers different types of resources such as technical guidance, outreach tools, and news to facilitate access to information on COVID-19 to small-scale fisheries actors.

Coalition for Fair Fisheries Agreements (CFFA) created a micro-blog to publish news on the impacts of COVID-19 upon African small-scale fisheries and the response from authorities and fishing communities.

ABALOBI from South Africa uses its 'ABALOBI Marketplace' app to facilitate the continuation of fishers' livelihoods by connecting fishers and consumers. ABALOBI also supports the creation of Community-Supported Fisheries, which provide fish to those in need because of the pandemic.

The South African Small-scale Fisheries Collective (SASSFC) in coordination with Department of Fisheries ensures that the regulations to control the spread of the virus have the minimum possible negative impact on fishermen, in relation to both harvest and post-harvest activities.

The **Fisheries Areas Network (FARNET)** collects initiatives in EU Member States that tackle the socio-economic challenges of COVID-19 faced by workers from the fisheries and aquaculture sectors.

EFFECT OF COVID-19 ON FISHERIES AND AQUACULTURE IN ASIA

Disruption of Capture Fisheries Livelihoods

In Asia during 2018, FAO estimated that 30.8 million people were engaged in the primary sector of marine and inland capture fisheries (FAO, 2018). Millions more were involved in secondary activities, such as post-harvest processing and marketing, in which women predominate. The pandemic has directly impacted almost all of these people. Many countries in Asia are among the top producers, exporters and importers of fish and fishery products. All countries in Asia have instituted a range of mitigation measures in varying degrees to stop the spread of COVID-19. However, in some countries, fishing is considered an essential activity and has been allowed to continue, as long as fishers and the public comply with the mitigation measures.

Stay-at-home orders, curfews, bans, and travel restrictions for national and international air and land travel have led to complete or partial closures of hotels, restaurants, catering services, local markets, as well as disruption of transportation and cold-chain systems. To enforce physical distancing measures, authorities have banned gatherings and festivities. Sudden and massive job losses have greatly reduced local demand, particularly for high-value foods. Reduced demand and closures of major global markets have stopped or greatly reduced international trade in fish products. On the other hand, some businesses support demand by stocking up their freezers, preparing for the eventual end of the pandemic when demand for canned fish products, particularly tuna, should increase. Countries that have kept their ports open continue to import whole frozen tuna from the Pacific and Indian Oceans to supply their canneries.

People working in the fisheries sector are plagued by fear and uncertainty. They fear catching the virus, and are uncertain when or if their livelihoods will return to normal. In general, small-scale fishers and women vendors have been the hardest hit. Although they can catch, eat and sell fish, they are highly dependent on daily income to buy basic necessities such as food, fuel and medicine. Reduced demand for fish has negatively impacted their income and health.

Many migrants working on vessels and in processing plants that have ceased operations have been unable to go back to their villages, or their countries. They remain stranded, often in cramped conditions, increasing their potential to become infected or spread the virus. Without work, they are more likely to fall into debt. Border closures also create gaps in domestic supplies of fish and fish products normally filled by cross-border trade. Land border closures (e.g. Cambodia–Vietnam and Singapore–Malaysia) have the potential to weaken food security. The closures also deprive fishers of markets for their catch (FAO, 2020).

Coping strategies and support to the sector

China launched the National Fish Demand and Supply Information Platform. Developed by the Chinese Aquatic Products Processing and Marketing Alliance (FAO, 2020b) under the guidance of the Ministry of Agriculture and Rural Affairs, the Alliance invited many leading companies and small-scale enterprises to register on the platform. By mobilizing resources, the platform has successfully helped thousands of fishers to sell their products.

In **Indonesia**, fishers targeting blue swimming crab for export have switched to other species and started selling them in the domestic market. Their alternative catch includes squid, shrimp and mixed finfish species (SFP, 2020).

In the **Philippines**, the Department of Agriculture introduced the Food Lane Conduct Pass to ensure the unimpeded supply and flow of food commodities, agriculture and fishery products, and inputs (Department of Agriculture, Philippines, 2020). Accredited holders of the Food Lane Vehicle Pass Card can travel through quarantine checkpoints. Moreover, the Bureau of Fisheries and Aquatic Resources initiated the "Seafood Kadiwa ni Ani at Kita on Wheels" (Datu, 2020). It is a rolling store that brings fresh fish products to communities affected by the lockdown. In addition, some local governments bought the catch of their small-scale fishers for inclusion in food packs and distributed to families affected by the enhanced community quarantine (Cabico, 2020).

Thailand (Manager, 2020; MCOT, 2020; ThaiPBS, 2020) has shut down its lucrative tourism sector. Local non-government organizations (NGOs) assist artisanal fishers, such as the Urak Lawoi group in Phuket, who cannot sell their fish, have no income from bringing tourists on their fishing boats, and no money to buy rice. The NGOs help them barter their dried fish products for rice from other vulnerable groups in the North and Northeast of Thailand. Government and donors have absorbed the logistics costs. The fishers also share the rice with their relatives and friends in other areas along the Andaman coast who do not produce processed fish. Villagers also store some of the rice in communities' rice buffer funds.

Impacts on Aquaculture

Impacts of Covid-19 on aquaculture businesses were immediate. Lockdowns subjected the value chain to severe domestic and international transport disruptions for production inputs, raw materials for processing, and finished products for domestic consumption and export. Strict enforcement of restrictions on the movement of materials and people, including workers, made farm inputs, such as feed and seed, unavailable. Small farmers suffered business losses because they could not sell their harvests, or were forced to sell at low prices. Many farmers cannot keep feeding their stock for too long without revenue as they must repay loans. There is also a significant effect on the value-chain actors, a large portion of whom are women, and their commodities that are strongly reliant on export and tourism. Segments that are labour intensive, specifically industrial-scale processing, are most affected by labour shortages.

In Bangladesh, fish farmers were unable to harvest and therefore, could not start a new production cycle. The result will be a decrease in fish supply in the coming months. It will also mean that upstream and downstream sources of employment will be lost or reduced (WorldFish, 2020). In Indonesia, hatcheries were unable to supply seed because they could not obtain specific pathogen-free broodstock from their sources abroad for lack of flights. In India, shrimp processing units have not been operating at optimal capacity due to an acute shortage of labour. The workers went home to their respective states and could not return because of locally imposed restrictions (Press Trust of India, 2020). In Indonesia, processing plants in cities struggled to obtain raw materials because of transport disruptions (Indonesian Traditional Fisherfolk Union, 2020). In the Philippines, logistics problems curtailed the fish supply. Consequently, the prices of two staples, tilapia and milkfish, have increased despite a price ceiling set by the government. The government later relaxed the restrictions to ease the movement of food supplies and essential services (FAO, 2020c).

In India and Thailand (Thai Department of Fisheries, 2020), shrimp exports have been most affected by reduced, delayed or cancelled orders from major markets such as China, the EU, Japan and the USA. The drop in demand forced processing plants to scale down, resulting in an oversupply of raw materials. Indonesia did not experience cancelled orders, but plants slowed down operations anyway because of a lack of workers. However, in some countries such as China, India and Thailand, processors continue to operate at reduced capacity with fewer workers per shift to comply with physical-distancing requirements.

The severity of impacts on commodities for domestic markets depends on local consumers' purchasing power and food substitution for farmed products. Unemployment has affected incomes and consumption patterns. People may find products such as eggs and canned fish more affordable than farmed fish.

Closures of offices and schools have also affected demand for fresh fish. The busy lifestyle of people in urban centers makes them rely on prepared meals. These meals often include fish products and served by small vendors and restaurants for take-away or on-site consumption. The shutdown of tourism and related businesses, such as hotels and restaurants also decreased demand, especially for high-value fresh or live fish. E-commerce has helped absorb the products to some extent. Meanwhile, underlying problems have exacerbated the pandemic's impacts on small farmers. These problems include the high cost of production, low farm-gate prices, diseases, persistent household debt, and natural hazards such as this year's drought in the Mekong Delta.

Recovery could be slowed down by constrained cashflow and low liquidity along the value chain and the likely debt burden incurred by all actors. Financial assistance to the sector will be much reduced because of severely strained public resources. Microfinance institutions also face liquidity problems, as their investors are wary that borrowers cannot pay back loans.

Ongoing coping measures and potential adaptation strategies

Responses by governments and the aquaculture industry include three outstanding measures for coping with and relieving the pandemic's impacts on the aquaculture value chain:

- Governments have allowed – while maintaining necessary precautions – the resumption of activities and services. This is to ensure uninterrupted production, input availability, and accessibility and access to market and facilitating services.
- Government and industry have been promoting local consumption of export products that have lost their markets, and have been developing online marketing platforms and assisting farmers in using them.

- Governments have been providing assistance to producers and processors/ exporters. This includes a produce price-stabilization scheme and enabling farmers to restock by facilitating access to farm inputs.

The pandemic can serve as a catalyst for the introduction, wider promotion, and faster adoption of plans and programmes to make the sector innovative, and socially and environmentally responsible. Responses to the pandemic can drive the resolution of the persistent social issues of gender inequity, social inequality, and lack of social protection for small farmers and workers. It also presents another opportunity: the integration of pandemic-related objectives and strategies into all existing work programmes and ongoing projects (FAO, 2020).

The country experiences suggest five key adaptation strategies. Integrating these strategies can achieve synergy and infuse the sector with greater resilience and adaptive capacity to this and other types of disasters that may occur alone or simultaneously.

- **Strengthen self-reliance**. While maintaining the export market, strengthen domestic markets by improving productivity, quality and safety standards. Diversify product forms and market channels and keep them affordable for local consumers.
- **Encourage multi-stakeholder investment in services to farmers**. Services rendered mainly by governments include research and development, extension, and even marketing, credit and insurance. With strained public resources, the private sector, NGOs and civil society organizations should now partner with the government to invest in, initiate and provide these services to hasten recovery and increase resilience.
- **Enhance human capital**. Equip small farmers and small processors, who are mostly women, with new knowledge and entrepreneurial skills. Give them the tools to anticipate and meet changing requirements for production and marketing and changes in consumer preferences.
- **Introduce social protection for all**. This includes a range of financial products such as health insurance bundled with mutual savings, and credit with crop insurance. Organizing producers can facilitate delivery.
- **Enhance social capital**. Promote the formation and professionalization of farmers' associations, women's associations, and youth associations. Link these to other actors along the value chain and with other associations to form a network. Frequent and close interaction with others through linkages and alliances promote the exchange of ideas and experiences that tend to generate innovation. Strong partnerships based on trust help small operators to better manage production and market risks.

Indian fishery Industry amid COVID-19

With a sudden lockdown of all fishing, fish trading, storage, refrigeration, and selling activities, the entire fishing industry that is the source of livelihood to nearly 16 million people faced the prospect of severe health risk, stranding, loss of income, and starvation. The unavailability of ice along with storage facilities being forcibly shut pushed fishers to take a measure they would not have otherwise resorted to i.e. dumping their catch back into the sea. The common image of what constitutes fishermen or fisherwomen is that of people who go out into the sea to catch fish. While they are crucial to the trade for obvious reasons, the fishing industry depends on multiple forms of personnel involved. This includes unloaders, migrant workers in storage facilities, boat suppliers, middlemen, market handlers, exporters, fish sellers, aquaculture farmers, and others involved in this complex trade. The abrupt lockdown put every single one of them in different stages of peril – some with regard to the future of the trade and many with regard to their very survival.

Amid this chaos, activists and NGOs have been working to provide fishers access to food, ration, safety, and healthcare. Dakshin Foundation, an NGO based in Bangalore, put together a task force to advocate for fishers' conditions and provide aid packages. Its efforts were also directed at resolving stranded cases across coastal India. It created a fundraiser to bring relief to the families of fishers in partnership with United Artists Association (UAA) in Odisha and SNEHA in Tamil Nadu. A community radio station "Kadal Osai" (Sound of the Ocean) for fishers in Tamil Nadu has been acting as a bridge between the government and fishers. The radio station began operations in 2016 and is said to be India's first community radio for fisherfolk. The station has at least 15000 active listeners and now it also provides information about the lockdown and measures to contain the spread of COVID-19 (Hemalatha, 2020). Officials from the fisheries department, police and panchayats (village councils) use the radio station to announce new protocols, physical distancing measures and other updates. The radio station invites local municipal staff, doctors, and police officers to respond to questions and concerns from fishers about the pandemic.

ICAR-Central Institute of Fisheries Technology (CIFT) prepared COVID-19 Advisories on fishing harbour, fishing boat, fishermen, fish transport, fish market and seafood processing plant while ICAR-Central Inland Fisheries Research Institute (CIFRI) released COVID-19 advisories for fishers of rivers, estuaries, creeks, reservoirs and wetlands.

Main Impacts of Covid-19 on fishing communities

Limited availability of cash reserves and/or rations in fishing hamlets: Fishing is the only source of daily income for many, and the sudden lockdown made it difficult for communities to provide for themselves and their families, leaving their women, children and elderly especially vulnerable. Unable to purchase medicines, emergency rations and basic supplies, and the inability to pay existing debts left people extremely vulnerable to indebtedness and open to exploitation by moneylenders, and open to abuse by unscrupulous merchants.

Stranded fishers: Tens of thousands of migrant fish workers were stranded in various parts of the country and abroad, with no money or means to get back home. Many were forced to remain on their boats or in temporary migrant camps in crowded quarters with limited access to food, medicine, water and hygiene. The present condition is highly counterproductive in the light of the COVID-19 spread, as they are unable to practice any physical distancing measures.

Sudden & unplanned halt in seafood supply: Even though exempt from lockdown restrictions, escalating fears amongst the public and enforcement bodies led to the closure of fisheries rather than creation of safeguards and measures to ensure productivity and seafood-based protein supply. Limited awareness of coronavirus coupled with lack of sanitation, soap supplies and access to personal protective equipment (PPE) created fear amongst some fishers to go fishing. Rising local demand and lack of fresh catch results in the sale of decayed fish that is highly hazardous to public health.

Declining profits from declining demand: The COVID-19 pandemic began affecting seafood exports from India as early as January, driving down prices well before the lockdown. The widespread myth that eating fish or meat would lead to catching COVID-19 also led to poor domestic sales of seafood leaving fishers with little money to tide through the current lockdown period and the monsoon ban period. Lack of seafood exports led to farm fishery products flooding the market, posing a direct threat to the marine fishing sector.

Mental health concerns: Livelihood and health related uncertainties raise anxiety levels and threaten the mental and physical well-being of these individuals and those involved in relief efforts (government, NGO and civil society members).

REBUILDING FISHING SECTOR AMID COVID-19

Public health & safety

The fishing community is currently plagued with misinformation on COVID-19. Government should support by sharing accurate information and providing Personal

Protective Equipment (PPE). Government endorsement is required to revive fisheries and consumer trust. National and state level fisheries authorities should create awareness of fish and other seafood products not being carriers of the coronavirus to help dispel the myth.

Continuing marine capture fisheries in the small-scale sector

State and district authorities should prioritize the re-opening of small-scale fisheries (SSF). SSF in territorial waters can help meet the food security needs of their own families and supply of seafood to hamlets and local coastal communities. While large-scale mechanised fishing operations may not be able to comply with COVID-safe norms without greater active engagement from the state, current lack of fish landings pose significant threat to the health and well-being of coastal communities accustomed to this protein source. Small-scale fishers, including traditional motorised and non-motorised boats, can easily comply with government's precautionary health and safety measures, including social distancing as they tend to involve fewer people (2-3 per boat), often land on beaches rather than crowded jetties, support lower volume of catch and consumer base. Government should actively assist the sector in setting up markets and supply chains to implement the same. With little support, small-scale fisheries can further reduce public interaction by conducting fish sales over the phone as well as by practicing door-to-door delivery. Women fish vendors can be empowered to utilize new modes of sales and distribution that maintain physical distance. Ice factories and refrigeration facilities across nation should also be opened up, albeit at lower capacities. Fish auctions controlled by state cooperative societies that help ensure fishers viable prices irrespective of market fluctuations in supply and demand are essential in reviving the fish economy.

Standard Operating Procedures (SOPs) in Fisheries and Aquaculture of India imposed by Central Department of Fisheries amid COVID-19

1. **Model Standard Operating Procedures (SOPs) related to movement of aquaculture farmers, fish workers; transport and marketing of fish, fish seed, feed and inputs for aquaculture activities; Fishing in larger inland water bodies; Fish Processing plants in COVID-19 Scenario**

   SOPs should be implemented through District Administration including District Fisheries Administration and need to be strictly followed by Fish farmers and farm workers/ Hatchery operators, and operating hands, Marginal fishers and

Fish marketing personnel (Wholesaler/ Retailer) and processing plant workers/ owners. The District Administration including Fisheries Department should coordinate with the local police, Farmer associations, hatchery associations and processors etc. Relaxations allowed for Inland fishery activities can be withdrawn at any time by the Government in case of non-compliance of SOPs or for such other reasons as may be decided by the Government of India.

The SOPs of Inland Fisheries may consist of 5 parts as follows (Annexure 10)

Annexure 10: SoPs of inland fisheries in COVID-19 Scenario

- Fish Production Units or Fish farm
- Fish seed Production Centre or Fish / Shrimp Hatchery
- Fish production from Inland Water bodies such as Reservoirs, Wetland, Rivers, Creeks and Estuaries
- Fish markets (Wholesale and Retail)
- Processing Plants

Sr.No	Particulars
1.	**Fish Production Units or Fish farm**
	✓ The Fish/shrimp farms registered or licensed by the States/Union Territories (UTs) will be allowed to carry out preparatory and Fish production activities like pond preparation, Seed segregation, Feeding, Water management, Pond management, Sampling, Fish harvesting etc. subject to the compliance of the Guidelines issued by the MHA and the SOPs issued by the concerned State/UT Government. ✓ For overall coordination and smooth implementation of farm operations during the COVID-19, the States/UTs may constitute a "Committee" headed by District Collector or his representative with relevant district officers including District Fisheries Officers as members. Representatives of local fish/shrimp farming associations and Progressive fish/shrimp farmers may be included in the Committee. ✓ The State/UTs Government shall establish a system for proper monitoring of the Fish/shrimp farms and their activities. A 'Nodal Office'/'Surveillance Booth' should be set up and nodal officer(s) be appointed at each Block/ Tehsil for monitoring of the farming activities and to address the related hindrance/ problems thereof. ✓ The contact number(s) of such Booth/ nodal officer(s) to remain functional at all times (24x 7) and displayed properly in the Block office / Tehsil etc. and made available to fish farmers. ✓ The State/UT Government and District Administration in consultation with the local Fish/shrimp farmer Associations/ Bodies shall plan and execute the fishing activities in such a manner, so as to maintain social distance during the farming activities. Further, if any fish farmer / worker/farm hands develop

Sr.No	Particulars
	symptoms like cough, headache, body ache, shortness of breath, nasal congestion, runny nose and sore throat etc., he should immediately communicate with the nodal officer(s) of the block/tehsil/district for further action. ✓ The owner of the farm shall ensure arrangement of all protective gears to avoid spread of Covid- 19 virus (Sanitizer, Soap etc.) for the workers/hands. The workers while entering in the farm should follow the sanitary norms and may change the dress they have used while travelling and should use the disinfected cloths while working inside the farm. ✓ Social distancing and proper hygiene practices shall be maintained during the activities such as pond preparation, Feeding, Fish/shrimp sampling, fish/shrimp catching or harvesting. Disinfectant spraying and cleanliness will be carried out at the pond sites/farm and to be continued as a regular practice till situation becomes normal. Visitors shall not be permitted inside the farm till the situation becomes normal. ✓ To avoid inconvenience to the farm owners, workers or farm hands, the local Fisheries Department and District Administration shall facilitate the issue of pass/permit for their travel to the farm from their place of residence. ✓ The State Government /UT Administration will set up a 'Facilitation Centre' in the Fisheries Department (for overall coordination work in the State/UT) and the details of this Centre to be conveyed to Disaster or Control Room Centre or Helpline Numbers of each district. ✓ The team deployed at 'Surveillance Booth' / nodal offices at Block/Tehsil level will monitor the movement of input loaded vehicles, farm implements, inorganic fertilizers etc for farm use so that the farmers or owners do not face any difficulty in procuring the inputs, materials and implements etc. ✓ Adequate facilities for Medical Check Up/Screening of fish farmers/workers shall be arranged by the local fishery offices in co-ordination with the Health Department of each District. Medical Screening Team should also make aware the fish farmers / farm hands/ workers about the symptoms related to COVID-19 and preventions (social distancing etc.) along with procedure to be followed related to it. ✓ Farm owner/ Farmers shall declare and share the list of the workers/ personnel to the nodal fishery officer(s) posted at the Surveillance Booth for maintaining the details in the register. Fisheries Department can devise their own format of recording details. ✓ District Administration should also earmark designated vehicles (in consultation with fish farmer Associations) exclusively for supply of farm inputs such as fish/shrimp seed, fish/shrimp feed, diesel, ice etc. and collection and transportation of Fish from the fish /shrimp farms. ✓ Disinfection of farm & implements such as nets / equipment for each pond has to be carried out before they are put into use. ✓ The Farm owner should procure seed, feed and other farm inputs locally as far as possible, and movement of workers should be preferably limited within the district only, to avoid any possible spread of COVID-19 infection.

Sr.No	Particulars
	✓ Fish farmers/ farm workers must provide contact details of their family members/relatives and AAROGYA SETU Mobile App to be downloaded on their phone (if they have compatible mobile device). ✓ Fish farmers/workers shall be subjected to thermal scanning at the entrance of farm / monitoring booth and they shall be allowed to enter into the farm after the thermal scanning and disinfection of their hands. ✓ The farmers/workers should be advised that no physical contacts or exchanges of items to be made by them with workers of adjacent farms, if any. Proper bio-security measures like crab/dog fencing, bird fencing, tyre-wash/vehicle wash, hand and leg dip should be provided in order to avoid cross contaminations. ✓ The fish transport vehicles should enter their proper movement details in the farm movement book and details may also be shared on daily basis with the nodal fishery officer(s). ✓ After completion of the work all the farm implements like nets, check trays, feeding buckets, scoop nets, water sampling bottles etc should be disinfected properly and sundried. ✓ The State/UT Fisheries Department/ Administration should ensure that the input shops like feed go-downs other farm inputs are kept open so that the Farmers or Farm owners don't face any difficulty in procuring the items for farm operations.
2.	**Fish seed Production Centre or Fish / Shrimp Hatchery**
	✓ The Fish / Shrimp hatchery registered or licensed by the States/Union Territories (UTs) will be allowed to carry out seed production activities, harvesting, packing and selling of the seed subject to the compliance of the Guidelines issued by the MHA and the SOPs issued by concerned State/UT Government. ✓ For overall coordination, and smooth implementation of hatchery operations during the COVID- 19, the States/UTs may constitute a "Committee" headed by District Collector or his representative with relevant district officers including District Fisheries Officers as members. Representatives of local fish/shrimp hatchery associations and Progressive fish/shrimp hatchery operators may be included in the Committee. ✓ The State/UT Government and District Administration in consultation with the local Fish/shrimp hatchery operators Associations/ Bodies shall plan and execute the seed production activities in such a manner, so as to maintain proper monitoring and social distancing. ✓ The owner of the hatchery shall ensure arrangement of all protective gears to avoid spread of Covid-19 virus (Sanitizer, Soap etc.) for the workers/hands. The worker while entering in the farm should ensure following necessary sanitary norms and change the dress they have used while travelling and should use the disinfected cloths while working inside the hatchery. If possible, every section worker should be kept in respective sections till the situation of COVID- 19 improves.

Sr.No	Particulars
	✓ Disinfectant spraying and cleanliness will be carried out at the hatchery before and after every cycle on regular basis. The SPF brood-stocks movement should also follow the CAA guidelines as in force. The seed once shown to the buyer should not be taken back to hatchery. Disinfection of Hatchery and related implements such as buckets, bins, nets / equipment for each tank has to be carried out before they are put into use. No visitors shall be permitted inside the hatchery till the situation becomes normal. ✓ The hatchery owner should procure seed, feed and other hatchery inputs locally as far as possible and movement of worker should be preferably limited within the district only. All the operators/workers shall be subjected to thermal scanning at the entrance of hatchery /monitoring booth and they shall be allowed to enter into the hatchery after the scanning. ✓ Subject to clearance from the health department, permission will be granted to the seed transport vehicles from the hatchery to the destination. The District Fishery Officer/ local administration should ensure that the live seed vehicle should not be detained enroute. The details of seed movement along with vehicle and staff should be maintained. ✓ After completion of one cycle all the hatchery complex and the implements should be disinfected and dried.
3.	**Fish production from Inland Water bodies such as Reservoirs, Wetland, Rivers, Creeks and Estuaries**
	✓ The fish harvesting from large water bodies (viz. reservoir, floodplain wetlands, Lakes, and Lagoons), rivers, creeks and estuaries can be carried out by using smaller country crafts, coracles with minimum number of fishers and by using cast nets, gill nets, shoot nets or by hook and lines only. ✓ As far as possible, the community fishing should be put on hold and only marginal fishers should be allowed to fish from the larger water bodies till the situation improves to normal. The boat and fishers will be allowed to carry out fishing have to return to the designated fish landing site only. They should not be allowed to land their fish at different landing centres of the lakes or reservoirs. The fishers should be advised that no physical contacts or exchanges of items to be made between fishing crafts or coracles while fishing. ✓ The boat owners/fishers should adhere to the guidelines issued by MoHFW, GoI to maintain social distance and also to maintain appropriate distance between the fishing crafts, coracles or logs. ✓ In case any fisher is reported sick and develop any symptoms like coughing, headache, fever, body ache, shortness of breath, nasal congestion, runny nose, and sore throat etc. while fishing in large water bodies, the fishing activity should be abandoned and the fishing crafts should return back to the designated landing site immediately. They should also not consume or touch their catch while returning from fishing. Such catch must be disposed of as per advise of health department and it must not be transferred from one fishing craft to others.

Sr.No	Particulars
	✓ The fishers should maintain social distance while on board, take all sanitary precautions and avoid contact with the fishers having symptoms. In no case they should come in contact with that symptomatic person. ✓ The smaller crafts, coracles or logs used for fishing in these water bodies along with nets should be disinfected before every fishing activity. The fishers should be screened at shore on daily basis as far as possible as they are exposed to adverse environments. ✓ The fishers should be advised to minimize the handling of catch onboard as well as at landing centre till the fishes are disposed off to the aggregator or trader.
4.	**Fish markets (Wholesale and Retail)**
	✓ Social distancing should be maintained in the fish/shrimp markets and cold storages. Dealers or workers of fish/shrimp collection sites/markets should adhere to preventive guidelines and social distancing as prescribed by MoHFW, GoI. ✓ All fish collection centers/markets (Whole sale or retail) are to be sanitized on daily basis as per the advice of the Health Authorities. ✓ Fish/shrimp landings shall be transported to the nearby markets, stalls, shops etc. through the sanitized vehicles for which necessary passes/ permission shall be issued by district authorities. ✓ The auctioning/selling of fish/shrimp may be carried out at the designated sites/areas as decided by the district authorities in consultation with the Fisheries Department /stakeholders and drawing of circles/line shall be made at appropriate distances, so as to maintain social distance among the bidders and sellers. ✓ The retail vending of fish shall not be permitted at the fish landing centers/ points/fish farms to avoid overcrowding. ✓ Suitable arrangements may be made by the district administration in consultation with Fisheries Department, Fishermen Cooperative Society/ Associations/bulk purchasers to sell the fish consumed by the locals in such a manner so that to maintain social distancing and hygienic conditions. In the case of door to door delivery of fish, appropriate steps may be taken for maintaining social distancing. ✓ As far as possible minimum handling of fish/shrimp has to be carried out and the persons engaged in catching, marketing, selling of fish should adhere proper sanitation/hygienic methods as prescribed from time to time by the local food safety department. ✓ The fish storage boxes, balance or the buckets or other handling implements should be thoroughly sanitized after every use. ✓ The fish vendors / retailer should restrict their movement from one area to other distant areas so as to avoid contamination and unhygienic conditions. ✓ The retail shops fish sale platform should be disinfected with Sodium Hypochlorite solution on daily basis.

Sr.No	Particulars
5.	**Processing Plants** ✓ Social distancing should be maintained in the processing plants while processing the fish/shrimp. The workers should also adhere to preventive guidelines and social distancing as prescribed by MoHFW, GoI. ✓ All collection trays or vehicle to be sanitized on daily basis as per the advice of the Health Authorities. ✓ The designated transport vehicles should be given permission by the Distt. Fishery Deptt./ District Administration. ✓ The Processing Plants should also follow the prescribed standard guidelines on food safety. ✓ Suitable arrangements may be made by the district administration in consultation with Fisheries Department, MPEDA to facilitate exports of the processed product. ✓ As far as possible minimum handling of fish/shrimp has to be carried out and the persons engaged in catching, marketing, selling of fish should adhere proper sanitation methods as prescribed from time to time by the food safety department. ✓ The processing activities have to be planned and executed in such a manner that if any worker develops symptoms like cough, headache, body ache, shortness of breath, nasal congestion, runny nose and sore throat etc., should not be allowed to the plant at any cost. ✓ On daily basis, the workers should be checked or screened thoroughly before they enter the plant. The plant workers or supervisors should restrict their movement from plant area to other areas so as to avoid contamination and unhygienic conditions. ✓ The processing plant should be disinfected regularly with Sodium Hypochlorite solution. The above list is not exhaustive and the District Administration in consultation with Fisheries Department and stakeholders may include such other appropriate measures to ensure compliance with the advisories issued by the Ministry of Health and Family Welfare, and Ministry of Home Affairs, Government of India.

2. Model Standard Operating Procedures (SOPs) related to movement of Fisherman and Fishing Boats in COVID-19 Scenario

These SOPs should be implemented through District Administration including District Fisheries Administration and need to be strictly followed by all fishermen/ owners and operators of fishing vessels, be it motorized, mechanized or non-motorized vessels/boats. The District Administration including Fisheries Department will coordinate with the Coast Guard and Coastal Security personnel for any help, including helping any distress call by fisherman or crew member while fishing in the sea. Relaxations allowed for marine fishing activities can be withdrawn at any time by the Government in case of non-compliance of SOPs or for such other reasons as may be decided by the Government of India.

The SOPs are categorized into four : (Annexure11)

Annexure 11: SOPs related to movement of Fisherman and Fishing Boats in COVID-19 Scenario

A. Before Departure of Fishing Vessels/Boats

B. During Fishing at Sea

C. Arrival of Vessels/Boats at Jetty/Shore/Fish Landing Centre

D. Procedure regarding Post Harvest and Transportation of Fish Catch

Sr.No	Particulars
1.	**Before Departure of Fishing Vessels/Boats**
	✓ The fishing boats/vessels authorized and permitted by the coastal States/ Union Territories (UTs) will be allowed to do fishing in the Territorial waters and the Exclusive Economic Zone (EEZ) subject to the compliance of uniform seasonal fishing ban period. The operation of such fishing boats/vessels will be subject to compliance of the Guidelines issued by the MHA and the SOPs issued by concerned State/UT Government. ✓ For overall coordination, oversight and smooth implementation of marine fishing operations during the COVID-19, the coastal States/UTs may constitute a "Committee" headed by District Collector with relevant district officers including District Fisheries Heads as members. Representatives of local fishermen associations and fishing vessels' owners/operators associations may be included in the Committee. ✓ The State/UT Government will establish a system for surveillance in the Fishing Harbors/fish landing sites. A 'Surveillance Booth' should be set up and nodal officer(s) be appointed at each Fishing Harbour (FH)/Fish Landing sites. ✓ The fishing vessels/boats with necessary permission will be allowed to sail from a designated location and return after completion of voyage to the same location. ✓ The contact number(s) of such Booth/ nodal officer(s) to remain functional at all times (24x 7)and displayed properly in the FH/FLCs etc. and made available to each operator(s) of fishing boats/vessels, which are permitted to fish. ✓ The State/UT Government and District Administration in consultation with the local Fishermen Associations/ Bodies will plan and execute the fishing voyages in such a manner, so as to maintain proper monitoring, control and surveillance including connectivity with the fishing boats. Further, the voyages have to be planned and executed in such a manner that if any crew member of a fishing boat/vessel reports symptoms like cough, headache, body ache, shortness of breath, nasal congestion, runny nose and sore throat etc., he should immediately be able to communicate with the nodal officer(s) of the district of departure and such fishing boat/vessel shall return to the designated fishing harbour/fish landing site within shortest possible time.

Sr.No	Particulars
	✓ The owner/operator of the boats/vessels shall ensure arrangement of all protective gears to avoid spread of Covid-19 virus (Sanitizer, Soap etc.) for the crew members. ✓ The crew members while venturing to sea for fishing activities to be advised to wear hand gloves, mask and ensure all precautions. ✓ The Owner/operator of the boat/vessel shall also ensure sanitization of the boat/vessel before and after each voyage. ✓ Social distancing and proper hygiene practices shall be maintained while loading of ice, water, ration, diesel and unloading of fish catch at the fishing harbours/fish landing sites. These operations shall be carried out with minimum operational staff. ✓ Disinfectant spraying and cleanliness has to be carried out at all jetties/ landing centres and has to be continued as a regular practice till situation becomes normal. ✓ Entry into Jetty/Fish landing sites should be restricted only for fishermen and crew members of the permitted vessels/boats (i.e. only for fishing related activities). ✓ To avoid inconvenience to the fishermen, the local Fisheries Department shall facilitate the issue of permit for the fishermen venturing for fishing on submission of their details (e.g., Name of Fishers, Registration No. of vessels, Fishers ID No./Aadhaar No., Date & duration of Voyage etc.) at least 24 hrs in advance. ✓ The State Government /UT Administration will set up a 'Facilitation Centre' in the Fisheries Department (for overall coordination work in the State/UT) and the details of this Centre to be conveyed to Disaster or Control Room Centre or Helpline Numbers of each district. ✓ The team deployed at 'Surveillance Booth' will monitor the movement of fishing boats/ vessels including upkeep and maintenance of boat/vessel movement books /records. ✓ Medical Check Up/Screening Camps to be setup in advance at FH/FLCs in co-ordination with Health Department of each District. Health Department should provide a separate team of doctors/medical practitioners/attendants so that somebody is always available for medical screening of fisherman and crew members as per their movement timings. Medical Screening Team should also make fisherman & crew members aware about the symptoms related to COVID-19 and preventions (social distancing etc.) along with procedure to be followed related to it. ✓ Boat operators and owners shall declare the list the crew members to the personnel posted at the Surveillance Booth /Fishing Department and are also directed to fill all details in the movement register including the date of sailing for fishing. Department can devise their own format of recording details. ✓ Disinfection of Fishing Boat & equipment for each vessel/boat has to be carried out before their departure.

Sr.No	Particulars
	✓ District Administration should also earmark some vehicles (in consultation with Fishermen Association) exclusively for supply of diesel, ice etc. and collection and transportation of Fish catch after arrival on fishing harbor / fish landing centre. ✓ The information of crew is to be given to Medical Screening Team by Fisheries Dept. ✓ The crew members should be fishermen, who are currently residing/present within that district only. Movement of crew should be preferably limited within the district only, since there can be chances of spread if crew is moving from any COVID-19 Hotspot or COVID – 19 Positive Case District or Locality. ✓ The boat operators and crew members allowed to venture out into sea for fishing activities should carry a valid document for identification purposes and the same should be ensured by local administration. ✓ All crew members must provide contact details of their family members/ relatives and AAROGYA SETU Mobile App to be downloaded on their phone (if they have compatible mobile device). ✓ All the fishing boat crew members will be subjected to thermal scanning at the monitoring booth and they will be allowed to embark into the boat only after the scanning. ✓ Sufficient provisions to be made for frequent hand washing and supply of soap / hand wash, Face Masks and sanitizers on board for all the crew members. Department will ensure that all the boat owners have complied these instructions. Subject to clearance from the health department, permission will be granted to depart for fishing. The details should be entered in the boat movement book and also in the records maintained at surveillance booth.
2.	**During Fishing at Sea**
	✓ The boat and crew members will be allowed to carry out fishing have to return to the designated fishing harbor/fish landing site within the stipulated date and time as permitted by the designated district authority and informed to Surveillance Booth/Fisheries Department. ✓ The fishing crew members to be advised that no physical contacts or exchanges of items to be made by them with crew members of other boats/ vessels while fishing out at sea. ✓ To track the location, monitoring and surveillance of fishing vessels and for maintaining connectivity with fishing boats/vessels, the concerned State/ UT Government and District Administration should put in place suitable system and extend help, if any in distress. Close coordination with Coastguard and Coastal Security Personnel shall be maintained in this regard by the concerned State/UT and District Administration. ✓ When any crew member is reported sick and develop any symptoms like coughing, headache, fever, body ache, shortness of breath, nasal congestion, runny nose, and sore throat etc. while fishing in sea, the fishing activity

Sr.No	Particulars
	should be abandoned and the fishing boat/vessel should return back to the designated jetty/landing site immediately. They should also not consume or touch their catch while returning from sea. Also, the catch must not be discharged in the sea. ✓ The crew members to maintain social distance while on board, take all sanitary precautions and avoid contact with the fellow crew members having symptoms. In no case they should come in contact with that symptomatic person.
3.	**Arrival of Vessels/Boats at Fishing Harbour/Fish Landing sites**
	✓ Fishing boats/vessels shall be allowed to return to the designated Fishing Harbour/Fish Landing site in a staggered manner. ✓ The operations and movements of persons for unloading of fish catch and transportation shall be planned properly and carried out with bare minimum operational staff. ✓ The boat and crew members should not land in any of the fishing harbor/ landing site other than the designated fishing harbor/fish landing site from where they were permitted to depart for fishing. ✓ The owner of the fishing vessel shall also ensure that the fishing vessel shall not enter any other ports/landing site. ✓ The boat should be anchored at an anchorage point near the jetty and kept in isolation. The crew members should contact the 'surveillance booth'/health authorities. After thorough checking and thermal scanner by the health authorities, the crew members will be allowed to disembark and the advice of health authorities and district authorities to be followed. ✓ The Owner of the boat/vessel shall ensure sanitization of the boat after each voyage. Fishing Boat & equipment should be sanitized as per the advice of the Health Authorities and shall be carried out immediately on arrival. Adequate care should be taken to avoid any contamination of the fish catch. ✓ On arrival, all the crew members will be scanned using hand held thermal scanner and thereafter, they will be allowed to move out the landing area / jetty. ✓ In case a crew member shows symptoms of COVID-19, all of the crew members have to undergo mandatory quarantine as per procedure. ✓ The fish catch received from the boat/vessel will be destroyed as per Bio-medical waste disposal norms prescribed by Ministry of Health and Family Welfare, GoI.
4.	**Procedure regarding Post Harvest and Transportation of Fish Catch**
	✓ Social distancing should be maintained in the landing points, fish markets and cold storages. Dealers or workers of fish collection sites/markets should also adhere to preventive guidelines and social distancing as prescribed by MoHFW. ✓ All fish collection centre/markets to be sanitized on daily basis as per the advice of the Health Authorities.

Sr.No	Particulars
	✓ Fish landings shall be transported to the nearby markets, stalls, shops etc. through the vehicles for which necessary Identification Cards shall be issued by district authorities. ✓ The auctioning of fish may be carried out systematically at the designated sites / areas as decided by the district authorities in consultation with the Fisheries Department /stakeholders and drawing of circles/line shall be made at appropriate distances, so as to maintain social distance among the bidders and sellers. ✓ The retail vending of fish shall not be permitted at the fish landing centers/ points to avoid overcrowding. ✓ Suitable arrangements may be made by the district administration in consultation with Fisheries Department, Fishermen Cooperative Society/ Associations/bulk purchasers to sell the fish consumed by the locals in such a manner so that to maintain social distancing and hygienic conditions. ✓ Appropriate steps may be taken for door to door delivery of fish to people by maintaining social distancing. The fish seller shall use the hand gloves and masks while selling the fish. The above list is not exhaustive and the District Administration in consultation with Fisheries Department and stakeholders may include such other appropriate measures to ensure compliance with the advisories issued by the Ministry of Health and Family Welfare, and Ministry of Home Affairs, Government of India.

Conclusion

COVID-19 has badly impacted global fisheries and aquaculture sector that is likely to affect growth of the sector in the coming years. The positive outcome might be the reduction in carbon emission across the countries and at the global level. The consumers preference may undergo changes post-covid probably towards sustainable and traceability-oriented seafood. Their preferences for home deliveries may offer opportunities in that segment (e-commerce platform). Standard Operating Procedures (SOPs) in Fisheries and Aquaculture of India have been imposed by the Central Department of Fisheries amid COVID-19. The SOPs of Inland Fisheries have been categorised into five: Fish Production Units or Fish farm; Fish seed Production Centre or Fish / Shrimp Hatchery; Fish production from Inland Water bodies such as Reservoirs, Wetland, Rivers, Creeks and Estuaries; Fish markets (Wholesale and Retail); and Processing plants. Model Standard Operating Procedures (SOPs) related to movement of Fisherman and Fishing Boats in COVID-19 Scenario have been categorised into four: Before Departure of Fishing Vessels/Boats; During Fishing at Sea; Arrival of Vessels/Boats at Fishing Harbour/ Fish Landing sites; and Procedure regarding Post Harvest and Transportation of Fish Catch.

CHAPTER 13

WAY FORWARD

Capture Fisheries

Capture fisheries play a significant role in meeting the nutritional requirements of the population and achieving the Millennium Development Goals. Marine fisheries around the world remain seriously threatened by overfishing, overcapacity and range of environmental problems. There has been global concern on the increasing fishing capacity in fisheries in different regions of the world. Cap on introduction of new fishing craft & gear in coastal sector, and diversification of excess fleet into deep-sea sector would reduce the fishing pressure in the coastal region and help in equitable distribution of fishing capacity in marine sector. Guidelines of CCRF caution against over-fishing and excess fishing capacity. For this purpose, states should devise and supplement measures to ensure that fishing effort is commensurate with the productive capacity of the fishery resources and their utilization.

In developed countries, sustainability (overall and for particular species) is maintained through catch quota or TAC (Total Allowable Catch) and the same principle is made easier by implementation of Individual Transferable Quotas (ITQs). This management approach is based on various elements such as economic and financial information; and integration of available data and the biological processes which determines the productivity of an exploited fish stock are usually not considered. As it is multi-species multi-gear fisheries in India, the relevance and feasibility of adopting such management system for India requires further studies. Hence, a combination of input controls and technical interventions is a more logical basis for management. Impose input controls or fishing effort management: Input controls are restrictions put on the intensity of use of gear that fishers use to catch fish. These refer to restrictions on the number and size of fishing vessels (fishing capacity controls), the amount of time fishing vessels is allowed to fish (vessel usage controls) or the product of capacity and usage (fishing effort control).

The measures to limit inputs require some form of restrictive licensing which will limit the total number of vessels engaged in a particular fishery together with their fishing power. Thus, if despite licensing, the fleet is too large for the particular fishery then it will be necessary to reduce its capacity. This can be achieved by removing vessels from the fleet; by reducing the amount of time fished by making all vessels fish for shorter periods; by limiting the amount or size of gear that a vessel can carry; or by reducing the efficiency of fishing effort. Monitoring and enforcement services may be combined since they need similar access to the fishery, and monitoring data provide the essential information on non-compliance with fishing effort.

Exploitation of juvenile fish results in considerable economic loss and also causes serious damage to the fish stock in terms of long-term sustainability of the resources. A minimum legal size (MLS) is seen as a fisheries management tool with the ability to protect juvenile fish, maintain spawning stocks and control the sizes of fish caught. Implementation of mesh size regulation and use of bycatch reduction devices is very crucial in reducing the bycatch and small fishes. Optimum mesh size reduces catch of small fishes and bycatch. Juvenile excluder devices should be installed in trawl to reduce juvenile catch. Awareness should be created among fisher community with regard to use of appropriate cod end mesh size as stipulated in MFRA to reduce bycatch.

Data on distribution of juveniles based on experimental fishing should be generated using GIS as mapping of juvenile-abundant areas would help to suggest restrictions on fishing grounds and fishing seasons to avoid catch of juveniles. Spatial and seasonal restrictions on fishing effort assist in avoiding biological overfishing that would be useful in the sustainable management of fishery resources.

Implementation of management plans with active stakeholder participation (co-management) & creation of awareness of CCRF principles at grass-root levels by active involvement of local fishermen communities, NGOs, IGOs and regional and international organizations is extremely important. Relevant training programs have to be organized for creating awareness about the conservation and management of fishery resources, environmental aspects, responsible fishing operations, hygienic fish handling and preservation and processing of fish among the fishers and allied segments.

Considering the need for a unified national fishery policy and the fact that the draft Marine Fishery regulation and management bill is already in the public domain, an Act is required to be put in place without too much delay. As there is no Act yet in place in India to regulate Indian Fishing Vessels or manage fishery resources beyond territorial waters up to the boundary of EEZ of India, and fisheries within the territorial waters are under tremendous pressure, it is essential to finalize the deep-sea fishing policy without further delay.

Monitoring, control and surveillance (MCS) system is one of the key components of fisheries management. It constitutes a powerful tool that can significantly contribute towards preventing, deterring and eliminating illegal, unregulated and unreported (IUU) fishing, ensuring sea safety and sustainable harvest of marine resources. It is recommended that the concerned state governments and the central government formulate a policy to establish the required system and also create awareness of the benefits of MCS among the fisher communities.

Considering the fact that only slightly over 50% of potential inland fisheries resources are currently harnessed, there is huge scope for increasing inland fish production and productivity from untapped inland fisheries resources. Inland and marine waters of India possess high diversity of ornamental fishes with over 195 indigenous varieties from northeast and western Ghats; and nearly 400 species from marine ecosystem. Indian ornamental fish trade consists of about 90% freshwater fish, most of which are cultured. Remaining 10% are marine ornamental fish, of which majority of varieties are captured. Strategies should be developed to enhance inland fish production from inland states/UTs, particularly less tapped resources from north-eastern states, cold-water fisheries in the Himalayan states/ UTs through region-specific management plans and technical measures.

Aquaculture

Aquaculture is one of the most promising food sectors of India. However, there are large resources, potential which is yet to be tapped. There is also a need to adopt new approaches of fish farming in addition to the existing ones. The Good Aquaculture Practices (GAqP) typically address environmental impacts, food safety, animal welfare and social aspects of aquaculture operations. Aquaculture in India is largely dependent on IMC and shrimps. However, national institutes have developed various technologies for culture of variety of fishes like pangasius, tilapia, cobia, Asian seabass, pearl spot, milk fish, pompano, groupers etc. Among these, pangasius and tilapia culture practices are picking up speed in different parts of the country and encouragement needs to be given for rest of the species. Farming filter feeders, such as bivalves, can have a positive impact on marine pollution and is another good option for further development.

Integrated Multi-Trophic Aquaculture (IMTA) is the farming of selected species in a way that allows uneaten feed, waste, nutrients and by-products of one species to be recaptured and converted into feed and energy for other crops and/or marine animals farmed in the system. This is one of the new systems developing gradually across the globe. It is possible to farm multiple complementary species within the same production area that allows farm operators to follow a more ecologically sound approach with better resilience to risks like disease outbreaks and climate change.

Offshore aquaculture systems in open sea areas cover less-valuable habitats and are less fragile than coastal zones, resulting in greater carrying capacities for large-scale aquaculture operations. Though it is one of the new options for aquaculture, it is necessary to study thoroughly both advantages and disadvantages of the system before actually starting operations. Cluster approach could be used for marine cage culture.

When businesses are unable to implement IMTA or offshore aquaculture, cage farming operations can significantly reduce effluent discharge and disease risk by employing bag nets instead of open nets.

Aquamimicry is the most advanced technology in the shrimp farming industry that provides natural live diets "Copepod" for post larvae prior to stocking. Aquamimicry is the intersection of aquatic biology and technology (synbiotics) mimicking the nature of aquatic ecosystems to create living organisms for the well-being of aquatic animals. It is a relatively new technique and is spreading fast especially in Asia. Aquaculture entrepreneurs in India have also started adopting this technique for shrimp aquaculture. However, it is in the initial stage and efforts in the right direction will help setting the trend.

Biofloc technology is an emerging method of preventing early mortality syndrome (EMS) infection. These systems convert toxic chemicals such as nitrate and ammonia into useful products such as fish feed. Studies show that shrimp infected with the EMS causing virus are more likely to survive in biofloc systems than clear seawater systems. Several experiments are being conducted currently in the country by different set of researchers and commercialization of this technology is possible in the coming times.

Recirculation Aquaculture Systems (RAS) is essentially a technology for farming fish or other aquatic organisms by reusing the water in the production. The technology is based on the use of mechanical and biological filters, and the method can in principle be used for any species grown in aquaculture such as fish, shrimps, clams, etc.

Digitalisation can boost production, curb diseases and enhance profitability. Facial recognition technology is one of the very new innovative techniques few companies practise for their aquaculture business. This generates predictive analytics to help operators make better decisions. One of the best examples is Cargill & Umitron.

There are other new techniques developed like automated video surveillance which is being used to determine when and how much to feed caged fishes & self-guided laser systems that kill the parasites e.g. sea lice without harming fish inside cage systems. These laser systems are placed in the centre of fish cages. When a camera detects the parasites, the laser targets the parasite without harming the caged fish.

To eliminate cases of food fraud, blockchain technology is one of the best solutions which is being used to improve the traceability and transparency of seafood supply chains. This technology is gaining importance in the aquaculture industry. Eruvaka Technologies, firm based at Vijayawada in Andhra Pradesh, India is one of the best examples for the use of IoT in aquaculture industry. The firm develops on-farm diagnostic equipment for aquaculture farmers to reduce their risk and increase productivity. They integrate sensors, mobile connectivity and decision tools for affordable aquaculture monitoring and automation. More such companies in the field will help develop the sector faster. Aquaconnect is another best example of using Artificial Intelligence in aquaculture. The firm was founded in 2017 as a full-stack aquaculture technology venture to offer data-driven farm advisory solutions and market place solutions to shrimp and fish farmers. 'FarmMOJO' is AI driven decision support dashboard that helps in understanding Farmer relationship management and potential sales discovery in real-time. The company claims FarmMOJO's data intelligence improves farm productivity and connects farmers with aquaculture industry stakeholders under its market place. Aquaconnect works with about 3000 farmers in Indian states of Andhra, Tamilnadu, and Gujarat to improve their farm revenues through technology intervention.

National Inland Fisheries and Aquaculture Policy, 2019 has been drafted and the government is in the process of initiating further steps for notification of the National Inland Fisheries & Aquaculture Policy (NIFAP). This too needs to be taken up at the earliest.

Realizing that the additional seafood requirement of the country in future years cannot be met by capture fisheries and inland aquaculture alone, though the draft National Mariculture Policy (NMP), 2019 is ready, an Act is yet to be put in place. This step would ensure sustainable farmed seafood production for the benefit of food and nutritional security of the nation and to provide additional livelihood and entrepreneurial opportunities to the coastal communities for better living.

The feed sector is actively developing alternative protein sources that are more efficient and environmentally friendly. Developing alternative feed ingredients that provide nutrition comparable to fish meal or soy but at lower environmental and financial costs is a priority and getting success will boost the development of aquaculture sector without impacting environment. Inroads should be made in the use of algae, particularly as a replacement for fish oil and high-grade hatchery feeds. There are pioneering ways to produce omega-3 fatty acids from heterotrophic algae, which require no light and can be produced efficiently and in greater quantities than light-dependent algae. More of such algal applications for using as a feed will help to reduce the burden on using the fish oil/meal. While still in the early stages of development, several companies are in the process of innovating to produce single-cell proteins for aquafeed. Single-cell proteins are proteins from the cells of microorganisms such as yeast, fungi, bacteria and algae.

Innovative concepts/approaches

In addition to the technologies in the aquaculture sector, new plant-based products are being developed that imitate seafood. As in the meat and dairy sectors, alternatives to seafood are being launched and cell-cultured seafood is in the process of development. These innovations could disrupt the protein industry by providing attractive, viable products through significantly less resource-intensive production methods. The production of plant-based and cultivated seafood is not limited by considerations like wild population productivity or geographical restrictions. Notable emerging plant-based brands and companies include Good Catch, Terramino Foods, and Ocean Hugger, all of which were formed since 2016. Good Catch has developed flaked fish products such as tuna, crab cakes, and fish burgers. India is in the initial stages of development of this sector; there are examples of few entrepreneurs however those are yet experimenting it in market and down the line at least 4-5 years may be required for this sector to start shaping up.

Counter measures for COVID-19

COVID-19 has badly impacted global fisheries and aquaculture sector that is likely to affect growth of the sector in the coming years. In general, it is being viewed that seafood production was lower than normal which disrupted the trade, strong decline in seafood prices at many places, low demand for luxurious seafood products and supply diversification. On the positive side, good growth was observed in retail sales through e-commerce/online deliveries. Model Standard Operating Procedures (SOPs) in Fisheries and Aquaculture of India have been imposed by the Central Department of Fisheries, GoI amid COVID-19 for undertaking various fisheries and aquaculture related activities as counter measures.

Sustainability and Development

The Pradhan Mantri Matsya Sampada Yojana (PMMSY) is a revolutionary scheme of Government of India that aims to bring about blue revolution in India through sustainable and responsible fisheries development. The scheme was launched with total allotment of Rs 20050 Crores for five-year period 2020-21 to 2024-25. Other important schemes are Fisheries & Aquaculture Infrastructure Development Fund (Rs 7522 Crores). The PMMSY scheme has set massive targets to achieve by 2024-25 through strategies such as enhancing fish production of 220 lakh tonnes, with 9% average annual growth of fish production, average aquaculture productivity of 05 tonnes/ha, fisheries exports of Rs 1,00,000 Crores, restricting the post-harvest losses to 10% and generating about 55 lakhs employment.

Doubling of exports by 2025 is one of the top priorities. Key organizations to play lead role in this endeavour are the Department of Fisheries & MPEDA. Some of the strategies towards this direction include aggressive marketing & brand promotion, market access, 100% adoption of traceability-certification-standards, species diversification, expansion of culturable areas under inland saline & brackish waters and technology infusion to improve productivity. The focus of PMMSY for developing domestic trade directs towards encouraging youth to become entrepreneurs; support for development of integrated cold storage cum retail chain; buy-back system from Fish Farming Producer Organizations (FFPO); live-fish marketing; and branding of Himalayan trout & promotion of domestic consumption of fish.

The Govt of India through PMMSY scheme has given high priority for traceability, certification and standards. It emphasizes on 100% compulsory accreditation of hatcheries, international certification & brand development, incentivize farmers for traceability, promoting sustainability through ecolabels which support minimal impact on environment, responsible sourcing, TED and marine mammal protection. FAO's recent flagship report "State of World Fisheries and Aquaculture" (SOFIA) 2020 highlights the importance of "Sustainability in Action". Possibilities of collaboration with international sustainable certification schemes in fisheries and aquaculture could be explored.

The present market size of Rs 500 crores of ornamental fisheries sector in India is expected to reach target of Rs 1500 crores by 2024-25 with the support of schemes under PMMSY. The new concept of integrated aquaparks has been launched with themes like dedicated parks for seaweed, shrimp, ornamental fishes, marine tuna, tilapia and pangasius.

'Fisheries' is a state subject and at the operational level, is a concurrent subject. All stake holders need to interact with many state and central ministries for any new developmental activity. Creating a 'Single-Window System' for all stake holders in the sector will certainly help in getting necessary information and preliminary clearances from one source. This platform will ensure proper coordination and create common understanding within inter- ministerial personnel (States and Centre) and facilitate faster growth of value chain not only at the domestic level but also boost exports.

Entering into joint venture (JV) for utilizing potential fish resources from the unutilized EEZ with large fishing vessel companies of international repute could be one of the profitable opportunities for India. Suitable policy with "single-window facility" will certainly be beneficial for earning huge forex.

Climate Change

Climate change-induced disasters in fisheries and aquaculture communities require an immediate humanitarian response in terms of relief, and robust infrastructure such as shelters. Communities must possess the capacity (assets, knowledge and skills) for climate related collective action. Support should be provided to community-based fisheries management organizations to enable their direct involvement in assessment of vulnerability and of measures to secure climate resilience.

Climate change adaptation strategies must be developed and implemented in cooperation with affected communities and their organizations through transparent processes, allowing for meaningful participation of fisheries stakeholders, including the poor and vulnerable in the fisheries and aquaculture sector. Climate adaptation strategies must emphasize the need for poverty eradication and food security, in accordance with the Paris Agreement, the United Nations 2030 Agenda for Sustainable Development.

Alternate fishing systems which have low impact and are fuel efficient should be adopted to improve energy efficiency and to reduce climate impact on fisheries. Energy efficient fishing systems should be developed.

Infrastructure development

Support for inputs and facilities like national brood banks, hatcheries, rearing facilities, quality seed units including specific pathogen-free or resistant seed facilities are required in freshwater, brackish water aquaculture and mariculture for enhancing production and productivity. Infrastructure and systems for seed and feed certification, input quality testing, aquatic animal health management including quarantine, and disease diagnostics laboratories and referral laboratories, capacity building and establishment of extension support services to be enhanced.

For optimum harnessing of potential deep sea and offshore fisheries resources in the Indian EEZ and High seas and reduce fishing pressure in inshore areas, support to be provided to traditional fishers for acquiring technologically advanced deep-sea fishing vessels along with support for training and capacity building in deep sea fishing.

Infrastructure support for development of fishing harbours and fish landing centres will go a long way in ensuring safe landing, berthing and hygienic fish handling. Strengthening Post-harvest infrastructure including cold chain would reduce post-harvest losses. Provision for more Patrol boats, Sea safety kits, Communication and tracking devices etc is a prerequisite for strengthening Monitoring, Control and Surveillance (MCS).

As India is an important member country of FAO and keeping in mind the sustainable development goals (SDG's) & Blue Growth Initiative, development of integrated modern coastal villages along the Indian coast is a top priority. Fishers of these villages would be empowered for securing their livelihood. Also, development of start-ups and innovation centres/incubations that are considered as the stepping stones for the businesses.

SOME ADVANCED TECHNOLOGIES FOR FISH PRODUCTION ACROSS THE GLOBE

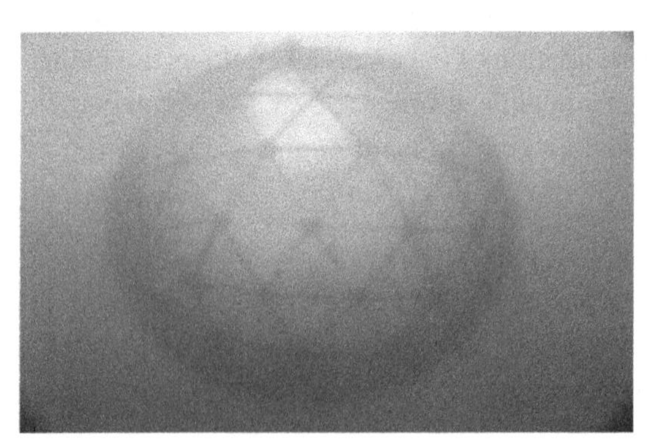

Aquapod is a unique containment system for marine aquaculture, suited for open ocean conditions and a diversity of species. The Aquapod is constructed of individual triangle net panels fastened together in a spheroid shape. This is invented by Ocean Farm Technologies.

The **Ocean Arks Technology** invented the concept which is based on usage of ship for salmon production. The vessel - designed for farming salmon, trout, tuna, seriola or cobia - has an approximate production capacity of 3900 tonnes. The vessel can move from one location to other to get better water for fish production. The concept is patented (Patent 201601709).

Salmon farming on solar energy (invented by Ocean Sun)

This innovation has more efficient energy production, low investment costs and longevity. Solar modules are installed on flexible and fluid plastic modules. The solar modules are cooled by direct heat transfer to water. Power outputs of 10% and more compared to conventional solar panels. The dimensions of floating power plants are determined based on power requirements and sea conditions at the location. All components included in the power plant are manufactured from environmentally friendly materials with a minimum carbon footprint.

Ocean Farm 1

This is world's first offshore fish farm manufactured in China. Ocean Farm 1 is a full-scale pilot facility for testing, learning, research and development. It is equipped for R&D activities, with particular focus on biological conditions and fish welfare. The structure is 110-metre wide, 70 metres in height, and contains 250,000 cubic metres in volume and can withstand magnitude 12 earthquakes. Around 20,000 sensors allow the marine site to achieve complete automation in monitoring and feeding the fish. The farm can mature up to 1.5 million fish in 14 months. It is also equipped with a 360-degree revolving gate for cleaning fish nets and driving fish shoals.

REFERENCES

Aadland, C. 2018. Norway approves permits for SalMar's offshore smart farm. IntraFish. Source: https://www.intrafish.com/aquaculture/1663517/norway-approves-permits-for-salmarsoffshore-smart-farm

Abraham, T. J., & Sasmal, D., 2009. Influence of salinity and management practices on the shrimp (Penaeus monodon) production and bacterial counts of modified extensive brackishwater ponds. *Turkish Journal of Fisheries and Aquatic Sciences*, *9*(1), 91-98.

ADB, 2005. An Evaluation of Small-scale Freshwater Rural Aquaculture Development for Poverty Reduction. Asian Development Bank, Manila.

Ahmed, N. and Thompson, S. 2019. "The blue dimensions of aquaculture: A global synthesis". In: Science of the Total Environment 652: 851-862

Ahmed, N., Thompson, S. and Glacer, M. 2018. "Integrated mangrove-shrimp cultivation: Potential for blue carbon sequestration". In: Ambio (A Journal of Human Environment by Springer). 47:441-452.

Ail, S. S., Krishnan, M., Jayasankar, J., Landge, A. and Shenoy, L., 2014. Evaluation of Compliance of Marine Fisheries of Kerala with Article 8 of FAO CCRF. *Fishery Technology*, 51:167-172.

Alongi, D.M., 2014. Carbon cycling and storage in mangrove forests. Annu. Rev. Mar. Sci. 6, 195–219.

Alverson, D. L. and Hughes, S. E., 1996. Bycatch: from emotion to effective natural resource management. *Reviews in Fish Biology and Fisheries*, 6(4):443-462.

Annual Report 2019-2020 of World Sustainability Organization, 2019. 27p.

Anon, 2000. The sustainable livelihood approach (SLA) and the Code of Conduct for Responsible Fisheries (CCRF) Liaison Bull. FAO sustain. Fish livelihood programme west Afr./bull. Liaison FAO programme Moyens existence durable penche Afr. Quest. No.1 pp. 1-2.

Anonymous, 2011. Report of the Working Group for Revalidating the Potential of Fishery Resources in the Indian Exclusive Economic Zone. New Delhi. DAHDF.

Aparna Roy, 2019. "Blue Economy in the Indian Ocean: Governance Perspectives for Sustainable Development in the Region", *ORF Occasional Paper No. 181,* January 2019, Observer Research Foundation. 34p.

Asian Development Bank (ADB), 2020. Asian Development Outlook (ADO) Series, 3 April 2020[online]. Manila. [Cited 15 May 2020]. https://www.adb.org/publications/series/asian-development-outlook

Asokan, P. K., Mohamed, K. S., & Sasikumar, G. 2013. Mussel farming and hatchery

Asokan, P. K., Mohamed, K. S., & Sasikumar, G. 2013. Mussel farming and hatchery http://eprints.cmfri.org.in/9730/1/Asokan.pdf

Ayer, N. W., & Tyedmers, P. H. 2009. Assessing alternative aquaculture technologies: life cycle assessment of salmonid culture systems in Canada. Journal of cleaner production, 17(3), 362-373. Source: https://www.sciencedirect.com/science/article/pii/S0959652608001820

Ayyappan, S., Moza Usha., Gopalakrishnan,A., Meenakumari,B., Jena, J.K. & Pandey, A.K., 2011. Handbook of Fisheries and Aquaculture, ICAR, 1116p.

Barange, M., Bahri, T., Beveridge, M.C.M., Cochrane, K.L., Funge-Smith, S. & Poulain, F., eds. 2018. *Impacts of climate change on fisheries and aquaculture: synthesis of current knowledge, adaptation and mitigation options.* FAO Fisheries and Aquaculture Technical Paper No. 627. Rome, FAO. 628 pp. 28: 621-623.

Baraniuk, C. 2018. How lasers and robo-feeders are transforming fish farming. BBC News. Source: https://www.bbc.com/news/business-43032542

Bartholomew, D. 2018. Rising Tide: Catalina Sea Ranch looks to expand, spins off research arm. Los Angeles Business Journal. Source: http://labusinessjournal.com/news/2019/jan/04/rising-tide/

Béné, C., Arthur, R., Norbury, H., Allison, E.H., Beveridge, M., Bush, S., Campling, L., Leschen, W., et al., 2016. Contribution of ûsheries and aquaculture to food security and poverty reduction: assessing the current evidence. World Dev. 79, 177–196.

Bhat, B. V. and Chembian, A. J., 2012. Marine fisheries policy in India-challenges for implementation. In: Shenoy L, Lakra, W. S, editors. Proceedings of national workshop on Code of Conduct for Responsible Fisheries. February, 1–2, 2012, Mumbai, India. CIFE; 2012, pp. 9-21.

BIM Guidance Note for Seafood Retailers – Labelling Requirement for Sale of Seafood, Issue 4, January 2018.

Blue revolution: a new initiative in fishery sector – A Parliamentary Reference Note No 39/RN/Ref./October/2017

Bondad-Reantaso, M.G., MacKinnon, B., Hao, B., Huang, J.,Tang-Nelson, K., Surachetpong, W., Alday-Sanz, V., Salman, M., Brun, E., Karunasagar, I., Hanson, L., Sumption, K., Barange, M., Lovatelli, A., Sunarto, A., Fejzic, N., Subasinghe, R., Mathiesen, Á.M. & Shariff, M. 2020. Viewpoint: SARS-CoV-2 (the cause of COVID-19 in humans) is not known to infect aquatic food animals nor contaminate their products. *Asian Fisheries Science*, 33: 74–78 [online]. [Cited 22 May 2020]. https://doi.org/10.33997/j.afs.2020.33.1.009

Bondad-Reantaso, M.G., Subasinghe, R.P. (Eds.), 2013. Enhancing the Contribution of Small-scale Aquaculture to Food Security, Poverty Alleviation and Socio-economic Development. FAO Fisheries and Aquaculture Proceedings No. 13, Rome

Bone, J., Clavelle, T., Ferreira, J. G., Grant, J., Ladner, I., Immink, A., Stoner, J and N. Taylor (2018). Best practices for aquaculture management guidance for implementing the ecosystem approach in Indonesia and beyond. Conservation International and the Sustainable Fisheries Partnership. Source: https://www.sustainablefish.org/Media/Files/Aquaculture/2018-Best-Practices-for-Aquaculture-Management

Boopendranath, M.R.,2012. Biodiversity conservation technologies in fisheries. Journal of Aquatic Biology & Fisheries Vol. 1, No. 1, 2012.

Bouillon, S., Borges, A.V., Castañeda-Moya, E., Diele, K., Dittmar, T., Duke, N.C., Kristensen, E., Lee, S.Y., et al., 2008. Mangrove production and carbon sinks: a revision of global budget estimates. Glob. Biogeochem. Cycles 22, GB2013.

Bregnballe, Jacob., 2015. A Guide to Recirculation Aquaculture. The Food and Agriculture Organization of United Nations (FAO) and EUROFISH International Organisation, 95 pp.

Brianna Cameron., Shannon O'Neill., Caroline Bushnell., Zak Wston., Elizabeth Derbes., Keri Szejda., (June) 2019. (Good Food Institute) State of Industry Report – Plant Based Meat Egg Dairy. 34p

Brianna Cameron., Shannon O'Neill., Liz Specht., Elizabeth Derbes., Keri Szejda., (June) 2019. (Good Food Institute) State of Industry Report – Cell Based Meat. 22p

Broadhurst, M. K., 2000. Modifications to reduce bycatch in prawn trawls: a review and framework for development. *Reviews in Fish Biology and Fisheries*, 10(1):27-60.

Bryant C, Szejda K, Parekh N, Deshpande V and Tse B., 2019. A Survey of Consumer Perceptions of Plant-Based and Clean Meat in the USA, India, and China. Front. Sustain. Food Syst. 3:11. doi: 10.3389/fsufs.2019.00011

Buck, B. H., Troell, M. F., Krause, G., Angel, D., Grote, B., & Chopin, T. 2018. State of the art and challenges for offshore integrated multi-trophic aquaculture (IMTA). Frontiers in Marine Science, 5, 165. Source: https://www.frontiersin.org/articles/10.3389/fmars.2018.00165/full

Cabico, G. K. 2020. How coastal towns can feed constituents by helping small-scale fishers. PhilStar News, 16 April 2020. (also available at https://www.philstar.com/nation/2020/04/16/2007734/howcoastal-towns-can-feed-constituents-helping-small-scale-fishers)

Chapagain, A.K., Hoekstra, A.Y., 2011. The blue, green and grey water footprint of rice from production and consumption perspectives. Ecol. Econ. 70, 749–758.

China Dialogue Ocean 2018. Marine Ranching in China. Source: https://chinadialogue-production.s3.amazonaws.com/uploads/content/file_en/10841/Marine_ranching_in_China_E_S.pdf

Chopin, T. 2013. Aquaculture, integrated multi-trophic (IMTA). In book: Encyclopedia of sustainability science and technology chapter: aquaculture, integrated multi-trophic (IMTA). Publisher: Springer, Dordrecht. Source: https://www.researchgate.net/publication_Aquaculture_Integrated_Multi-Trophic_IMTA

Chopin, T., Robinson, S., Sawhney, M., Bastarache, S., Belyea, E., Shea, R., & Fitzgerald, P. 2004. TheAquaNet integrated multi-trophic aquaculture project: rationale of the project and development of kelp cultivation as the inorganic extractive component of the system. Bulletin-Aquaculture Association of Canada, 104(3), 11. Source: https://www.researchgate.net/publication/269990513. The AquaNet integrated multi-trophic aquaculture project Rationale of the project and development of kelp cultivation as the inorganic extractive component of the system

Clavelle, T. 2020. Global fisheries during COVID-19. In: *Global Fishing Watch* [online]. [Cited 22 May 2020]. https://globalfishingwatch.org/data-blog/global-fisheries-during-covid-19/

Clements, J.C., Chopin, T., 2017. Ocean acidiûcation and marine aquaculture in North America: potential impacts and mitigation strategies. Rev. Aquac. 9, 326–341.

Coalition for Fair Fisheries Arrangements (CFFA), 2020. Hard hit by the Covid-19 crisis, Ivorian women in artisanal fisheries also see it as an opportunity to address long postponed issues. In: *CFFA* [online]. [Cited 22 May 2020]. www.cffacape.org/news-blog/hard-hit-by-the-covid-19-crisis-ivorian-women-in-artisanal-fisheries-also-see-it-as-an-opportunityto-address-long-postponed-issues

Costa-Pierce, B.A., 2002. Ecological Aquaculture: The Evolution of the Blue Revolution. Blackwell Science Ltd, Oxford.

Cubitt, F., Butterworth, K., and McKinley, R. S. 2008. A synopsis of environmental issues associated with salmon aquaculture in Canada. In Aquaculture, Innovation and Social Transformation (pp.123-162). Springer, Dordrecht. Source: https://link.springer.com/chapter/10.1007/978-1-4020-8835-3_10

DADF, 2017. *Annual report 2016-17*. Department of Animal Husbandry, Dairying and Fisheries. Ministry of Agriculture, Government of India, 162 pp.

Datu, C.L.J. 2020. 'Seafood Kadiwa Ni Ani at Kita on Wheels' visits CSFP barangays. Philippine Information Agency [online]. [Cited 13 April 2020]. https://pia.gov.ph/news/articles/1038864

De Silva, S.S., Soto, D., 2009. Climate change and aquaculture: potential impacts, adapta- tion and mitigation. In: Cochrane, K., De Young, C., Soto, D., Bahri, T. (Eds.), Climate Change Implications for Fisheries and Aquaculture: Overview of Current Scientiûc Knowledge. FAO Fisheries and Aquaculture Technical Paper No. 530, Rome, pp. 151–212.

Deo, A. D., Anand, P. S., & Ravichandran, P. 2013. BMP in shrimp farming with special reference to West Bengal.

Department of Animal Husbandry Dairying and Fisheries, Ministry of Agriculture and Farmers Welfare, Government of India, September 2017. "Guidelines in Brief on Centrally Sponsored Scheme on Blue Revolution: Integrated Development and Management of Fisheries". 08p.

Department of Animal Husbandry Dairying and Fisheries, Ministry of Agriculture and Farmers Welfare, Government of India, January-August 2018. "Guidelines in Brief on Centrally Sponsored Scheme on Blue Revolution: Integrated Development and Management of Fisheries". 08p.

Department of Animal Husbandry Dairying and Fisheries, Ministry of Agriculture and Farmers Welfare, Government of India, Feb 2019. Revised Guidelines CSS on Blue Revolution: Integrated Development & Management of Fisheries. 54p

Department of Fisheries, 2019. Ministry of Fisheries, Animal Husbandry and Dairying, Govt of India, "Handbook on Fisheries Statistics 2018". 190p

Department of Fisheries, 2020. Ministry of Fisheries, Animal Husbandry and Dairying, Govt of India, Operational Guidelines for Pradhan Mantri Matsya Sampada Yojana, 237p.

Department of Fisheries, Govt of India, 2017. National Policy on Marine Fisheries. 22p.

Department of Fisheries, Govt of India, 2019. Draft National Inland Fisheries and Aquaculture Policy. 12p.

Department of Fisheries, Govt of India, 2019. National Mariculture Policy – Revised Draft. 14p.

Department of Fisheries, Govt of India, 2019. National Marine Fisheries Bill – Revised Draft. 10p.

Department of Fisheries, Ministry of Agriculture & Farmers Welfare, Government of India, February 2019. Guidelines on Fisheries and Aquaculture Infrastructure Development Fund (FIDF). 15p

Diana, J.S., Egna, H.S., Chopin, T., Peterson, M.S., Cao, L., Pomeroy, R., Verdegem, M., Slack, W.T., et al., 2013. Responsible aquaculture in 2050: valuing local conditions and human innovations will be key to success. Bioscience 63, 255–262.

Dineshbabu, A. P., Thomas, S. and Radhakrishnan, E. V., 2010. Bycatch from trawlers with special reference to its impact on commercial fishery, off Mangalore. 327-334.

Duarte, C.M., Holmer, M., Olsen, Y., Soto, D., Marbà, N., Guiu, J., Black, K., Karakassis, I., 2009. Will the oceans help feed humanity? Bioscience 59, 967–976.

Duarte, C.M., Losada, I.J., Hendriks, I.E., Mazarrasa, I., Marbà, N., 2013. The role of coastal plant communities for climate change mitigation and adaptation. Nat. Clim. Chang. 3, 961–968.

Economist Intelligence Unit. 2015. Investing in the Blue Economy—Growth and Opportunity in a Sustainable Ocean Economy. Briefing Paper

Economy Profile – India, Doing Business 2020 Report, 125p

Edwards, P., 2015. Aquaculture environment interactions: past, present and likely future trends. Aquaculture 447, 2–14.

Emerenciano, M., Gaxiola, G., Cuzon, G., 2013. Bioûoc technology (BFT): a review for aquaculture application and animal food industry. In: Matovic, M.D. (Ed.), Biomass Now – Cultivation and Utilization. InTech, Rijeka, Croatia, pp. 301–328.

Erickson-Davies, M. 2018. New study finds mangroves may store way more carbon than we thought. Mongabay. Source: https://news.mongabay.com/2018/05/new-study-finds-mangroves-maystore-way-more-carbon-than-we-thought/

Erwin, K. L. 2009. Wetlands and global climate change: the role of wetland restoration in a changing world. Wetlands Ecology and management, 17(1), 71. Source: https://www.wetlands.org/wp-content/uploads/2015/11/Wetlands-and-Global-Climate-Change.pdf

Eurofish, 2012. *Overview of the world's anchovy sector and trade possibilities for Georgian anchovy products*. Retrieved from http://www.fao.org/fileadmin/user_upload /Europe/documents/Publications/Anchovies_report_2.03.2012.pdf

Euronews, 2020. Coronavirus: supporting Europe's battered fishing industry. In: *Euronews* [online]. [Cited 22 May 2020]. www.euronews.com/2020/05/19/coronavirus-supporting-europe-s-battered-fishingindustry

FAIRR in collaboration with KKS Advisors and VB Consultancy, 2019. FAIRR Initiative. Shallow returns? ESG risks and opportunities in aquaculture. 146p.

Falkenmark, M., Rockström, J., 2006. The new blue and green water paradigm: breaking new ground for water resources planning and management. J. Water Resour. Plan. Manag. 132, 129–132.

FAO, 1995. Code of Conduct for Responsible Fisheries, 41 pp.

FAO, 1998. Food and Agriculture Organization of the United Nations (FAO). (1998). Part 4: Outlook: Expected trends in supply and demand. In *The state of world fisheries and aquaculture.* Retrieved from http://www.fao.org/docrep/w9900e/w9900e05.htm

FAO, 1999. Indicators for sustainable development of marine capture fisheries. *FAO Technical Guidelines for Responsible Fisheries. No.8. FAO,* Rome, 68 pp.

FAO, 2006. Building Adaptive Capacity to Climate Change: Policies to Sustain Livelihoods and Fisheries (FAO, Rome) New Directions in Fisheries, A Series of Policy Briefs on Development Issues, No 8. *2006*

FAO, 2007. Building adaptive capacity to climate change, policies to sustain livelihoods and fisheries. New directions in fisheries- A series of policy briefs on development issues. No.8. FAO, Rome, 86 pp.

FAO, 2009. Responsible fish trade. "*FAO Technical Guidelines for Responsible Fisheries*" No. 11. Rome, 23p.

FAO, 2009. *State of the World's Fisheries and Aquaculture,* Food and Agriculture Organization of the United Nations, Rome. 176pp.

FAO, 2009. Guidelines for the ecolabelling of fish and fishery products from marine capture fisheries. Revision 1.108p

FAO, 2011. India and FAO Achievement Success Stories. 16p

FAO, 2011. Guidelines for the Ecolabelling of Fish and Fishery Products from Inland Capture Fisheries. 106p.

FAO, 2014 B. *The state of world fisheries and aquaculture.* Food and Agriculture Organization of the United Nations (FAO), Rome Food and Agriculture Organization of the United Nations (FAO), Fish and Aquaculture Department. (2014b). *Global capture production statistics 2012.* Retrieved from ftp://ftp.fao.org/FI/STAT/ Overviews/CaptureStatistics2012.pdf

FAO, 2014b. Global Blue Growth Initiative and Small Island Developing States. Rome.

FAO, 2014c. SAVE FOOD: Global Initiative on Food Loss and Waste Reduction. Background paper on the economics of food loss and waste. Working paper. Rome.

FAO, 2016. *Food Outlook: Biannual Report on Global Food Markets, June 2016.* Rome. 130p

FAO, 2016. *Food Outlook: Biannual Report on Global Food Markets, October 2016.* Rome. 132p

FAO, 2016. The State of World Fisheries and Aquaculture 2016. Contributing to food security and nutrition for all. Rome, 200 pp.

FAO, 2016. *Fishery and aquaculture stastistics 2014.* Food and Agricultural Organisation, Rome, 204 pp.

FAO, 2016 A. Brief on fisheries, aquaculture and climate change in the Intergovernmental Panel on Climate Change Fifth Assessment Report. Source: http://www.fao.org/3/a-i5871e.pdf

FAO, 2016 B. Fisheries and Aquaculture and Climate Change. Source: http:// www.fao.org/ fishery/climatechange/en

FAO, 2017. A leaflet on "India and FAO 2017", 02p

FAO, 2017. *Food Outlook: Biannual Report on Global Food Markets, June 2017.* Rome. 142p

FAO, 2017. Sustainable Development Goals. FAO working for SDG 14, 2017. 35p (http://www.fao.org/3/a-i7298e.pdf)

FAO, 2017. FAO inputs in relation to resolution a/res/69/245 concerning "Oceans and the law of the sea". For the report of the Secretary-General to the seventieth session of the United Nations General Assembly. Source: https:// www.un.org/depts/los/general_assembly/ contributions_2017/FAO.pdf

FAO, 2018 *Food Outlook: Biannual Report on Global Food Markets, July 2018*. Rome. 161p

FAO, 2018. Country Programming Framework for India. 24p

FAO, 2018. *Food Outlook - Biannual Report on Global Food Markets – November 2018*. Rome. 104 pp.

FAO, 2018. *The State of World Fisheries and Aquaculture 2018 - Meeting the sustainable development goals*. Rome.

FAO, 2018. The State of World Fisheries and Aquaculture: Meeting the Sustainable Devel- opment Goals. Food and Agriculture Organization of the United Nations, Rome.

FAO, 2019. A leaflet on "India and FAO 2019", 02p

FAO, 2019. *FAO yearbook. Fishery and Aquaculture Statistics 2017/FAO annuaire. Statistiques des pêches et de l'aquaculture 2017/FAO anuario.* Estadísticas de pesca y acuicultura 2017. Rome/Roma.

FAO, 2019. Fishery and Aquaculture Country Profiles, The Republic of India. Food and Agricultural Organisation, Rome. Source: http://www.fao.org/fishery/facp/IND/en

FAO, 2019. *Food Outlook - Biannual Report on Global Food Markets – November 2019*. Rome.

FAO, 2019. *Food Outlook - Biannual Report on Global Food Markets*. Rome.

FAO, 2019. *GLOBEFISH Highlights October 2019 ISSUE, with Jan. – Jun. 2019 Statistics – A quarterly update on world seafood markets*. Globefish Highlights no. 4-2019.

FAO, 2019. National Aquaculture Legislation Overview of India. Food and Agricultural Organisation, Rome. Source: http://www.fao.org/fishery/legalframework/nalo_india/es

FAO, 2019. Statistical Capacity Assessment for FAO relevant SDG Indicators – India, 2018/19. 07p

FAO, 2020. How is COVID-19 affecting the fisheries and aquaculture food systems. Rome. https://doi.org/10.4060/ca8637en

FAO, 2020. Summary of the impacts of the COVID-19 pandemic on the fisheries and aquaculture sector: Addendum to the State of World Fisheries and Aquaculture 2020. Rome. https://doi.org/10.4060/ca9349en

FAO, 2020. *The effect of COVID-19 on fisheries and aquaculture in Asia.* Bangkok. https://doi.org/10.4060/ca9545en

FAO. 2020a. *How is COVID-19 affecting the fisheries and aquaculture food systems?* [online]. [Cited 22 May 2020]. www.fao.org/3/ca8637en/CA8637EN.pdf

FAO. 2020b. *Migrant workers and the COVID-19 pandemic* [online].[Cited 22 May 2020]. www.fao.org/3/ca8559en/CA8559EN.pdf

FAO, 2020. *The impact of COVID-19 on fisheries and aquaculture – A global assessment from the perspective of regional fishery bodies: Initial assessment, May 2020*. No. 1. Rome. https://doi.org/10.4060/ca9279en

FAO, FAO Global Record (http://www.fao.org/global-record/en/)

FAO,2011. Technical guidelines on aquaculture certification, 122 p

Ferreira, J. G., and Bricker, S. B. 2016. Goods and services of extensive aquaculture: shellfish culture and nutrient trading. Aquaculture international, 24(3), 803-825. Source: https://link.springer.com/article/10.1007/s10499-015-9949-9

Ferreira, J. G., Sequeira, A., Hawkins, A. J. S., Newton, A., Nickell, T. D., Pastres, R., and Bricker, S. B. 2009. Analysis of coastal and offshore aquaculture: application of the FARM model to multiple systems and shellfish species. Aquaculture, 292(1-2), 129-138. Source: https://www.academia.edu/12758998/ Analysis of coastal and offshore aquaculture Application of the FARM model to multiple systems and shellfish species

FICCI-KAS,2019. Report on blue economy: Global best practices, Takeaways for India and partner nations,142p.

Flanders Investment and Trade Market Survey, June 2018. "Growing Opportunities Food Market in India". 37p.

Food Business News, 2018. Poultry producer invests in maker of plant-based seafood alternatives. Source:https://www.foodbusinessnews.net/articles/12296-poultry-producer-invests-in-makerof-plant-based-seafood-alternatives

FRAD, CMFRI, 2019., Marine Fish Landings in India - 2018. Technical Report. CMFRI, Booklet No.15/2019, Kochi. pp.1-9

Fromberg, A. 2017. Anti-fish farm flotilla protests Tassal expansion at Okehampton Bay. Australian Broadcasting Corporation. Source: https://www.abc.net.au/news/2017-09-17/okehampton-bayfish-farm-protest-takes-to-the-water/8954512

Ghosh, S., Rao, M. H., Kumar, M. S., Mahesh, V. U., Muktha, M. and Zacharia, P. U., 2014. Carbon footprint of marine fisheries: life cycle analysis from Visakhapatnam. *Current science*, pp.515-521.

Good Food Institute, January 22, 2019. "An Ocean of Opportunity: Plant based and cell-based seafood for sustainable oceans without sacrifice". 38p

Gopakumar, G. 2010. "Mariculture Technologies for Augmenting Marine Resources, In: Coastal Fishery Resources of India: Conservation and Sustainable Utilisation (Meenakumari, B., Boopendranath, M.R., Edwin, L., Sankar, T.V., Gopal, N. and Ninan, G., Eds.), p. 39-58, Society of Fisheries Technologists (India), Cochin.

Gopakumar, G. 2016. Overview of Mariculture; Winter School on Technological Advances in Mariculture for Production Enhancement and Sustainability, 1: 1-10. Central Marine Fisheries Research Institute

Government of India, 2015. Order No. 30035/15/97-Fy (T-1) vol. IV. Department of Animal Husbandry, Dairying and Fisheries, Ministry of Agriculture, Government of India. Krishi Bhavan, New Delhi: Dated: 03rd June, 2015.

Government of India, 2016. Sustainable Development Goals (SDGs) – Draft Mapping, Development Monitoring and Evaluation Office, NITI Aayog, New Delhi. 2016, 30p

Government of India, 2017. Press Note on "Blue Revolution" by Press Information Bureau (PIB), Ministry of Agriculture & Farmers Welfare, Government of India, dated 11.04.2017

Government of India, 2017.National Policy on Marine Fisheries, 2017, Ministry of Agriculture, Department of Animal Husbandry, Dairying and Fisheries. F. No. 21001/ 05/2014-FY (Ind) Vol. V

Government of India, 2018. Order No. 30035/15/97-Fy (T-1) vol. V. Department of Animal Husbandry, Dairying and Fisheries, Ministry of Agriculture, Government of India. Krishi Bhavan, New Delhi: Dated: 22 Feb., 2018.

Government of India, 2018. Press Note on "Creation of Fisheries and Aquaculture Infrastructure Development Fund" by Press Information Bureau (PIB), Ministry of Agriculture & Farmers Welfare, Government of India dated 24.10.2018

Guidelines for Seafood Retailers by Department of Fisheries, Government of Western Australia, 2002. 42p.

Hall, S.J., Delaporte, A., Phillips, M.J., Beveridge, M., O'Keefe, M., 2011. Blue Frontiers: Man- aging the Environmental Costs of Aquaculture. The WorldFish Center, Penang.

Halwart, M., van Dam, A.A. (Eds.), 2006. Integrated Irrigation and Aquaculture in West Africa: Concepts, Practices and Potential. Food and Agriculture Organization of the United Nations, Rome.

Hanchard, B., 2004. The implementation of the 1995 FAO Code of Conduct for Responsible Fisheries in the Pacific Islands. *FAO,* Rome. October, 2004.

Handisyde, N.T., 2008. The effects of Climate Change on World Aquaculture: A Global Prospective. 1-151p.

Harkell, L. 2018 B. In pictures: China's offshore pen designs. Undercurrent News. Source: https://www.undercurrentnews.com/2018/10/02/why-china-is-a-hub-for-building-offshore-aquaculturepens/

Harrington, M. J., Ransom, A. M. and Andrews, A. R., 2005. Wasted fishery resources: Discards bycatch in the USA. *Fish and Fisheries,* 6: 350-361.

Hemalatha, K. 2020. Sounds of the sea give fishermen COVID-19 updates. Livemint. 12 May 2020. (also available at https://www.livemint.com/news/india/sounds-of-the-sea-give-fishermen-covid-19- updates-11589277117590.html)

Hoff, H., Falkenmark, M., Gerten, D., Gordon, L., Karlberg, L., Rockström, J., 2010. Greening the global water system. J. Hydrol. 384, 177–186.

http://cifa.nic.in/

http://cift.res.in/cift-about

http://cift.res.in/uploads/userfiles/CIFT-COVID-19%20Advisory-English.pdf

http://cofcau.nic.in/aboutcof.html

http://infofish.org/v3/index.php/component/k2/item/165-fao-calls-for-new-vision-for-fisheries?utm_source=GLOBEFISH+Newsletter&utm_campaign=4d070d133c-EMAIL_CAMPAIGN_2017_11_29_COPY_01&utm_medium=email&utm_term=0_30087880c0-4d070d133c-587748341

http://kufos.ac.in/home/

http://nfdb.gov.in/

http://ris.org.in/sdg/india-and-sustainable-development-goals-way-forward

http://www.caa.gov.in/

http://www.ciba.res.in/

http://www.cifri.res.in/

http://www.cifri.res.in/advisory/1%20Advisory%20on%20Rivers,%20estuaries%20(Covid%2019)(Eng).pdf

http://www.cifri.res.in/advisory/2%20Advisory%20on%20reservoir%20&%20Wetland%20(Covid%2019)(Eng).pdf

http://www.cift.res.in/covid-19-advisory-note

http://www.cmfri.org.in/

http://www.dof.gov.in/sites/default/filess/Model%20SOPs%20for%20Marine%20Fishries%20dated%2018.4.2020%20by%20DoF,%20GoI.pdf

http://www.dof.gov.in/sites/default/filess/Model%20SOPs%20of%20DoF,%20GoI%20on%20Inland%20Fisheries%20&%20Aquaculture.pdf

http://www.fao.org/2019-ncov/q-and-a/impact-on-fisheries-and-aquaculture/en/

http://www.fao.org/3/ca8959en/ca8959en.pdf

http://www.fao.org/about/fao-and-the-un/en/

http://www.fao.org/about/meetings/cofi-sub-committee-on-aquaculture/en/

http://www.fao.org/about/who-we-are/departments/en/

http://www.fao.org/about/who-we-are/worldwide-offices/en/

http://www.fao.org/apfic/en/

http://www.fao.org/aquaculture/en/

http://www.fao.org/faoterm/collection/aquaculture/en/

http://www.fao.org/faoterm/collection/fisheries/en/

http://www.fao.org/fisheries/blue-growth/en/

http://www.fao.org/fisheries/en/

http://www.fao.org/fishery/about/cofi/aquaculture/en

http://www.fao.org/fishery/about/cofi/en

http://www.fao.org/fishery/area/search/en

http://www.fao.org/fishery/capture/en

http://www.fao.org/fishery/countrysector/naso_india/en

http://www.fao.org/fishery/legalframework/nalo_india/es

http://www.fao.org/fishery-aquaculture/en/

http://www.fao.org/global-record/en/

http://www.fao.org/india/programmes-and-projects/en/

http://www.fao.org/UNFAO/struct-e.htm

http://www.fao.org/voluntary-guidelines-small-scale-fisheries/news-and events/detail/en/c/1272868/

http://www.mospi.nic.in/

http://www.nbfgr.res.in/

http://www.rgca.org.in/

http://www.unsceb.org/

https://aquaconnect.blue/

https://dcfr.res.in/

https://economictimes.indiatimes.com/news/politics-and-nation/andhra-pradesh-announces-one-time-aid-of-rs-2000-to-6000-fishermen-struck-in-gujarat-due-to lockdown/articleshow/75292848.cms?from=mdr

https://in.one.un.org/page/sustainable-development-goals/

https://in.one.un.org/page/sustainable-development-goals/sdg-14/

https://mumbaimirror.indiatimes.com/coronavirus/news/15000-tonnes-of-fish-dumped/articleshow/75090040.cms

https://nacsampeda.org.in/

https://sanctuarynaturefoundation.org/article/how-the-lockdown-threw-16-million-fishers-into-chaos

https://stats.oecd.org/glossary/detail.asp?ID=2713#:~:text=The%20total%20allowable%20catch%20(TAC,terms%20of%20numbers%20of%20fish.

https://timesofindia.indiatimes.com/city/chennai/4-85l-of-fishing-community-get-rs-1k-covid-aid/articleshow/75260935.cms

https://timesofindia.indiatimes.com/city/kochi/lockdown-fisheries-sector-lost-rs-11652cr-in-40-days/articleshow/76718383.cms

https://www.business-standard.com/article/news-ani/kerala-cm-announces-financial-assistance-for-fishermen-beedi-workers-others-120040901594_1.html

https://www.cife.edu.in/index.html

https://www.dakshin.org/wp-content/uploads/2020/04/Petition-Requesting-Attn-to-Fisher-Health-Livelihoods_13.04.20.pdf

https://www.livemint.com/news/india/sounds-of-the-sea-give-fishermen-covid-19-updates-11589277117590.html

https://www.nationsencyclopedia.com/United-Nations-Related-Agencies/The-Food-and-Agriculture-Organization-of-the-United-Nations-FAO-STRUCTURE.html

https://www.pminewyork.gov.in/

https://www.thehindubusinessline.com/economy/agri-business/covid-19-causes-a-daily-loss-of-224-crore-to-indias-fishery-sector/article31388582.ece

https://www.tnjfu.ac.in/vision2020

https://www.un.org/ecosoc/en/home

https://www.un.org/en/sections/about-un/overview/index.html

https://www.un.org/en/sections/un-charter/chapter-i/index.html

https://www.un.org/en/sections/un-charter/chapter-ii/index.html

https://www.un.org/en/sections/un-charter/chapter-iii/index.html

https://www.un.org/en/sections/un-charter/chapter-x/index.html

https://www.un.org/en/sections/un-charter/chapter-xv/index.html

ICAR funded short course on 'World Trade Agreements and Indian Fisheries Paradigms: A Policy Outlook', Kochi; CMFRI, 17-26 September 2012. 357-363p

Indonesian Traditional Fisherfolk Union (DPP KNTI), 2020. Covid-19 Outbreak: Socio-economic Impact on Small-scale Fishery and Aquaculture in Indonesia. Focus on the Global South, 17 April 2020. (also available at https://focusweb.org/covid-19-outbreak-socio-economic-impact-on-small-scale-fisherand-aquaculture-in-indonesia/)

International Institute for Sustainable Development, 2016. “State of Sustainability Initiatives Review: Standards and Blue Economy”. 207p

International Monetary Fund (IMF), 2020. World Economic Outlook, April 2020: The Great Lockdown [online]. Washington, DC. [Cited 15 May 2020]. https://www.imf.org/en/Publications/WEO/Issues/2020/04/14/weo-april-2020

IPCC, 2014. Climate Change 2014: Synthesis Report – Summary for Policymakers. Inter- governmental Panel on Climate Change, Valencia.

Jacquet, J., Franks, B., Godfrey-Smith, P. and Sánchez-Suárez, W. The Case Against Octopus Farming. Issues in Science and Technology 35, no. 2 (Winter 2019): 37–44. Source: https://issues.org/the-case-against-octopus-farming/

Jayasankar, P., 2018. “Present Status of Freshwater Aquaculture in India: A Review”, Indian J Fish. 65 (4), 157-165.

Jiménez Cisneros, B.E., Oki, T., Arnell, N.W., Benito, G., Cogley, J.G., Döll, P., Jiang, T. & Mwakalila, S.S. 2014. Freshwater resources. *In* V.R. Barros, C.B. Field, D.J. Dokken, M.D. Mastrandrea, K.J. Mach, T.E. Bilir, M. Chatterjee *et al.*, eds. *Climate change 2014: Impacts, adaptation, and vulnerability. Part B: Regional aspects. Contribution of Working Group II to the Fifth Assessment Report of the Intergovernmental Panel on Climate Change.* Cambridge, UK and New York, Cambridge University Press. pp. 229–2690. (also available at https://www.ipcc.ch/pdf/assessment-report/ar5/wg2/WGIIAR5-Chap3_FINAL.pdf).

Kapetsky, J.M., Aguilar-Manjarrez, J., Jenness, J., 2013. A Global Assessment of Offshore Mariculture Potential From a Spatial Perspective. FAO Fisheries and Aquaculture Technical Paper No. 549, Rome.

Karnatak, G., & Kumar, V. 2014. Potential of cage aquaculture in Indian reservoirs. *International Journal of Fisheries and Aquatic Studies*, *1*(6), 108-112

Kauffman, J.B., Arifanti, V.B., Trejo, H.H., Garcia, M.C.J., Norfolk, J., Cifuentes, M., Hadriyanto, D., Murdiyarso, D., 2017. The jumbo carbon footprint of a shrimp: carbon losses from mangrove deforestation. Front. Ecol. Environ. 15, 183–188.

Kauffman, J.B., Heider, C., Norfolk, J., Payton, F., 2014. Carbon stocks of intact mangroves and carbon emissions arising from their conversion in the Dominican Republic. Ecol. Appl. 24, 518–527.

Kaushik, S.J., 1999. Nutrient requirements, supply and utilization in the context of carp culture. Aquaculture, 129: 1–4, 225-241. ISSN 0044-8486,https://doi.org/10.1016/0044-8486(94)00274-R.

Kelleher, K., 2004. Discards in the World’s Marine Fisheries: An Update, FAO Fisheries Technical Paper. No. 470, FAO, Rome

Kelleher, K., 2005. Discards in the world's marine fisheries. An update. 2005. FAO Fisheries Technical Paper, 470, pp.131.

Kharatmol, B.R, 2018. Study on compliance of trawl net fishery of Maharashtra coast, India with provisions of FAO CCRF. PhD thesis, Central Institute of Fisheries Education, Mumbai. 173 pp.

Kumar M., Gurjar, U.R., Keer, N.R., Chandravanshi, K. S., Shukla, A., Kumar, S., Gupta,S., and Pal, P.S. 2018. Professional Fisheries Education in India: History, Current Status and Future – A Review, In: Int.J.Curr.Microbiol.App.Sci (2018) 7(6): 3395-3409. Source: https://www.ijcmas.com/7-6-2018/Manmohan%20Kumar,%20et%20al.pdf

Kumar, A. B. and Deepthi, G. R., 2006. Trawling and by-catch: Implications on marine ecosystem. *Current Science*, 90(8):922-931.

Kumar, D., Shyam, S. S., Shenoy, L., Biradar, R. S., Ananthan, P. S., and Kamat, S., 2007. Third Zonal and Aquaculture Policy: Responsible Fisheries and Sustainable aquaculture perspective for West Coast States Goa, Gujarat, Karnataka, Kerala, Maharashtra and Union Territories of daman and Diu, Dadra and Nagar Haveli and Lakshadweep. In: *3rd* Zonal and Aquaculture Policy:, 21-23 June 2007, Goa.

Kumawat, T., Shenoy, L., Chakraborty, S. K., Deshmukh, V. D. and Raje, S. G., 2015. Compliance of bag net fishery of Maharashtra coast, India with Article 7 of the FAO Code of Conduct for Responsible Fisheries. *Marine Policy,* 56:9–15.

Lakshmi, A. and Ramya, R., 2009. "Fisheries and Climate change". *Coasttrack* Vol.8 (1) J6. une 2009.1-5p.

Louis R. D'Abramo & Shyn Shin Sheen., 1994. Nutritional requirements, feed formulation, and feeding practices for intensive culture of the freshwater prawn *Macrobrachium rosenbergii.* Reviews in Fisheries Science, 2:1, 1-21, DOI: 10.1080/10641269409388551

Maloney, 2018. Eat More Bivalves! Source: https://www.fishfarming.com/blog/eat-more-bivalves.html

Manager, 2020. Rice barters with fish programme: rice of more than 7 tonnes from Karen community in Northern Thailand arrived Andaman. Manager [Online]. Bangkok. [Cited 28 April 2020]. https://mgronline.com/south/detail/9630000044237 McGinn, A.P., 1998. Blue revolution: the promises and pitfalls of ûsh farming. World Watch 11 (2), 10–19.

McGinn, A.P., 1998. Blue revolution: the promises and pitfalls of ûsh farming. World Watch 11 (2), 10–19.

Mcleod, E., Chmura, G.L., Bouillon, S., Salm, R., Björk, M., Duarte, C.M., Lovelock, C.E., Schlesinger, W.H., et al., 2011. A blueprint for blue carbon: toward an improved un- derstanding of the role of vegetated coastal habitats in sequestering CO_2. Front. Ecol. Environ. 9, 552–560.

MCOT, 2020. Fish barters with rice: opportunity during pandemic for sea gypsy in Rawai Beach. Thai News Agency [Online]. Bangkok. [Cited 18 April 2020]. https://www.mcot.net/viewtna/5e9b08e5e3f8e40af443ae14

Mele, Gianluca. 2014. "Mauritania Economic Update: 1." World Bank, Washington, DC. http://documents.worldbank.org/curated/en/2014/07/20133610/mauritania-economic-update

Menzel, L., Matovelle, A., 2010. Current state and future development of blue water availability and blue water demand: a view at seven case studies. J. Hydrol. 384, 245–263.

Ministry of Commerce and Industry, Government of India, September 2019. IBEF Report on "Agriculture and Allied Industries"

Ministry of Commerce and Industry, Government of India, September 2019. IBEF Report on "E-commerce Industry in India"

Ministry of Commerce and Industry, Government of India, September 2019. IBEF Report on "Retail Industry in India"

Mohamed, K. S., Vijayakumaran, K., Zacharia, P. U., Sathianandan, T.V., Maheswarudu, G., Kripa, V., Narayanakumar, R., Prathibha Rohit, Joshi, K.K., Sankar, T. V., Leela Edwin, Ashok Kumar, K., Bindu J, Nikita Gopal and Pravin Puthra, 2017. Indian Marine Fisheries Code: Guidance on a Marine Fisheries Management Model for India. CMFRI *Marine Fisheries Policy Series,* 4: 120 pp.

Mohamed, K. S., Zacharia, P. U., Maheswarudu, G., Sathianandan, T. V., Abdussamad, E. M., Ganga, U., Pillai, S. L., Sobhana, K. S., Nair, R. J., Josileen, J. and Chakraborty, R .D., 2014. Minimum Legal Size (MLS) of capture to avoid growth overfishing of commercially exploited fish and shellfish species of Kerala. *Marine Fisheries Information Service; Technical and Extension Series*, No. 220, pp. 3-7.

Mohamed, K.S., CMFRI, 2013. "Ecolabelling in Fisheries: Boon or Bane in Improving Trade". In: Winter School on Structure and Function of Marine Ecosystems: Fisheries. 39: 332-340.

Mohanty, S.K., Priyadarshi Dash, Aastha Gupta and Pankhuri Gaur, 2015. "Prospects of Blue Economy in the Indian Ocean by RIS (Research and Information System for Developing Countries)". 87p

Molden, D., 2007. Water for Food, Water for Life: A Comprehensive Assessment of Water Management in Agriculture. Earthscan, London, and International Water Manage- ment Institute, Colombo.

Movik, S., Mehta, L., Mtisi, S., Nicol, A., 2005. A "blue revolution" for African agriculture? Int. Dev. Stud. Bull. 36 (2), 41–45.

Murdiyarso, D., Purbopuspito, J., Kauffman, J.B., Warren, M.W., Sasmito, S.D., Donato, D.C., Manuri, S., Krisnawati, H., et al., 2015. The potential of Indonesian man- grove forests for global climate change mitigation. Nat. Clim. Chang. 5, 1089–1092.

NAAS, 2015. Aquaculture Certification in India: Criteria and Implementation Plan. Policy Paper No. 77, National Academy of Agricultural Sciences, New Delhi: 16 p.

NABARD, 2018. Sectoral Paper on Fisheries and Aquaculture. 75p.

National Account Statistics, 2019. National Statistical Office, Ministry of Statistics and Programme Implementation (MoSPI), Government of India. 378p.

National Consultation on SDGs Sustaining Life: Integrating Biodiversity Concerns, Ecosystems Values and Climate Resilience in India's Planning Process Focus on SDG 13, 14 and 15, 08-09 February 2017

National Fisheries Development Board, 2019. A report on "Aquaculture Technologies Implemented by NFDB". 58p

National Fisheries Development Board, 2020. Operational Guidelines of PMMSY, Ready Reckoner, Government of India, Ministry of Fisheries, Animal Husbandry and Dairying, Dept of Fisheries, 25p.

National Fisheries Development Board. A leaflet on "Ushering Blue Revolution in India". 02p. Source: http://nfdb.gov.in/PDF/E%20Publications/13%20NFDB%20%20Ushering%20Blue%20Revolution%20in%20India.pdf

National Plan for Conservation of Aquatic Ecosystems: A presentation to PM by NPCA officials, 14 July 2016

Naylor, R.L., Goldburg, R.J., Primavera, J.H., Kautsky, N., Beveridge, M.C.M., Clay, J., Folke, C., Lubchenco, J., et al., 2000. Effect of aquaculture on world ûsh supplies. Nature 405, 1017–1024.

Nellemann, C., Corcoran, E., Duarte, C.M., Valdés, L., De Young, C., Fonseca, L., Grimsditch, G. (Eds.), 2009. Blue Carbon: The Role of Healthy Oceans in Binding Carbon. GRID- Arendal, United Nations Environment Programme.

Neori, A., Troell, M., Chopin, T., Yarish, C., Critchley, A., Buschmann, A.H., 2007. The need for a balanced ecosystem approach to blue revolution aquaculture. Environ. Sci. Pol. Sustain. Dev. 49 (3), 36–43.

Niti Aayog, Government of India, 2017. Voluntary National Review Report on Implementation of Sustainable Development Goals in India. 31p

Niti Aayog, Government of India, 2019. Localising SDG's Early Lessons from India. 121p.

NOFIMA, 2014. Nofima Atlantic salmon in closed containment systems research update. Source:http://www.tidescanada.org/wp-content/uploads/2015/03/Bendik-Fyhn-Terjesen-Nofima-Atlantic-Salmon-in-Closed-Containment-Systems-Research-Update.pdf

NRAI Technopak India Food Service Report, 2016.

OECD/FAO, 2019. *OECD-FAO Agricultural Outlook 2019-2028*, OECD Publishing, Paris/Food and Agriculture Organization of the United Nations, Rome. *https://doi.org/10.1787/agr_outlook-2019-en*

Olivier, 2016 A global review of the ecosystem services provided by bivalve aquaculture. https://onlinelibrary.wiley.com/doi/full/10.1111/raq.12301

Parker, R. W., Blanchard, J. L., Gardner, C., Green, B. S., Hartmann, K., Tyedmers, P. H. and Watson, R. A., 2018. Fuel use and greenhouse gas emissions of world fisheries. *Nature Climate Change*, 8(4):333.

Partos, L (2010). FAO: fish feed costs to remain high. Seafood Source. Source: https://www.seafoodsource.com/news/aquaculture/fao-fish-feed-costs-to-remain-high

Pauly, D., Christensen, V., Guénette, S., Pitcher, T. J., Sumaila, U. R., Walters, C. J.and Zeller, D., 2002. Towards sustainability in world fisheries. *Nature*, pp.689-695.

Pendleton, L., Donato, D.C., Murray, B.C., Crooks, S., Jenkins, W.A., Siûeet, S., Craft, C., Fourqurean, J.W., et al., 2012. Estimating global "blue carbon" emissions from conver- sion and degradation of vegetated coastal ecosystems. PLoS One 7 (9), e43542.

Penman, D. J., Gupta, M. V., & Dey, M. M. (Eds.). 2005. *Carp genetic resources for aquaculture in Asia* (Vol. 1727). WorldFish.

Pérez-Ramírez, M., Phillips, B., Lluch- Belda, D., & Lluch-Cota, S. (2012). Perspectives for implementing fisheries certification in developing countries. *Marine Policy, 36*: 297–302.

Pillay, T. V. R. & Kutty, M. N. 2005. *Aquaculture: principles and practices* (No. Ed. 2). Blackwell publishing.

Pitcher, T. J., Kalikoski, D. and Pramod, G. (eds), 2006. Evaluations of Compliance with the UN Code of Conduct for Responsible Fisheries. Fisheries Centre Research Reports 14(2): 1191.

Poore, J. and Nemecek, T. 2018. Reducing food's environmental impacts through producers and consumers. Science, 360 (6392), 987-992. Source: https://science.sciencemag.org/ content/360/6392/987

Portner, O.H. and Knust, R., 2007. "Climate Change Affects Marine Fishes Through the Oxygen Limitation of Thermal Tolerance" *Science,* Vol. 315. no. 5808, pp. 95-97

Pradhan Mantri Matsya Sampada Yojana, Operational Guidelines, Government of India – Ministry of Fisheries, Animal Husbandry and Dairying – Department of Fisheries, June 2020, 237 pp

Prasad Amita , 2017. MoEFCC. A presentation on National Consultation on SDG 13, 14 and 15.

Public Report: Coller FAIRR Protein Producer Index, 2019. 104p

Rabobank, 2015. *Rabobank world seafood trade map 2015.* Retrieved from https://far.rabobank.com/en/sectors/animalprotein/world-seafood-trade-map.html

Ramachandran, C., 2002. FAO Code of Conduct for Responsible Fisheries (Malayalam translation). Project Report. CMFRI, Kochi.

Ramsar. 2015. State of the World's Wetlands and Their Services to People: A Compilation of Recent Analyses. Ramsar Briefing Note 7. Ramsar Convention on Wetlands.

Rasenberg, M. 2013. GHG emissions in aquatic production systems and marine fisheries. Source:https://core.ac.uk/download/pdf/29222560.pdf

Rockström, J., Falkenmark, M., Karlberg, L., Hoff, H., Rost, S., Gerten, D., 2009. Future water availability for global food production: the potential of green water for increasing re- silience to global change. Water Resour. Res. 45, W00A12.

Rockström, J., Lannerstad, M., Falkenmark, M., 2007. Assessing the water challenge of a new green revolution in developing countries. PNAS 104, 6253–6260.

Rustomjee, C. 2016. Financing the Blue Economy in Small States. Center for International Governance Innovation.

Sachs, J., Schmidt-Traub, G., Kroll, C., Lafortune, G., Fuller, G. 2019. *Sustainable Development Report 2019.* New York: Bertelsmann Stiftung and Sustainable Development Solutions Network (SDSN).

Sainsbury, K. 2010. Review of ecolabelling schemes for fish and fishery products from capture fisheries. FAO Fisheries and Aquaculture Technical Paper. No. 533. 93p

Sajali, U. S. B. A., Atkinson, N. L., Desbois, A. P., Little, D. C., Murray, F. J., & Shinn, A. P., 2019. Prophylactic properties of biofloc-or Nile tilapia-conditioned water against Vibrio parahaemolyticus infection of whiteleg shrimp (Penaeus vannamei). Aquaculture, 498, 496-502. Source: https://www.sciencedirect.com/science/article/pii/S0044848618307725

Salgaonkar, A. A., 2016. "Sustainable Certifications in Global Seafood Trade". In: Progressive Grocer. 64-68p

Salgaonkar, A. A., 2018. "An Overview of Indian Fisheries and Seafood Business". In: The India Food Report. Chapter 2.5: 82-87p

Salgaonkar, A. A., and Prasad, R. K., 2017. "Food Processing Sector and Major Segments". In: Progressive Grocer. 34-38p.

Sanyukta Samaddar, 2019. Niti Aayog, Government of India. A ppt on "Localising SDGs in India"

Sehgal, K. L.,1999. Coldwater fish and fisheries in the Indian Himalayas: Culture. *Fish and fisheries at Higher Altitudes. Asia*, (385).

Shenoi S. S. C., 2017. Life below water: Sustainable Management of Coastal and Marine Ecosystems. National Consultation on SDGs, Integrating Biodiversity Concerns, Ecosystem Values and Climate Resilience for India's Sustainable Development, WWF, New Delhi.

Shenoy, L. and Biradar, R. S., 2005. Marine Fishing Regulation Acts of Coastal States of India-A Compendium. Central Institute of Fisheries Education, Mumbai, India, 73 pp.

Shenoy, L. and Lakra, W. S., (ed.), 2012. Proceedings of National workshop on Code of Conduct for Responsible Fisheries, February 1-2, 2012, Mumbai, India. CIFE; 102 pp.

Shenoy, L., 2011. Safety in fishing operation. In: Manual on Code of Conduct for Responsible Fisheries and sea safety. Short-term training program for fishers in collaboration with state Fisheries Department, CIFE, 20-22 October, 2011, 27-30 pp.

Siikamäki, J., Sanchirico, J.N., Jardine, S., McLaughlin, D., Morris, D.F., 2012. Blue Carbon: Global Options for Reducing Emissions from the Degradation and Development of Coastal Ecosystems. Resources for the Future, Washington DC.

Simpson, S., 2011. The blue food revolution: making aquaculture a sustainable food source. Scientiûc American February, pp. 54–61.

SOFIA, 2014. State of World Fisheries and Aquaculture, Food and Agriculture Organization of the United Nations, Rome. pp.1-192.

SOFIA, 2016. State of World Fisheries and Aquaculture, Food and Agriculture Organization of the United Nations, Rome. pp.1-190.

Sustainable Fisheries Partnership (SFP), 2020. Impacts of COVID-19 in Target 75 Fisheries. Summary of preliminary findings. Honolulu: Sustainable Fisheries Partnership Foundation [online]. San Francisco, CA. [Cited 20 May 2020] https://globalmarinecommodities.org/en/publications/impacts-of-covid-19- in-target-75-fisheries/

Tan, R.R. and Culaba, A.B., 2009. Estimating the carbon footprint of tuna fisheries. *WWWF Binary Item*, 17870, 14pp.

Thai Department of Fisheries, 2020. Impact of Covid-19. March 2020. Bangkok.

Thai PBS, 2020. Fish barters with rice. Thai PBS [Online]. Bangkok. [Cited 28 April 2020]. https://program.thaipbs.or.th/WeFightCovid19/episodes/6899

The Fish Site, 2016. Early Mortality Syndrome | Disease guide. Source: https://thefishsite.com/diseaseguide/early-mortality-syndrome

The India Food Report, 2016. 429p

The India Food Report, 2018. 233p

Thong, P. Y., 2014. Use of Bioflocs in Shrimp Farming. Source: https://www.researchgate.net/ publication/305905355 Use of Bioflocs in Shrimp Farmig

Thrane, M., 2004. Energy consumption in the Danish fishery: identification of key factors. *Journal of Industrial Ecology*, *8*(1 2):223-239.

Troell, M., Eide, A., Isaksen, J., Hermansen, Ø., & Crépin, A. S., 2017. Seafood from a changing Arctic. Ambio, 46(3), 368-386. Source: https://link.springer.com/article/10.1007/s13280-017-0954-2#citeas

Troell, M., Joyce, A., Chopin, T., Neori, A., Buschmann, A. H., & Fang, J. G., 2009. Ecological engineering in aquaculture—potential for integrated multi-trophic aquaculture (IMTA) in marine offshore systems. Aquaculture, 297(1-4), 1-9. Source: https://www.sciencedirect.com/science/article/abs/pii/S0044848609007856

Tyedmers, P.H., Watson, R. and Pauly, D., 2005. Fuelling global fishing fleets. *AMBIO: A Journal of the Human Environment*, 34(8), pp.635-639.

UN, 2015. "Transforming our world: Agenda 2030 on Sustainable Development". 35p

UN, 2016. The First Global Integrated Marine Assessment: World Ocean Assessment I, by the Group of Experts of the Regular Process. New York.

UN, 2019. Special edition: Progress towards the Sustainable Development Goals, 2019. 39p

UN, 2019. The Sustainable Development Goal Report, 2019. 60p

UN, Niti Aayog. "SDG India Index - Baseline Report 2018". 270p

UNCTAD, 2016. Review of Maritime Transport 2016. United Nations, Geneva.

UNDESA, 2014a. Blue Economy Concept Paper. United Nations, New York.

UNEP, 2013. Green Economy Definition. Nairobi.

UNEP, 2015. Blue Economy—Sharing Success Stories to Inspire Change. UNEP Regional Seas Report and Studies No. 195. Nairobi.

Various links of www.fao.org and UN organisation

Varkey, D., Pramod, G. and Pitcher, T. J., 2006. An Estimation of Compliance of the Fisheries of India with Article 7 (Fisheries Management) of the FAO (UN) Code of Conduct for Responsible Fisheries Centre Research Reports, 14(2):1191 pp.

Vincke, M. M. J., 1992. Integrated farming of fish and livestock: present status and future development. In *FAO/IPT Workshop on Integrated Livestock-Fish Production Systems, Kuala Lumpur (Malaysia), 16-20 Dec 1991.*

Vivekanandan, E., 2011. Climate Change and Indian Marine Fisheries, Central Marine Fisheries Research Institute. Special Publication, vol. 105, 97pp.

Vivekanandan, E., Singh, V. V. and Kizhakudan, J. K., 2013. Carbon footprint by marine fishing boats of India. *Current Science*, *105*(3):361-366.

Webster, C. D., & Jana, B. B. 2003. *Sustainable aquaculture: global perspectives*. CRC Press.

White Paper on "A Guide to Visual Merchandising in the Fresh Department" by METTLER TOLEDO, 2011.

White, K., O'Neill, B., Tzankova, Z., 2004. At a Crossroads: Will Aquaculture Fulûl the Promise of the Blue Revolution? A SeaWeb Aquaculture Clearinghouse Report, USA.

Wiedmann, T. and Minx, J., 2008. A definition of 'carbon footprint'. *Ecological economics research trends*, *1*, pp.1-11.

World Bank and FAO, 2009. The sunken billions. The economic justification for fisheries reform. Agriculture and Rural Development Department, The World Bank, Washington DC, 100 p.

World Bank and UNDESA, 2017. The Potential of the Blue Economy: Increasing Long-term Beneûts of the Sustainable Use of Marine Resources for Small Island Developing States and Coastal Least Developed Countries. United Nations Department of Eco- nomic and Social Affairs, and World Bank, Washington DC.

World Bank, 2012a. Inclusive Green Growth—The Pathway to Sustainable Development. Washington, DC.

World Bank, 2016a. Blue Economy Development Framework—Growing the Blue Economy to Combat Poverty and Accelerate Prosperity.

World Bank, 2016b. Sunken Billions Revisited. Progress and Challenges in Global Marine Fisheries. Washington, DC.

World Bank, 2020. World Bank predicts sharpest decline of remittances in recent history. In: *The World Bank* [online]. [Cited 22 May 2020]. www.worldbank.org/en/news/press-release/2020/04/22/world-bankpredicts-sharpest-decline-of-remittances-in-recent-history

WorldFish, 2020. Recommendations to safely open the aquaculture (carp, tilapia and pangasius) supply chain open during the COVID-19 Crisis. 7 May 2020. Bangladesh.

WWF, 2015. Living blue planet report. Species, habitats and human well-being. Eds. Tanzer, *J et al.* WWF, Gland, Switzerland, 68 pp.

www.bioshrimp.com

www.friendofthesea.org

Yadava, Y., 2000. Report of the National Workshop on the Code of Conduct for Responsible Fisheries. Bay of Bengal Programme, Chennai, India, 23 pp.

Yuan, J., Xiang, J., Liu, D., Kang, H., He, T., Kim, S., ... & Ding, W., 2019. Rapid growth in greenhouse gas emissions from the adoption of industrial-scale aquaculture. Nature Climate Change, 1. Source: https://www.nature.com/articles/s41558-019-0425-9

ACRONYMS

AHIDF: Animal Husbandry Infrastructure Development Fund
AIS: Automatic Identification System
AMR: Anti-Microbial Resistance
AOC: Aqua One Center
APFAMGS: Andhra Pradesh Farmer Managed Groundwater Systems
ASC: Aquaculture Stewardship Council
BAP: Best Aquaculture Practices
BIS: Bureau of Indian Standards
BMP: Best Management Practices
BOBLME: Bay of Bengal Large Marine Ecosystem
BOBP: Bay of Bengal Programme
BOBP-IGO: Bay of Bengal Programme-Intergovernmental Organization
BRICS: Brazil Russia India China and South Africa
CAA: Coastal Aquaculture Authority
CAC: Central Apex Committee
CAGR: Compound Annual Growth Rate
CAMC: Central Approval & Monitoring Committee
CAU: Central Agricultural University
CCRF: Code of Conduct for Responsible Fisheries
CFC: Consumption of Fixed Capital
CFFA: Coalition for Fair Fisheries Agreements
CIBA: Central Institute of Brackishshwater Aquaculture
CIFA: Central Institute of Freshwater Aquaculture
CIFE: Central Institute of Fisheries Education

CIFRI: Central Inland Fisheries Research Institute

CIFT: Central Institute of Fisheries Technology

CIG: Common Interest Groups

CITES: Convention on International Trade in Endangered Species

CMFRI: Central Marine Fisheries Research Institute

CMS: Convention on the Conservation of Migratory Species of Wild Animals

CMSP: Coastal and Marine Spatial Plans

COFI: Committee on Fisheries

CPF: Country Programming Framework

CRZ: Coastal Regulation Zone

CSS: Centrally Sponsored Scheme

DADF: Department of Animal Husbandry Dairying and Fisheries

DAHD: Department of Animal Husbandry and Dairying

DARE: Department of Agricultural Research & Education

DCFR: Directorate of Coldwater Fisheries Research

DIVA: Directorate of Incubation and Vocational Training in Aquaculture

DLC: District Level Committee

DoF: Department of Fisheries

DPR: Detailed Project Report

DSA: Directorate of Sustainable Aquaculture

EAF: Ecosystem Approach to Fisheries

EE: Eligible Entities

EEZ: Exclusive Economic Zone

EIA: Environmental Impact Assessment

EMS: Early Mortality Syndrome

e-NAM: (electronic) National Agricultural Market

EPC: Export Promotion Council

FAO: Food and Agriculture Organization

FARNET: Fisheries Areas Network

FCRI: Fisheries College and Research Institute

FDI: Foreign Direct Investment

FFDA: Fish Farmers' Development Agencies

FFPO: Fish Farming Producer Organization

FH: Fishery Harbour

FIDF: Fisheries and Aquaculture Infrastructure Development Fund

FIFO: First In First Out

FOS: Friend of the Sea

FRAD CMFRI: Fishery Resources Assessment Division (of) Central Marine Fisheries Research Institute

FRs: Feasibility Reports

FSI: Fishery Survey of India

FSSAI: Food Safety and Standards Authority of India

FWS: Farmers Water School

GAP: Good Aquaculture Practice

GAqP: Good Aquaculture Practice

GCF: General Collateral Financing

GDP: Gross Domestic Product

GEF: Global Environment Facility

GFSI: Global Food Safety Initiative

GIS: Geographical Information System

GNDP: Gross National Disposable Income

GNI: Gross National Income

GoI: Government of India

GSSI: Global Sustainable Seafood Initiative

GTFL: Garware Technical Fibres Ltd

GVA: Gross Value Added

HAB: Harmful Algal Bloom

HACCP: Hazard Analysis Critical Control Point

HDPE: High Density PolyEthylene

IAY: Indira Awaas Yojana

IBEF: Indian Brand Equity Organization

ICAR: Indian Council of Agricultural Research

ICMR: Indian Council of Medical Research

ICSF: International Collective in Support of Fishworkers

ICT: Information & Communication Technology

IFAD: International Fund for Agricultural Development
IMTA: Integrated Multi-Trophic Aquaculture
IoT: Internet of Things
IOTC: Indian Ocean Tuna Commission
IPOA: International Plans of Action
ITQ: Individual Transferable Quota
IUCN: International Union for Conservation of Nature
IUU Fishing: Illegal, Unreported and Unregulated Fishing
KBA: Key Biodiversity Areas
KCC: Kisan Credit Card
KMFRA: Kerala Marine Fisheries Regulation Act
KUFOS: Kerala University of Fisheries and Ocean Studies
LDCs: Least Developed Countries
LIFDCs: Low-Income Food-Deficit Countries
LLDCs: Landlocked Developing Countries
LOA: Length Overall
LVB: Low Value Bycatch
MCS: Monitoring Control and Surveillance
MDGs: Millennium Development Goals
MFRA: Marine Fishing Regulation Acts
MGNAREGA: Mahatma Gandhi National Rural Employment Guarantee Act
MGNREGS: Mahatma Gandhi National Rural Employment Guarantee Scheme
MoA & FW: Ministry of Agriculture and Farmers Welfare
MoEFCC: Ministry of Environment, Forest and Climate Change
MoES: Ministry of Earth Sciences
MoSPI: Ministry of Statistics and Programme Implementation
MPEDA: Marine Product Export Development Authority
MRL: Maximum Residue Limits
NABARAD: National Bank for Agriculture and Rural Development
NaCSA: National Centre for Sustainable Aquaculture
NBAP: National Biodiversity Action Plan
NBFGR: National Bureau of Fish Genetic Resources
NBT: National Biodiversity Targets

NCDC: National Centre for Disease Control

NCDC: National Cooperatives Development Corporation

NFDB: National Fisheries Development Board

NGOs: Non-Governmental Organizations

NICRA: National Initiative on Climate-resilient Agriculture

NIFAP: National Inland Fisheries and Aquaculture Policy

NITI Aayog: National Institution for Transforming India Aayog

NLE: Nodal Loaning Entities

NMP: National Mariculture Policy

NNI: Net National Income

NPMF: National Policy on Marine Fisheries

NRLM: National Rural Livelihood Mission

NVA: Net Value Added

OSPARC: Orissa Shrimp Seed Production Supply and Research Centre

PAC: Project Appraisal Committee

PFCE: Private Final Consumption Expenditure

PFRs: Preparation of Pre-Feasibility Reports

PMAY: Pradhan Matri Awaas Yojana

PMEU: Project Monitoring and Evaluation Unit

PMMSY: Pradhan Mantri Matsya Sampada Yojana

PMU: Project Monitoring Unit

PPE: Personal Protective Equipment

PVC: Poly Vinyl Chloride

QCI: Quality Council of India

RAS: Recirculation Aquaculture System

RAS: Recirculatory Aquaculture Systems

RFABs: Regional Fisheries Advisory Bodies

RFBs: Regional Fishery Bodies

RFMOs: Regional Fisheries Management Organizations

RGCA: Rajiv Gandhi Centre for Aquaculture

RKVY: Rashtriya Krishi Vikas Yojana

SARA: Sustainable Augmented Reality Audits

SCs: Scheduled Castes

SDGs: Sustainable Development Goals

SHG: Self-help Groups

SIDS: Small Island Developing States

SLAMC: State Level Approval & Monitoring Committee

SOFIA: State of World Fisheries and Aquaculture

SPF: Specific Pathogen Free

SSF: Small-Scale Fisheries

STs: Scheduled Tribes

SVC: Spring Viremia in Carp

TAC: Total Allowable Catch

TANUVAS: Tamil Nadu Veterinary and Animal Sciences University

TASPARC: The Andhra Pradesh Shrimp Seed Production Supply and Research Centre

TCP: Transmission Control Protocol

TED: Turtle Excluder Device

TEFRs: Techno Economic Feasibility Reports

TNAU: Tamil Nadu Agricultural University

TNJFU: Tamil Nadu Dr J Jayalalithaa Fisheries University

UAA: United Artists Association

UC: Utilization Certificate

UN DESA: United Nations Department of Economic and Social Affairs

UN: United Nations

UNCLOS: United Nations Convention on the Law of the Sea

UNCTAD: United Nations Conference on Trade and Development

UNEP: United Nations Environment Programme

USAID: United States Agency for International Development

UT: Union Territory

UVI: Unique Vessel Identifier

VG-SSF: Voluntary Guidelines on Sustainable Small-Scale Fisheries

WFFP: World Forum of Fisher Peoples

WFP: World Food Programme

WTO: World Trade Organization

WWF: World Wide Fund for Nature